Ernst Heinrich Hirschel

Basics of Aerothermodynamics

Ernst Heinrich Hirschel

Basics of Aerothermodynamics

With 147 Figures

Springer

Prof. Dr. E. H. Hirschel
Herzog-Heinrich-Weg 6
85604 Zorneding
Germany
e.h.hirschel@t-online.de

Jointly published with the American Institute of Aeronautics and Astronautics (AIAA)

ISBN 978-3-642-06050-2 e-ISBN 978-3-540-26519-1

Springer. Part of Springer Science+Business Media
springeronline.com

Softcover reprint of the hardcover 1st edition 2005

Cover design: deblik Berlin
Printed on acid free paper 62/3020/M - 5 4 3 2 1 0

Preface

The last two decades have brought two important developments for aerothermodynamics. One is that airbreathing hypersonic flight became the topic of technology programmes and extended system studies. The other is the emergence and maturing of the discrete numerical methods of aerodynamics/aerothermodynamics complementary to the ground-simulation facilities, with the parallel enormous growth of computer power.

Airbreathing hypersonic flight vehicles are, in contrast to aeroassisted re-entry vehicles, drag sensitive. They have, further, highly integrated lift and propulsion systems. This means that viscous effects, like boundary-layer development, laminar-turbulent transition, to a certain degree also strong interaction phenomena, are much more important for such vehicles than for re-entry vehicles. This holds also for the thermal state of the surface and thermal surface effects, concerning viscous and thermo-chemical phenomena (more important for re-entry vehicles) at and near the wall.

The discrete numerical methods of aerodynamics/aerothermodynamics permit now - what was twenty years ago not imaginable - the simulation of high speed flows past real flight vehicle configurations with thermo-chemical and viscous effects, the description of the latter being still handicapped by insufficient flow-physics models. The benefits of numerical simulation for flight vehicle design are enormous: much improved aerodynamic shape definition and optimization, provision of accurate and reliable aerodynamic data, and highly accurate determination of thermal and mechanical loads. Truly multidisciplinary design and optimization methods regarding the layout of thermal protection systems, all kinds of aero-servoelasticity problems of the airframe, et cetera, begin now to emerge.

In this book the basics of aerothermodynamics are treated, while trying to take into account the two mentioned developments. According to the first development, two major flight-vehicle classes are defined, pure aeroassisted re-entry vehicles at the one end, and airbreathing cruise and acceleration vehicles at the other end, with all possible shades in between. This is done in order to bring out the different degrees of importance of the aerothermodynamic phenomena for them. For the aerothermodynamics of the second vehicle class the fact that the outer surfaces are radiation cooled, is especially taken into account. Radiation cooling governs the thermal state of the surface, and hence all thermal surface effects. At the center of attention is the flight in the earth atmosphere at speeds below approximately 8.0 km/s and at altitudes below approximately 100.0 km.

The second development is taken into account only indirectly. The reader will not find much in the book about the basics of discrete numerical methods. Emphasis was laid on the discussion of flow physics and thermo-chemical phenomena, and on the provision of simple methods for the approximate quantification of the phenomena of interest and for plausibility checks of data obtained with numerical methods or with ground-simulation facilities. To this belongs also the introduction of the Rankine-Hugoniot-Prandtl-Meyer-(RHPM-) flyer as highly simplified configuration for illustration and demonstration purposes.

The author believes that the use of the methods of numerical aerothermodynamics permits much deeper insights into the phenomena than was possible before. This then warrants a good overall knowledge but also an eye for details. Hence, in this book results of numerical simulations are discussed in much detail, and two major case studies are presented. All this is done in view also of the multidisciplinary implications of aerothermodynamics.

The basis of the book are courses on selected aerothermodynamic design problems, which the author gave for many years at the University of Stuttgart, Germany, and of course, the many years of scientific and industrial work of the author on aerothermodynamics and hypersonic flight vehicle design problems. The book is intended to give an introduction to the basics of aerothermodynamics for graduate students, doctoral students, design and development engineers, and technical managers. The only prerequisite is the knowledge of the basics of fluid mechanics, aerodynamics, and thermodynamics.

The first two chapters of introductory character contain the broad vehicle classification mentioned above and the discussion of the flight environment. They are followed by an introduction to the problems of the thermal state of the surface, especially to surface radiation cooling. These are themes, which reappear in almost all the remaining chapters. After a review of the issues of transport of momentum, energy and mass, real-gas effects as well as inviscid and viscous flow phenomena are treated. In view of the importance for air-breathing hypersonic flight vehicles, and for the discrete numerical methods of aerothermodynamics, much room is given to the topic of laminar-turbulent transition and turbulence. Then follows a discussion of strong-interaction phenomena. Finally a overview over simulation means is given, and also some supplementary chapters.

Throughout the book the units of the SI system are used, with conversions given at the end of the book. At the end of most of the chapters, problems are provided, which should permit to deepen the understanding of the material and to get a "feeling for the numbers".

Zorneding, April 2004 *E. H. Hirschel*

Acknowledgements

The author is much indebted to several persons, who read the book or parts of it, and gave critical and constructive comments.

First of all I would like to thank G. Simeonides and W. Kordulla. They read all of the book and their input was very important and highly appreciated.

Many thanks are due also to A. Celic, F. Deister, R. Friedrich, S. Hein, M. Kloker, H. Kuczera, Ch. Mundt, M. Pfitzner, C. Weiland, and W. Staudacher, who read parts of the book.

Illustrative material was directly made available for the book by many colleagues, several of them former doctoral and diploma students of mine. I wish to thank D. Arnal, J. Ballmann, R. Behr, G. Brenner, S. Brück, G. Dietz, M. Fertig, J. Fischer, H.-U. Georg, K. Hannemann, S. Hein, A. Henze, R. K. Höld, M. Kloker, E. Kufner, J. M. Longo, H. Lüdecke, M. Marini, M. Mharchi, F. Monnoyer, Ch. Mundt, H. Norstrud, I. Oye, S. Riedelbauch, W. Schröder, B. Thorwald, C. Weiland, W. Zeiss. General permissions are acknowledged at the end of the book.

Special thanks for the preparation of the figures is due to H. Reger, S. Klingenfuss, B. Thorwald, and F. Deister, and to S. Wagner, head of the Institut für Aerodynamic und Gasdynamik of the University of Stuttgart, for sponsoring much of the preparation work.

Finally I wish to thank my wife for her support and her never exhausted patience.

E. H. H.

Table of Contents

1 Introduction

In this book basics of aerothermodynamics are treated, which are of importance for the aerodynamic and structural layout of hypersonic flight vehicles. It appears to be useful to identify from the begin classes of hypersonic vehicles, because aerothermodynamic phenomena can have different importance for different vehicle classes. This holds especially for what is usually called "heat loads". In this book we introduce the "thermal state of the surface", which encompasses (and distinguishes between) thermal surface effects on wall and near-wall viscous-flow and thermo-chemical phenomena, and thermal (heat) loads on the structure.

1.1 Classes of Hypersonic Vehicles and their Aerothermodynamic Peculiarities

The scientific and technical discipline "aerothermodynamics" is multidisciplinary insofar as aerodynamics and thermodynamics are combined in it. However, recent technology work for future advanced space transportation systems has taught that "aerothermodynamics" should be seen from the beginning in an even larger context.

In aircraft design, a century old design paradigm exists, which we call Cayley's design paradigm, after Sir George Cayley (1773 - 1857), one of the early English aviation pioneers [1]. This paradigm still governs thinking, processes and tools in aircraft design, but also in spacecraft design. It says, that one ought to assign functions like lift, propulsion, trim, pitch and yaw stabilization and control, et cetera, plainly to corresponding subsystems, like the wing, the engine (the propulsion system), the tail unit, et cetera. These subsystems and their functions should be coupled only weakly and linearly. Then one is able to treat and optimize each subsystem with its function, more or less independent of the others, and nevertheless treats and optimizes the whole aircraft which integrates all subsystems.

For space planes, either re-entry systems, or cruise/acceleration systems (see the classification below), Cayley's paradigm holds only partly. So far this was more or less ignored. But if future space-transportation systems (and also hypersonic aircraft) are to be one order of magnitude more cost-effective than now, it must give way to a new paradigm. This should be possible because of the rise of computer power, provided that proper multidisciplinary simulation and optimization methods can be developed and brought into practical use [2].

It is not intended to introduce such a new paradigm in this book. However, it is tried to present and discuss aerothermodynamics in view of the major roles of it in hypersonic vehicle design, which reflects the need for such a new paradigm.

Different hypersonic vehicles pose different aerothermodynamic design problems. In order to ease the discussion, four major classes of hypersonic vehicles are introduced[1]. These are, with the exception of class 4, classes of aeroassisted vehicles, i. e. vehicles, which fly with aerodynamic lift in the earth atmosphere at altitudes below approximately 100.0 km, and with speeds below 8.0 km/s, Fig. 1.1.

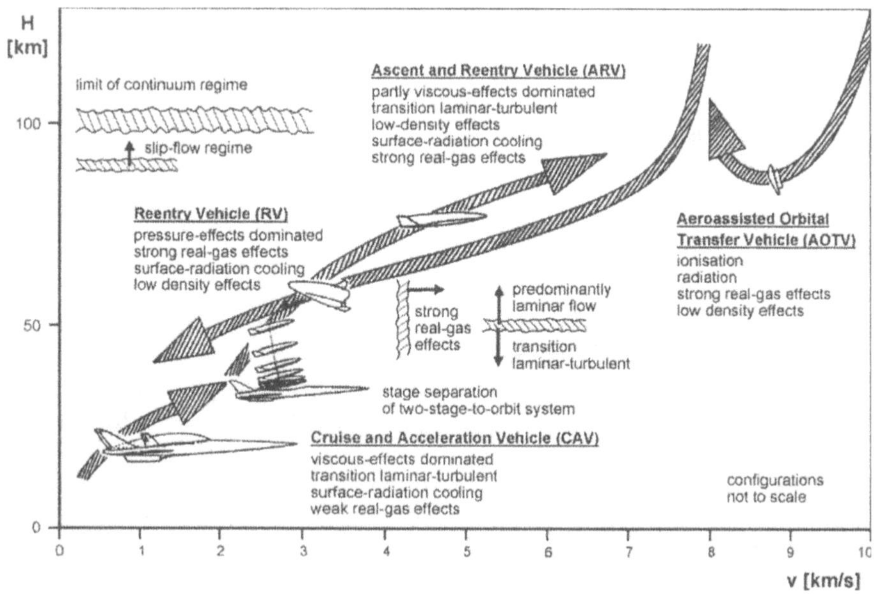

Fig. 1.1. The four major classes of hypersonic vehicles and some characteristic aerothermodynamic phenomena [4].

Of the mentioned vehicles so far only the Space Shuttle (and BURAN) actually became operational. All other are hypothetical vehicles or systems, which have been studied and/or developed to different degrees of completion [5]. The four classes are:

[1] A detailed classification of both civil and military hypersonic flight vehicles is given in [3].

1. Winged re-entry vehicles (RV), like the US Space Shuttle and the X-38[2], the Russian BURAN, the European HERMES, the Japanese HOPE. RV-type flight vehicles are launched typically by means of rocket boosters, but can also be the rocket propelled upper stages of two-stage-to-orbit (TSTO) space-transportation systems like SÄNGER, STAR-H, RADIANCE, MAKS.
2. Cruise and acceleration vehicles with airbreathing propulsion (CAV), like the lower stages of TSTO systems, e. g., SÄNGER, STAR-H, RADIANCE, but also hypothetical hypersonic air transportation vehicles (Orient Express, or the SÄNGER lower stage derivative). Flight Mach numbers would lie in the ram propulsion regime up to $M_\infty = 7$, and the scram propulsion regime up to $M_\infty = 12$ (to 14).
3. Ascent and re-entry vehicles (in principle single-stage-to-orbit (SSTO) space-transportation systems) with airbreathing (and rocket) propulsion (ARV), like the US National Aerospace Plane (NASP/X30), Oriflamme, HOTOL, and the Japanese Space Plane. The upper stages of TSTO-systems and purely rocket propelled vehicles, like Venture Star/X33, FESTIP FSSC-01, FSSC-15 et cetera are not ARV-type flight vehicles, because with their large thrust at take-off they do not need low-drag airframes.
4. Aeroassisted orbital transfer vehicles (AOTV), also called Aeroassisted Space Transfer Vehicles (ASTV), see, e. g., [6].

Each of the four classes has specific aerothermodynamic features and multidisciplinary design challenges. These are summarized in Table 1.1.

Without a quantification of features and effects we can already say, see also Fig.1.1, that for CAV- and ARV-type flight vehicles viscosity effects, notably laminar-turbulent transition and turbulence (which occur predominantly at altitudes below approximately 40.0 to 60.0 km) play a major role, while thermo-chemical effects are very important with RV-, ARV-, and AOTV-type vehicles. With the latter, especially plasma effects (ionization, radiation emission and absorption) have to be taken into account [6].

In Table 1.1 aerothermodynamic and multidisciplinary design features of the four vehicle classes are listed. The main objective of this list is to sharpen the perception, that for instance a CAV-type flight vehicle, i. e. an airbreathing aeroassisted system, definitely poses an aerothermodynamic (and multidisciplinary) design problem quite different from that of a RV-type vehicle. The CAV-type vehicle is aircraft-like, slender, flies at small angles of attack, all in contrast to the RV-type vehicle. The RV-type flight vehicle is a pure re-entry vehicle, which is more or less "only" a deceleration system,

[2] The X-38 is NASA's demonstrator of the previously planned crew rescue vehicle of the International Space Station.

Table 1.1. Comparative consideration of the aerothermodynamic features and multidisciplinary design features of four major classes of hypersonic vehicles.

Item	Re-entry vehicles (RV)	Cruise and acceleration vehicles (CAV)	Ascent and re-entry vehicles (ARV)	Aeroassisted orbital transfer vehicles (AOTV)
Mach number range	28 - 0	0 - 7(12)	0(7) - 28	20 - 35
Configuration	blunt	slender	opposing design requirements	very blunt
Flight time	short	long	long(?)/short	short
Angle of attack	large	small	small/large	head on
Drag	large	small	small/large	large
Aerodynamic lift/drag	small	large	large/small	small
Flow field	compressibility-effects dominated	viscosity-effects dominated	viscosity-effects/ compressibility-effects dominated	compressibility-effects dominated
Thermal surface effects: 'viscous'	not important	very important	opposing situation	not important
Thermal surface effects: 'thermo-chemical'	very important	important	opposing situation	very important
Thermal loads	large	medium	medium/large	large
Thermo-chemical effects	strong	weak/medium	medium/strong	strong
Rarefaction effects	initially strong	weak	medium/strong	strong
Critical components	control surfaces	inlet, nozzle/ afterbody, control surfaces	inlet, nozzle/ afterbody, control surfaces	control devices
Special problems	large Mach number span	propulsion integration, thermal management	propulsion integration, opposing design requirements	plasma effects

however not a ballistic one. Therefore it has a blunt shape, and flies at large angles of attack in order to increase the effective bluntness[3].

Thermal loads always must be considered together with the structure and materials concept of the respective vehicle, and its passive or active cooling concept. As will be discussed later, the major passive cooling means for outer surfaces is surface-(thermal-)radiation cooling [8]. The thermal management of a CAV-type or ARV-type flight vehicle must take into account all thermal loads (heat sources), cooling needs and cooling potentials of airframe, propulsion system, sub-systems and cryogenic fuel system.

1.2 RV-Type and CAV-Type Flight Vehicles as Reference Vehicles

In the following chapters we refer to RV-type and CAV-type flight vehicles as reference vehicles. They represent the two principle vehicle classes on which we, regarding aerothermodynamics, focus our attention. ARV-type vehicles combine their partly contradicting configurational demands, whereas AOTV-type vehicles are at the fringe of our interest. Typical shapes of RV-type and CAV-type vehicles are shown in Fig. 1.2.

Typical flight Mach number and angle of attack ranges as function of the altitude of the Space Shuttle, [11], and the SÄNGER space-transportation system, [12], up to stage separation are given in Fig. 1.3. During the re-entry flight of the Space Shuttle the angle of attack remains larger than 20° down to $H \approx 35.0\ km$, where the Mach number is $M_\infty \approx 5$. SÄNGER on the other hand has an angle of attack below $\alpha = 10°$, before the stage separation at $M_\infty \approx 7$ occurs.

The flight Mach-number, the flight altitude, and the angle of attack ranges govern many of the aerothermodynamic phenomena. We illustrate the determining characteristics with the help of the RHPM-flyer, Chapter 11, which is a sufficient good approximation of RV-type and CAV-type flight vehicles, Table 1.2 and 1.3. For a convenient restitution of the data we use triplets of M_∞, H and α which are not necessarily present in Fig. 1.3. For the same reason the ratio of specific heats was chosen to be $\gamma = 1.4$, and the exponent in the power law of the viscosity, Sub-Section 4.2.2, to be $\omega_\mu = 0.65$.

We observe from the Tables 1.2 and 1.3 the following tendencies, see also Section 2.1:

- RV-type flight vehicles are characterized by a strong flow compression on the windward side, resulting in $M_w = 1.76$. In reality we have even a large

[3] We note that, for instance, future RV-type flight vehicles may demand large down and cross range capabilities (see some of the FESTIP study concepts [7]). Then Aerodynamic lift/drag "small" for RV-type vehicles in Table 1.1 actually should read "small to medium".

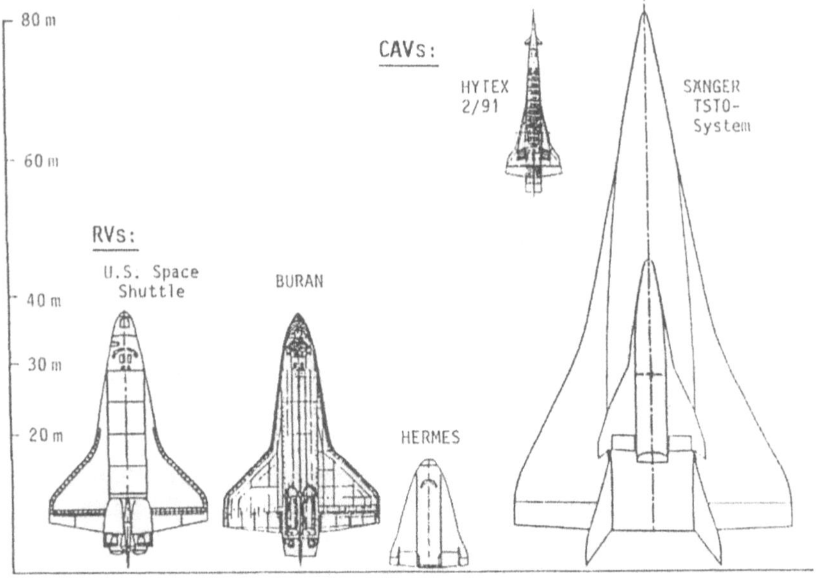

Fig. 1.2. Shape (planform) and size of hypersonic flight vehicles of class 1 (RV-type flight vehicles) and 2 (CAV-type flight vehicles) [9]. HYTEX: experimental vehicle studied in the German Hypersonics Technology Programme [10].

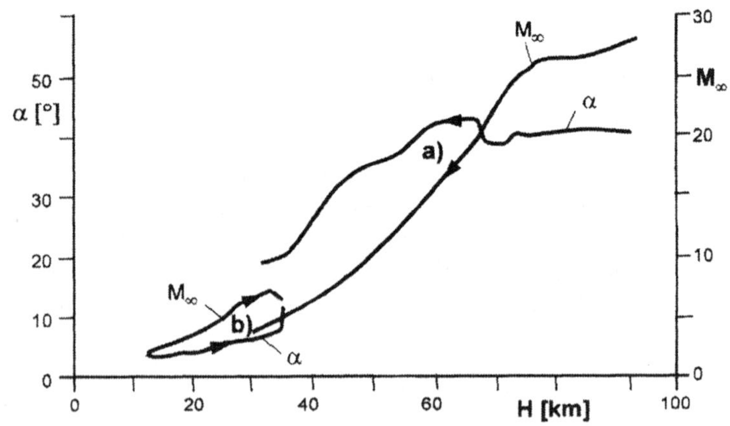

Fig. 1.3. Flight Mach number M_∞ and angle of attack α of **a**) the Space Shuttle, [11], and **b**) the two-stage-to-orbit space-transportation system SÄNGER up to stage separation, [12], as function of the flight altitude H.

Table 1.2. Flow parameters on the windward (w) and the lee (l) side of the RHPM-RV-flyer at 70.0 km altitude and an angle of attack $\alpha = 40°$, $\gamma = 1.4$, $\omega_\mu = 0.65$. The flow parameters are constant along the lower and the upper surface.

Location	M	T	p	ρ	Re^u_∞
∞	20	$219.69\ K$	$5.52\ Pa$	$8.75 \cdot 10^{-4}\ kg/m^3$	$3.62 \cdot 10^4\ m^{-1}$
w	1.76	$50.0\, T_\infty$	$295.0\, p_\infty$	$5.9\, \rho_\infty$	$0.29\, Re^u_\infty$
l	$\to \infty$	$\to 0\, T_\infty$	$\to 0\, p_\infty$	$\to 0\, \rho_\infty$	$\to 0\, Re^u_\infty$

Table 1.3. Flow parameters on the windward (w) and the lee (l) side of the RHPM-CAV-flyer at 30.0 km altitude and an angle of attack $\alpha = 7°$, $\gamma = 1.4$, $\omega_\mu = 0.65$. The flow parameters are constant along the lower and the upper surface.

Location	M	T	p	ρ	Re^u_∞
∞	6	$226.51\ K$	$1.20 \cdot 10^3\ Pa$	$1.842 \cdot 10^{-2}\ kg/m^3$	$2.26 \cdot 10^6\ m^{-1}$
w	5	$1.36\, T_\infty$	$2.72\, p_\infty$	$2.0\, \rho_\infty$	$1.59\, Re^u_\infty$
l	7.19	$0.72\, T_\infty$	$0.32\, p_\infty$	$0.45\, \rho_\infty$	$0.57\, Re^u_\infty$

subsonic pocket there. During a Space Shuttle re-entry one has typically at maximum $M_w \approx 2.5$, and mostly $M_w < 2$, [13]. The strong compression leads to large temperatures at still moderate densities, so that high-temperature real-gas effects are present[4], Chapter 5.
The unit Reynolds number Re^u is smaller than that at infinity. The boundary layer will be laminar at this altitude and it is at most a low supersonic boundary layer. Laminar-turbulent transition, Chapter 8, will happen only at altitudes below 60.0 to 40.0 km, where the boundary layer is also at most a low supersonic boundary layer. Due to the small unit Reynolds number the boundary layer is thick, Chapter 7, and hence radiation cooling is effective, Chapter 3.
On the lee-side it is indicated that the Prandtl-Meyer expansion limit, Chapter 6, has been reached. This does not match reality, but we can conclude that there no high-temperature real-gas effects are present, except for possible non-equilibrium frozen flow coming from the stagnation-point region. The boundary layer is extremely thick, radiation cooling is very effective.

- At CAV-type flight vehicles, due to the flight at small angle at attack, no large compression effects occur. We find them only in the blunt nose region and possibly at (swept) leading edges, ramps and control surfaces. High-temperature real-gas effects will essentially be restricted to these con-

[4] The use of $\gamma = 1.4$ of course gives much too high temperatures. With $\gamma = 1.25$ one gets $T_w \approx 25.0\ T_\infty$, which is more realistic, but does not change our conclusion.

figuration parts and to the boundary layers. They will increase of course with increasing flight speed.
On the windward side the Mach number is slightly below M_∞, on the lee side slightly above. The boundary layers are hypersonic boundary layers. The unit Reynolds numbers are large enough, so that laminar-turbulent transition will happen. The boundary layers are thick enough for an effective radiation cooling.

1.3 The Objectives of Aerothermodynamics

The aerothermodynamic design process is embedded in the vehicle design process. Aerothermodynamics has, in concert with the other disciplines, the following objectives:

1. Aerothermodynamic shape definition, which has to take into account the thermal state of the surface [14], if, for instance, it strongly influences the drag of the vehicle (CAV, ARV), or the performance of a control surface (all classes):
   a) Provision of aerodynamic performance, flyability and controllability on all trajectory elements (all vehicle classes).
   b) Aerothermodynamic airframe/propulsion integration for rocket propelled (RV, ARV) and especially airbreathing (CAV) vehicles.
   c) Aerothermodynamic integration of reaction control systems (RV, ARV, AOTV).
   d) Aerothermodynamic upper stage integration and separation for TSTO vehicles.
2. Aerothermodynamic structural loads determination for the layout of the structure and materials concept, the sizing of the structure, and the external thermal protection system (TPS) or the internal thermal insulation system, including possible active cooling systems for the airframe:
   a) Determination of mechanical loads (surface pressure, skin friction), both as static and dynamic loads, especially also acoustic loads.
   b) Determination of thermal loads for both external and internal surfaces/structures.
3. Surface properties definition (external and internal flowpath):
   a) In view of external surface-radiation cooling, the important "necessary" surface property is radiation emissivity. It governs the thermal loads of structure and materials, but also the thermal-surface effects on the near-wall viscous-flow and thermo-chemical phenomena.

b) "Permissible" surface properties are surface irregularities like roughness, waviness, steps, gaps et cetera in view of laminar-turbulent transition and turbulent boundary-layer flow. For CAV-type and ARV-type flight vehicles they must be "sub-critical" in order to avoid unwanted increments of viscous drag, and of the thermal state of the surface[5]. For RV-type vehicles surface roughness is an inherent matter of the layout of the thermal protection system (TPS). There especially unwanted increments of the thermal state of the surface are of concern on the lower part of the re-entry trajectory. In this context the problems of micro-aerothermodynamics on all trajectory segments are mentioned, which are connected to the flow, for instance, between tiles of a TPS or flow in gaps of control surfaces. All sub-critical, i. e. "permissible", values of surface irregularities should be well known, because surface tolerances should be as large as possible in order to minimize manufacturing cost.
Another "permissible" surface property is the surface catalytic behaviour, which should be as small as possible, in order to avoid unwanted increments of the thermal state of the surface, e. g., of the surface temperature. Usually surface catalytic behaviour, together with emissivity and anti-oxydation protection are properties of the surface coating of the airframe or the TPS material.

This short consideration shows that aerothermodynamics indeed must be seen not only in the context of aerodynamic design as such. It is an element of the truly multidisciplinary design of hypersonic flight vehicles, and must give answers and inputs to a host of design issues.

1.4 The Thermal State of the Surface and Radiation-Cooled Outer Surfaces as Focal Points

Under the "thermal state of the surface [14]" we understand the temperature of the gas at the surface, *and* the temperature gradient, respectively the heat flux, in it normal to the surface. As will be shown in Chapter 3, these are not necessarily those of and in the surface material. Regarding external surfaces we note, that these are, with some exceptions, in general only radiation cooled, if we consider aeroassisted space transportation vehicles or hypersonic aircraft flying in the earth atmosphere at speeds below approximately 8.0 km/s [8].

The thermal state of the surface thus is defined by

[5] Sub-critical means that laminar-transition is not triggered prematurely, and that in turbulent flow neither skin friction nor heat transfer are enhanced by surface irregularities, Chapter 8.

- the actual temperature of the gas at the wall surface, T_{gw}, and the temperature of the wall, T_w, with $T_{gw} \equiv T_w$, if low-density effects (temperature jump) are not present,
- the temperature gradient in the gas at the wall, $\partial T/\partial n|_{gw}$, in direction normal to the surface[6], respectively the heat flux in the gas at the wall, q_{gw}, if the gas is a perfect gas or in thermo-chemical equilibrium, Chapter 5,
- and the temperature gradient in the material at the wall surface, $\partial T/\partial n|_w$, in (negative) direction normal to the surface, respectively the heat flux q_w (tangential gradients are also neglected). The heat flux q_w is not equal to q_{gw}, if radiation cooling is employed, Section 3.1.

If one considers a RV-type flight vehicle, the thermal state of the surface concerns predominantly the structure and materials layout of the vehicle, and not so much its aerodynamic performance. This is because the RV-type vehicle flies a "braking" mission, where the drag on purpose is large (blunt configuration, large angle of attack, Table 1.1). Of course, if a flight mission demands large down-range or cross-range capabilities in the atmosphere, this may change somewhat.

The situation is different for an (airbreathing) CAV-type flight vehicle, which like any aircraft is drag-sensitive, and where viscous effects, which are affected strongly by the thermal state of the surface, in general play an important role, Table 1.1.

This book puts emphasis on this fact by distinguishing between the classical "thermal (heat) loads", which are of importance for the structure and materials concept of a vehicle, and "thermal-surface effects", which concern wall and near-wall viscous-flow and thermo-chemical phenomena, Fig. 1.4. This holds for both the external and the internal flowpath of a vehicle.

Both thermal-surface effects and thermal loads are coupled directly to the necessary and permissible surface properties, which were mentioned in the previous section.

In the following Table 1.4 wall and near-wall viscous-flow and thermo-chemical phenomena as well as structure and materials issues are listed, which are influenced by the thermal state of the surface. Partly our knowledge of these phenomena is still limited. We note in any case that there are viscous phenomena, especially laminar-turbulent transition, which are influenced by both T_w and $\partial T/\partial n|_{gw}$. This is also true for catalytic surface recombination.

Regarding materials and structures we note first of all that "thermal loads" encompasses both the wall temperature T_w and the heat flux into

[6] The temperature gradients, and hence the heat fluxes tangential to the surface, which in downstream direction appear especially with radiation cooled surfaces, are neglected in this consideration, but there might be situations, where this is not permitted. We neglect also slip-flow and non-equilibrium effects, which are discussed in Chapters 4 and 5.

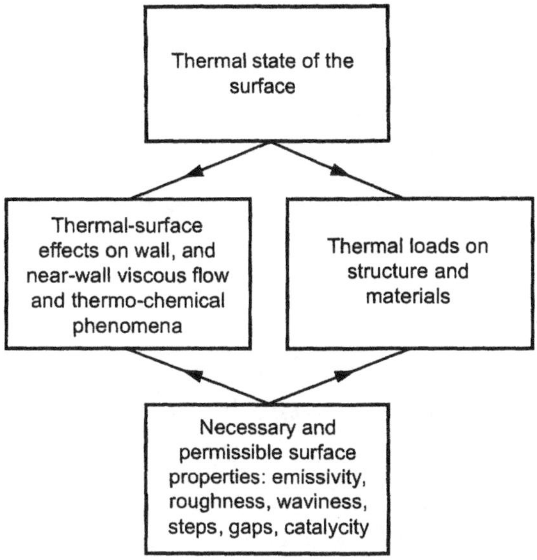

Fig. 1.4. The thermal state of the surface and its different aero-thermal design implications.

Table 1.4. Wall and near-wall viscous-flow/thermo-chemical phenomena, and structure and materials issues influenced by the thermal state of the surface (n is the direction normal to the surface, ( ) indicates indirect influence).

Item	T_w	$\partial T/\partial n\|_{gw}$	$\partial T/\partial n\|_w$
Boundary-layer thicknesses(δ, δ_1, ...)	X		
Skin friction	X		
Heat flux in the gas at the wall q_{gw}	X	X	
Surface-radiation heat flux q_r	X	(X)	(X)
Laminar-turbulent transition	X	X	
Turbulence	?	?	
Controlled and uncontrolled flow separation	X		
Shock/bondary-layer interaction	X		
Hypersonic viscous interaction	X		
Catalytic surface recombination	X	(X)	
Transport properties at and near the surface	X	X	
Heat flux into the wall q_w	X	(X)	X
Material strength and endurance	X		
Thickness of TPS or internal insulation (time integral of q_w)	X		X

the wall q_w. The wall temperature T_w governs the choice of surface material (and coating) in view of strength and endurance, and $\partial T/\partial n|_w$ or q_w, for instance, as function of flight time, the thickness of the thermal protection system (TPS) of a RV-type flight vehicle.

In the Chapters 5, 7 and 8 emphasis will be put on the discussion of the influence of the thermal state of the surface, especially in presence of radiation cooling, on wall, and near-wall thermo-chemical and viscous-flow phenomena. This influence, of course, is additional to that of the basic parameters Reynolds number, Mach number, stream-wise and cross-wise pressure gradients, et cetera.

1.5 Scope and Content of the Book

The aerodynamic and aerothermodynamic design of flight vehicles presently is undergoing large changes regarding the tools used in the design processes. Discrete numerical methods of aerothermodynamics find their place already in the early vehicle definition phases. This is a welcome development from the viewpoint of vehicle design, because only with their use the necessary completeness and accuracy of design data can be attained. However, as will be shown and discussed in this book, computational simulation still suffers, like that with other analytical methods, and of course also ground-facility simulation, under the insufficient representation of real-gas and of turbulence phenomena. This is a shortcoming, which in the long run has to be overcome.

It remains the problem that computational simulation gives results on a high abstraction level. This is similar with the application of, for instance, computational methods in structural design. The user of numerical simulation methods therefore must have very good basic knowledge of both the phenomena he wishes to describe and their significance for the design problem at hand.

Therefore this book has the aim to foster:

- the knowledge of aerothermodynamic phenomena regarding their significance in vehicle design,
- the understanding of their qualitative dependence on flight parameters, vehicle geometry, et cetera,
- and their quantitative description.

As a consequence of this the classical approximate methods, and also the modern discrete numerical methods, in general will not be discussed in detail. The reader is referred either to the original literature, or to hypersonic monographs, which introduce to their basics in some detail, for instance [15], [16]. However, approximate methods, and very simple analytical considerations will be employed where possible to give basic insights and to show basic trends. For this reason also a highly simplified flight-vehicle configuration, the

Rankine-Hugoniot-Prandtl-Meyer- (RHPM-) flyer is introduced. It basically is only an infinitely thin flat plate at angle of attack. The flow past it is determined by means of simple shock-expansion theory. In all chapters also results of computational or ground-facility simulation as well as flight data will be used to broaden the picture.

The following Chapter 2 treats the flight environment, which is in the frame of this book predominantly the earth's atmosphere below approximately 100.0 km altitude. Chapter 3 is devoted to the discussion of the thermal state of the vehicle surface. Especially the phenomena connected to radiation cooling of outer surfaces are considered.

Chapter 4 gives the basic mathematical formulations regarding transport of mass, momentum and energy. Emphasis is put on the presentation and discussion of similarity parameters, and of the boundary conditions at the vehicle surface. They govern, together with the free-flight parameters and the vehicle geometry, all what happens in the flow past the vehicle and on the surface of the vehicle, the latter being the major concern of the vehicle designer.

Chapters 5, 6, 7, 8, and 9 then treat the topics real-gas phenomena, inviscid phenomena, attached high-speed viscous flow, laminar-turbulent transition and turbulence, and strong-interaction phenomena. Concerning wall-near viscous-flow and thermo-chemical phenomena we put emphasis on the discussion of the influence of thermal-surface effects, i. e., the influence of the thermal state of the surface on these phenomena.

A discussion of computational and ground-facility simulation means follows in Chapter 10. Again we will not go into details, for instance of algorithms or of wind-tunnels and wind-tunnel measurement techniques. Instead basic potentials and deficits of the simulation means will be considered, including problems of in-flight simulation.

Chapter 11 introduces the RHPM-flyer, in Chapter 12 we collect the governing equations of hypersonic flows in general coordinates. Both chapters serve as reference chapters for the forgoing ones. The book closes with constants, functions et cetera, symbols, the author and the subject index, and permissions.

References

1. G. CAYLEY. "On Arial Navigation". *Nicholson's Journal.* Vol. 24, 1809, pp. 164-174, Vol. 25, 1810, pp. 81 - 87, 161 - 173.
2. E. H. HIRSCHEL. "Towards the Virtual Product in Aircraft Design?" *J. Periaux, M. Champion, J.-J. Gagnepain, O. Pironneau, B. Stoufflet, P. Thomas (eds.), Fluid Dynamics and Aeronautics New Challenges.* CIMNE Handbooks on Theory and Engineering Applications of Computational Methods, Barcelona, Spain 2003, pp. 453 - 464.

3. D. M. Bushnell. "Hypersonic Ground Test Requirements". *F. K. Lu, D. E. Marren (eds.), Advanced Hypersonic Test Facilities.* Vol. 198, Progress in Astronautics and Aeronautics, AIAA, Washington, DC, 2002, pp. 1 - 15.
4. E. H. Hirschel. "Viscous Effects". Space Course 1991, RWTH Aachen, 1991, pp. 12-1 to 12-35.
5. H. Kuczera, P. W. Sacher. "Reusable Space Transportation Systems". Springer, Berlin/Heidelberg/New York and Praxis Publishing, Chichester, 2005.
6. H. F. Nelson, ed. "Thermal Design of Aeroassisted Orbital Transfer Vehicles". *Vol. 96 of Progress in Astronautics and Aeronautics.* AIAA, Washington D. C., 1985.
7. H. Kuczera, P. W. Sacher, Ch. Dujarric. "FESTIP System Study - an Overview". *7th AIAA International Space Plane and Hypersonic System and Technology Conference 1996.* Norfolk, Virginia, 1996.
8. E. H. Hirschel. "Heat Loads as Key Problem of Hypersonic Flight". Zeitschrift für Flugwissenschaften und Weltraumforschung (ZFW), Vol. 16, No. 6, 1992, pp. 349 - 356.
9. E. H. Hirschel. "Hypersonic Aerodynamics". Space Course 1993, Vol. 1, Technical University München, 1993, pp. 2-1 to 2-17.
10. P. W. Sacher, R. Kunz, W. Staudacher. "The German Hypersonic Experimental Aircraft Concept". 2nd AIAA International Aerospaceplane Conference, Orlando, Florida, 1990.
11. S. D. Williams. "Columbia, the First Five Flights Entry Heating Data Series, Volume 1: an Overview". NASA CR - 171 820, 1984.
12. E. H. Hirschel. "The Technology Development and Verification Concept of the German Hypersonics Technology Programme". AGARD R-813, 1986, pp. 12-1 to 12-15.
13. W. D. Goodrich, S. M. Derry, J. J. Bertin. "Shuttle Orbiter Boundary-Layer Transition: A Comparison of Flight and Wind-Tunnel Data". AIAA-Paper 83-0485, 1983.
14. E. H. Hirschel. "Thermal Surface Effects in Aerothermodynamics". *Proc. Third European Symposium on Aerothermodynamics for Space Vehicles, Noordwijk, The Netherlands, November 24 - 26, 1998.* ESA SP-426, 1999, pp. 17 - 31.
15. J. D. Anderson. "Hypersonic and High Temperature Gas Dynamics". McGraw-Hill, New York, 1989.
16. J. J. Bertin. "Hypersonic Aerothermodynamics". AIAA Education Series, Washington, 1994.

2 The Flight Environment

Hypersonic flight either of space-transportation systems or of hypersonic aircraft in the earth atmosphere is in the focal point of the book. Hence the flight environment considered here is that which the earth atmosphere poses. The basic features and properties are discussed, and references are given for detailed information.

2.1 The Earth Atmosphere

The earth atmosphere consists of several layers, the troposphere from sea level up to approximately 10.0 km, the stratosphere between 10.0 km and 50.0 km, the mesosphere between 50.0 km and 80.0 km, and the thermosphere above approximately 80.0 km altitude, Fig. 2.1. The weather phenomena occur mainly in the troposphere, and consequently the fluctuations there mix and disperse introduced contaminants. These fluctuations are only weakly present at higher altitudes.

The stratosphere is characterized by a temperature plateau around 220.0 K to 230.0 K, in the mesosphere it becomes colder, in the thermosphere the temperature rises fast with altitude. Ecologically important is the altitude between 18.0 and 25.0 km with the vulnerable ozone layer.

The composition of the atmosphere can be considered as constant in the homosphere, up to approximately 100.0 km altitude. In the heterosphere, above 100.0 km altitude, it changes with altitude. This is important especially for computational simulations of aerothermodynamics. Note that also around 100.0 km altitude the continuum domain ends (Section 2.3).

It should be mentioned, that these numbers are average numbers, which partly depend strongly on the degree of geographical latitude of a location, and that they are changing with time (seasons, atmospheric tides, sun-spot activities). A large number of reference and standard atmosphere models is discussed in [1], where also model uncertainties and limitations are noted.

In aerothermodynamics we work usually with the U. S. standard atmosphere [2], in order to determine static pressure (p_∞), density (ρ_∞), static temperature (T_∞) et cetera as function of the altitude, Table 2.1. The 15°C standard atmosphere assumes a temperature of 15°C at sea level. A graphical view of some properties of the standard atmosphere is given in Fig. 2.1.

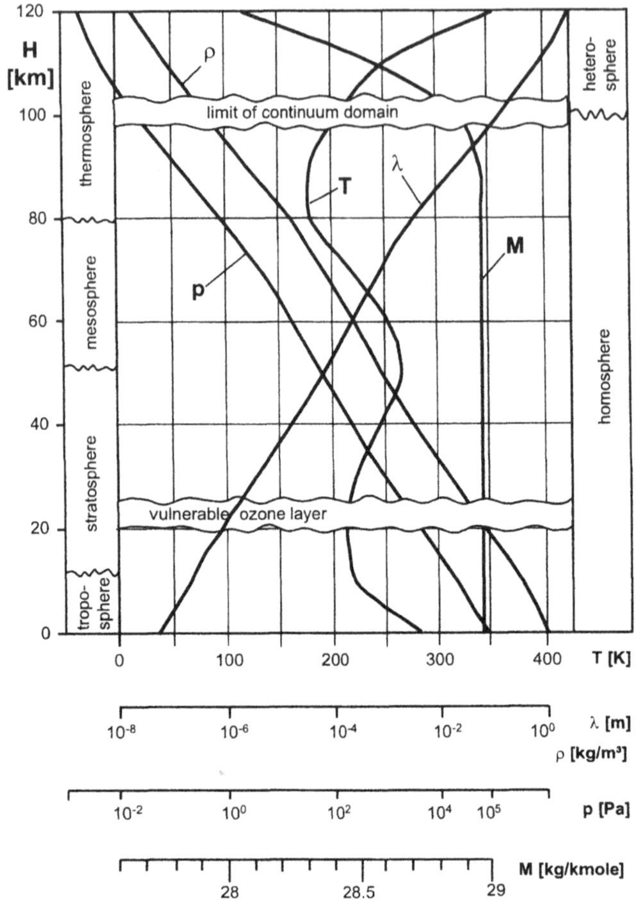

Fig. 2.1. Atmospheric layers and some properties of the atmosphere as function of the altitude, based on [2] (see also Table 2.1).

Uncertainties in atmospheric data influence guidance and control of a hypersonic flight vehicle. Large density fluctuations, which predominantly can occur in approximately 60.0 to 80.0 km altitude, must be compensated during, for example, a re-entry flight, otherwise a (down) range deviation would occur[1]. A 25 per cent smaller density than assumed at that altitude would lead, without correcting measures, to an approximately 100.0 km larger downstream range [3].

[1] This is a phenomenon which is primarily governed by the vehicle drag. The flight-path angle down to approximately 60.0 km altitude is 2° or less, so that the vehicle flies almost a circular trajectory. The drag then is an aerothermodynamic drag whereas the aerodynamic lift is initially only a fraction of the effective lift.

Table 2.1. Properties of the 15°C U. S. standard atmosphere as function of the altitude, [2].

Altitude H $[km]$	Temperature T $[K]$	Pressure p $[Pa]$	Density ρ $[kg/m^3]$	Mean free path λ $[m]$	Mean molecular weight M $[kg/kg\text{-}mole]$
0.0	288.150	$1.013 \cdot 10^5$	$1.225 \cdot 10^0$	$6.633 \cdot 10^{-8}$	28.9644
10.0	225.320	$2.641 \cdot 10^4$	$4.084 \cdot 10^{-1}$	$1.990 \cdot 10^{-7}$	28.9644
20.0	217.359	$5.531 \cdot 10^3$	$8.865 \cdot 10^{-2}$	$9.165 \cdot 10^{-7}$	28.9644
30.0	226.506	$1.198 \cdot 10^3$	$1.842 \cdot 10^{-2}$	$4.411 \cdot 10^{-6}$	28.9644
40.0	250.334	$2.874 \cdot 10^2$	$3.999 \cdot 10^{-3}$	$2.032 \cdot 10^{-5}$	28.9644
50.0	270.650	$7.978 \cdot 10^1$	$1.027 \cdot 10^{-3}$	$7.912 \cdot 10^{-5}$	28.9644
60.0	255.758	$2.244 \cdot 10^1$	$3.056 \cdot 10^{-4}$	$2.658 \cdot 10^{-4}$	28.9644
70.0	219.690	$5.518 \cdot 10^0$	$8.751 \cdot 10^{-4}$	$9.285 \cdot 10^{-4}$	28.9644
80.0	182.839	$1.036 \cdot 10^0$	$1.974 \cdot 10^{-5}$	$4.117 \cdot 10^{-3}$	28.9644
90.0	183.921	$1.648 \cdot 10^{-1}$	$3.121 \cdot 10^{-6}$	$2.603 \cdot 10^{-2}$	28.9618
100.0	212.504	$3.049 \cdot 10^{-2}$	$4.982 \cdot 10^{-7}$	$1.625 \cdot 10^{-1}$	28.8674
110.0	258.017	$7.379 \cdot 10^{-3}$	$9.823 \cdot 10^{-8}$	$8.156 \cdot 10^{-1}$	28.5570
120.0	360.325	$2.556 \cdot 10^{-3}$	$2.395 \cdot 10^{-8}$	$3.288 \cdot 10^0$	28.0673

For in-flight tests, for instance for vehicle-parameter identification purposes, or to obtain data on aerothermodynamic or other phenomena, it is mandatory to have highly accurate instantaneous "air data" on the trajectory in order to correlate measured parameters, Section 10.4. The air data are the thermodynamic data p_∞, T_∞, ρ_∞, and the vehicle speed vector $\underline{v}_\infty$ relative to the surrounding air space.

The atmosphere with its properties determines the free-stream parameters of a flight vehicle. These in turn govern the aerothermodynamic phenomena and the aerodynamic performance. We give a short overview in Figs. 2.2 to 2.4 [4] over the main free-stream parameters and some aerothermodynamic phenomena for the altitude domain 0.0 $km \leqq H \leqq 65.0$ km, and the flight-speed domain 0.0 $km/s \leqq v_\infty \leqq 7.0$ km/s. Indicated are nominal design points of the supersonic passenger aircraft Concorde, of four reference concepts LK1 to LK4 of a German hypersonic technology study [5], of the SÄNGER lower stage (staging condition) [6], and of the X-30 (cruise) [7]. Typical trajectory data of re-entry vehicles (US Space Shuttle, HERMES) are included, too.

Fig. 2.2 shows iso-Mach-number lines. These lines are more or less parallel to iso-speed lines. Turbo propulsion is possible up to $M_\infty \approx 3$, i. e. $v_\infty \approx 1.0$ km/s, ram-jet propulsion between $3 \lessapprox M_\infty \lessapprox 7$, i. e. 1.0 $km/s \lessapprox v_\infty \lessapprox 2.0$ km/s, and finally scram-jet propulsion between $4 \lessapprox M_\infty \lessapprox 12$ to 14, i. e. 1.2 $km/s \lessapprox v_\infty \lessapprox 4.0$ to 4.5 km/s.

Iso-unit Reynolds-number lines are at larger altitudes approximately parallel to iso-altitude lines. They show that at flight below approximately 50.0

km altitude boundary layers will be predominantly turbulent, Fig. 2.3, if the unit Reynolds number $Re^u_\infty = \rho_\infty u_\infty / \mu_\infty \approx 10^5\ m^{-1}$ is taken, with due reservations, as zero-order criterion. This means that the design of airbreathing flight vehicles (CAV-type vehicles) always has to cope with laminar-turbulent transition and turbulence. The laminar portion at the front part of a flight vehicle will be small at small altitudes and finally will extend over the vehicle length when $Re^u_\infty \approx 10^5\ m^{-1}$ is approached. On RV-type flight vehicles boundary layers are laminar during re-entry flight above approximately 60.0 to 40.0 km altitude, i. e. especially in the domain of large high-temperature real-gas effects and large thermal loads.

Included in Fig. 2.3 are also lines of constant total temperature (equilibrium real gas). They indicate that with increasing speed thermal loads indeed become a major design problem of airframes of hypersonic flight vehicles, and of airbreathing propulsion systems.

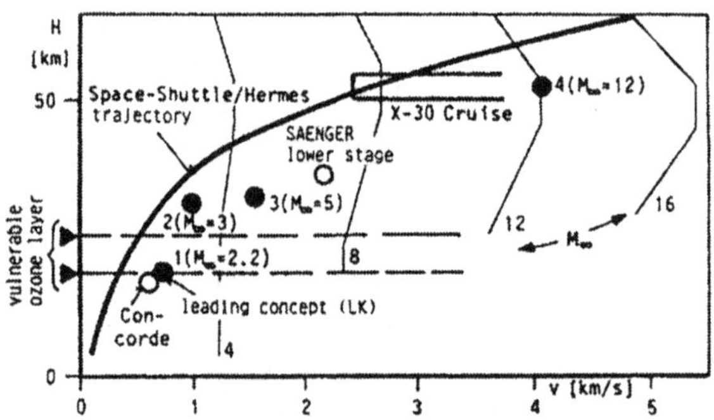

Fig. 2.2. Hypersonic flight-vehicle concepts in the velocity-altitude map ($v \equiv v_\infty$) with flight Mach numbers M_∞, [4] (based on [8]).

Fig. 2.4 finally shows that at flight speeds below $v_\infty \approx 0.8\ km/s$ air can be considered as a calorically and thermally perfect gas. For $0.8\ km/s \lessapprox v_\infty \lessapprox 2.6\ km/s$ vibration excitation must be taken into account. Above $v_\infty \approx 2.5\ km/s$ first dissociation of oxygen and then of nitrogen occurs. The dissociation of both gas constituents depends strongly on the flight altitude. At altitudes below $\approx 25.0\ km$ we can expect in general equilibrium, above it non-equilibrium real-gas behaviour. These statements are only of approximate validity, in reality flight vehicle form and size play a major role.

2.2 Atmospheric Properties and Models

The earth atmosphere is a gas consisting (if dry) of molecular nitrogen (N_2, 78.084 volume per cent), molecular oxygen (O_2, 20.946 volume per cent),

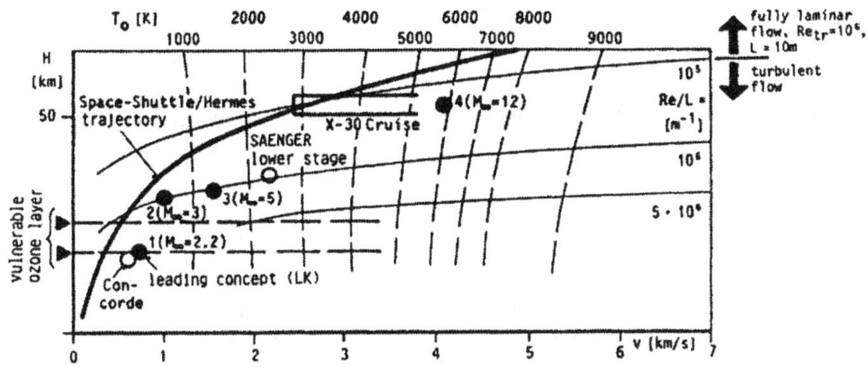

Fig. 2.3. Hypersonic flight-vehicle concepts in the velocity-altitude map with flight unit Reynolds numbers ($Re/L \equiv Re^u_\infty$), and equilibrium real-gas total temperatures T_o ($\equiv T_t$) [4] (based on [8]).

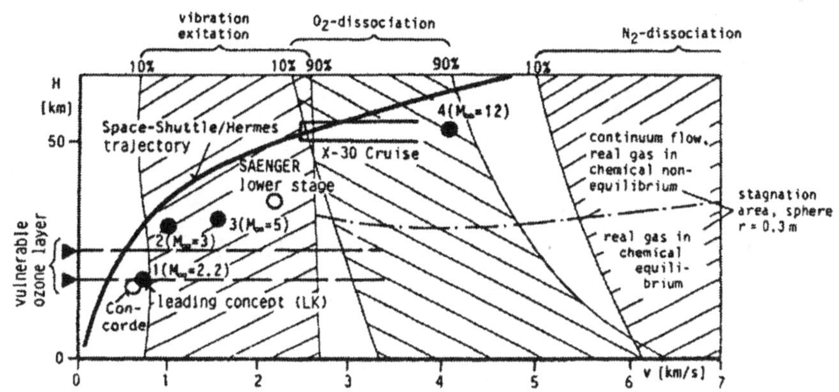

Fig. 2.4. Hypersonic flight-vehicle concepts in the velocity-altitude map with high-temperature real-gas effects [4] (based on [8]).

argon (Ar, 0.934 volume per cent), carbon dioxide (CO_2, 0.033 volume per cent), and some other spurious gases.

For our purposes we assume that the undisturbed air consists only of the molecules N_2 and O_2, with all three translational, and two rotational degrees of freedom of each molecule excited, Chapter 5. During hypersonic flight in the earth atmosphere this (model) air will be heated close to the flight vehicle due to compression and viscous effects. Consequently then first the (two) vibration degrees of freedom of the molecules become excited, and finally dissociation and recombination takes place. The gas is then a mixture of molecules and atoms. In aerothermodynamics at temperatures up to 8,000.0 K air can be considered as a mixture of the five species [9], [10]:

$$N_2,\ N,\ O_2,\ O,\ NO.$$

At high temperatures ionization may occur. Since it involves only little energy, which can be neglected in the overall flow-energy balance, we can disregard it in general. However if electromagnetic effects (radio-frequency transmission) are to be considered, we must take into account that additional species appear, see Fig. 2.5:

$$N_2^+,\ N^+,\ O_2^+,\ O^+,\ NO^+,\ e^-.$$

The air can, in general, be considered as a mixture of n thermally perfect gas species [9]. This holds for the undisturbed atmosphere, and for both the external and the internal flow path of hypersonic vehicles .

For such gases it holds:

- The pressure is the sum of the partial pressures p_i:

$$p = \sum_{i=1}^{n} p_i. \tag{2.1}$$

- The density is the sum of the partial densities ρ_i:

$$\rho = \sum_{i=1}^{n} \rho_i. \tag{2.2}$$

- The temperature T in equilibrium is the same for all species.
- The equation of state, with the universal gas constant R_0, is

$$p = \sum_{i=1}^{n} p_i = \sum_{i=1}^{n} \rho_i \frac{R_0}{M_i} T = \rho T \sum_{i=1}^{n} \omega_i \frac{R_0}{M_i} = \rho T R, \tag{2.3}$$

R being the gas constant of the mixture. The mass fraction ω_i of the species i is:

$$\omega_i = \frac{\rho_i}{\rho}, \tag{2.4}$$

and the mean molecular weight M:

$$\frac{1}{M} = \sum_{i=1}^{n} \frac{\omega_i}{M_i}. \tag{2.5}$$

- The mole fraction x_i is:

$$x_i = \frac{c_i}{c}, \tag{2.6}$$

with the molar concentration c_i

$$c_i = \frac{\rho_i}{M_i}, \tag{2.7}$$

which has the dimension $kg-mole/m^3$. The molar density of a gas mixture then is defined by:

$$c = \sum_{i=1}^{n} c_i. \tag{2.8}$$

The mean molecular weight of the undisturbed binary model air at sea level, as well as the two mass fractions, can be determined with the above eqs. (2.1) to (2.5). With the data from Table 2.1, and the molecular weights, Section 13.1, of molecular nitrogen (M_{N_2} =28.02 kg/kg-$mole$), and molecular oxygen (M_{O_2} = 32 kg/kg-$mole$), we obtain ω_{N_2} = 0.73784, ω_{O_2} = 0.26216, and M_{air} = 28.9644 kg/kg-$mole$. The latter being in perfect agreement with the value in Table 2.1.

For this model air we can determine the composition in volume per cent (volume fraction r_i), which is slightly different from that previously quoted, since the spurious gases are now neglected. Defining the fractional density ρ_i^* by:

$$\rho_i^* = \rho \frac{M_i}{M}, \tag{2.9}$$

the volume fractions of nitrogen and oxygen are given by

$$r_{N_2} = \frac{\omega_{N_2} \rho_{O_2}^*}{\omega_{N_2} \rho_{O_2}^* + \omega_{O_2} \rho_{N_2}^*}, \tag{2.10}$$

and

$$r_{O_2} = \frac{\omega_{O_2} \rho_{N_2}^*}{\omega_{N_2} \rho_{O_2}^* + \omega_{O_2} \rho_{N_2}^*}. \tag{2.11}$$

The volume fractions are finally r_{N_2} = 0.7627 (76.27 volume per cent), and r_{O_2} = 0.2373 (23.73 volume per cent).

The actual "equilibrium" composition of a mixture of thermally perfect gases is always a function of the temperature and the density or the pressure. Fig. 2.5, [9], gives an example.

At large pressures and low temperatures van-der-Waals effects can occur. In flight they usually can be neglected, but in high-enthalpy hypersonic ground-simulation facilities, especially in high-pressure/high-Reynolds number facilities they can play a non-negligible role. In Section 5.1 we come back to the van-der-Waals effects.

2.3 Flow Regimes

As can be seen in Fig. 2.1, density and pressure are decreasing rapidly with rising altitude, while the temperature is more or less constant up to 100.0

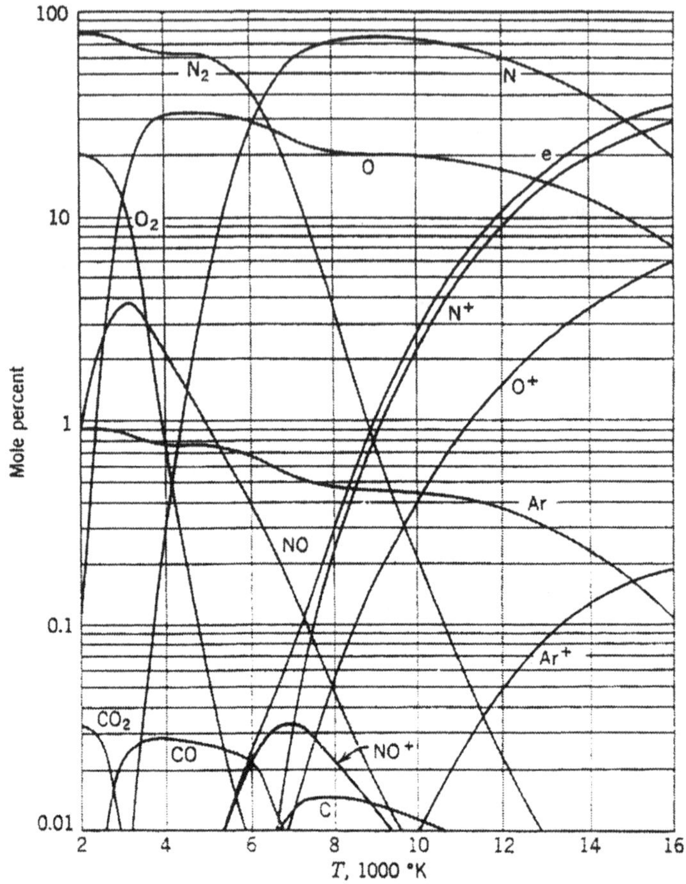

Fig. 2.5. Equilibrium composition of air at $p = 10^{-2}$ *atm* ($1.01325 \cdot 10^3$ *Pa*) as function of the temperature [9].

km altitude with $T \approx 200.0$ to $300.0\ K$. On the other hand the mean free path λ is increasing fast with altitude.

The mean free path is the average distance that a gas particle travels between successive collisions with other particles. Here λ is used for a qualitative characterization of the flow regimes, which we encounter in hypersonic flight, and hence the approximate relation [9]:

$$\lambda = \frac{16}{5}\frac{\mu}{\rho}\frac{1}{\sqrt{2\pi RT}}, \qquad (2.12)$$

can be used.

The ratio of the mean free path and a characteristic length L of a flow field is called the "overall" [11] Knudsen number Kn:

$$Kn = \frac{\lambda}{L}. \tag{2.13}$$

The characteristic length L can be a body length, a nose radius, a boundary-layer thickness, or the thickness of a shock wave, depending on the problem at hand.

Take for instance the thickness δ of a laminar, incompressible flat-plate (Blasius) boundary layer [12]:

$$\frac{\delta}{x} = c\frac{1}{\sqrt{Re_{\infty,x}}}, \tag{2.14}$$

as characteristic length, with $Re_{\infty,x} = \rho_\infty u_\infty x/\mu_\infty$, then:

$$Kn = \frac{\lambda}{\delta} = \frac{\lambda}{cx}\sqrt{Re_{\infty,x}}. \tag{2.15}$$

The Knudsen number can also be expressed in terms of flight Mach number M_∞ and Reynolds number $Re_{\infty,L}$, L being for instance the vehicle length:

$$Kn = \frac{16}{5}\sqrt{\frac{\gamma}{2\pi}}\frac{M_\infty}{Re_{\infty,L}} = 1.28\sqrt{\gamma}\frac{M_\infty}{Re_{\infty,L}}. \tag{2.16}$$

For a laminar boundary layer like above then it holds:

$$Kn \sim \frac{M_\infty}{\sqrt{Re_{\infty,x}}}. \tag{2.17}$$

The Knudsen number is employed to distinguish approximately between flow regimes:

- continuum flow:

$$Kn \lessapprox 0.01, \tag{2.18}$$

- continuum flow with slip effects (slip flow and temperature jump at body surfaces):

$$0.01 \lessapprox Kn \lessapprox 0.1, \tag{2.19}$$

- disturbed free molecular flow (gas particles not only collide with the body surface, but also with each other):

$$0.1 \lessapprox Kn \lessapprox 10, \tag{2.20}$$

- free molecular flow (gas particles collide only with the body surface, Newton limit [13]):

$$10 \lessapprox Kn. \tag{2.21}$$

No sharp limits exist between the four flow regimes[2]. In Fig. 2.6 we show the Knudsen number $Kn_\infty = \lambda_\infty/L$ of some hypersonic vehicles and some

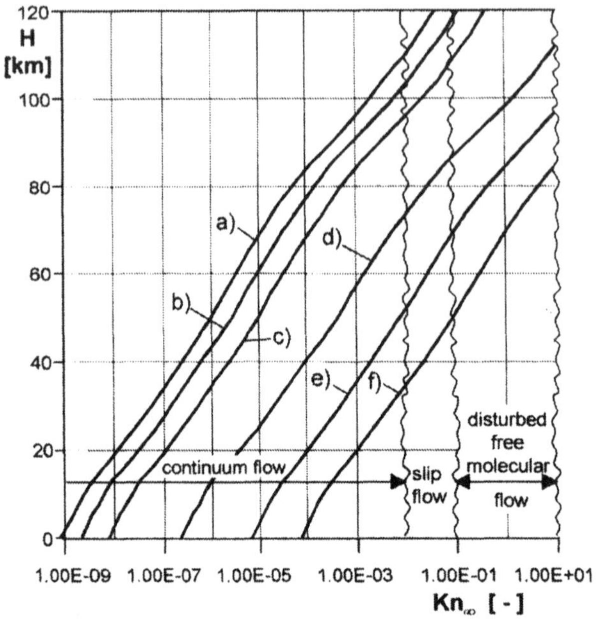

Fig. 2.6. Knudsen numbers Kn_∞ as function of altitude H for **a**) the SÄNGER lower stage, $L \approx 80.0\ m$; **b**) the Space Shuttle, $L \approx 30.0\ m$; **c**) the X-38, $L \approx 8.0\ m$; **d**) a nose cone, (D =) $L \approx 0.3\ m$; **e**) a pitot gauge, (D=) $L \approx 0.01\ m$; **f**) a measurement orifice, (D =) $L \approx 0.001\ m$.

components, with the characteristic length L taken as vehicle length or component diameter.

Globally the lower stage of SÄNGER (case a), with a maximum flight altitude of approximately 35.0 km [14], the Space Shuttle (case b), and also the X-38 (case c), remain fully in the continuum regime in the interesting altitude range below approximately 100.0 km. For the nose cone (case d) slip-flow effects can be expected above 80.0 km altitude. The pitot gauge (case e) is in the continuum regime only up to approximately 50.0 km because of its small diameter, and likewise the measurement orifice (case f) only up to 40.0 km altitude.

Regarding the aerothermodynamic simulation means, which will be discussed shortly in Chapter 10, we note that ground-test facilities are available for all flow regimes. However, in almost no case is a full simultaneous simulation of all relevant parameters possible for hypersonic flows.

Computational simulation for the continuum and for the slip-flow regime can be made with the classical tools, to which now the methods of numerical aerothermodynamics for the solution of the Euler equations, the Navier-

[2] In [11] it is shown, that with a "locally" defined Knudsen number more precise statements can be made.

Stokes equations, and the derivatives of the latter (boundary-layer methods, viscous shock-layer methods et cetera) are counted[3].

It is interesting to note that the Newton theory [13], which is exact for free molecular flow, can be used as inviscid computational tool in the continuum regime with sufficient accuracy down to $M \approx 2$ to 4, Section 6.7.

A major question that arises, is how shock waves, which are only a few mean-free paths thick, are treated in discrete numerical methods for the continuum regime. Shock waves, which as typical compressibility effects appear from transonic free-stream Mach numbers upwards, are found in the flow past all supersonic and hypersonic flight vehicles operating in this regime. We come back to this problem in Sub-Section 6.3.3.

2.4 Problems

Problem 2.1 During re-entry of a RV-type vehicle the density at 70.0 km altitude is 10 per cent lower than assumed. The actual drag hence is 10 per cent smaller than the nominal drag ($D = 0.5\rho_\infty v_\infty^2 C_D(\alpha) A_{ref}$). The vehicle flies with an angle of attack $\alpha = 45°$. We assume a constant velocity v_∞, the reference area A is constant anyway. How must $C_{D_{nom}}$ approximately be changed in order to recover the nominal drag?

Solution: Make a Taylor expansion around the actual drag coefficient and keep the linear term only:

$$\rho_{nom} C_{D_{nom}} = \rho_{actual} C_{D_{corr}} = \rho_{actual}\left(C_{D_{nom}} + \Delta\alpha \frac{dC_{D_{nom}}}{d\alpha} + O(\Delta\alpha^2)\right).$$

Assume that the RV-type vehicle can be approximated by a flat plate (the RHPM-flyer, Chapter 11), but determine the drag with the help of Newton's theory, Section 6.7:

$$C_D = 2\sin^3\alpha,$$

and find finally the result $\Delta\hat{\alpha} = 0.037$, respectively $\Delta\alpha = 2.12°$. Check the result with the corrected angle of attack $\alpha + \Delta\alpha$ and the Newton relation.

Problem 2.2 Take from Table 2.1 the density for a) 30.0 km and b) 80.0 km altitude. Compute for both cases the partial densities ρ_{N_2} and ρ_{O_2} and determine the partial pressures p_{N_2} and p_{O_2}. Compare the results with the density and pressure data in Table 2.1.

[3] It is noted already here, that flow-physics models (laminar-turbulent transition, turbulence, turbulent separation), which are required in viscous flow simulations for altitudes below approximately 50.0 km, have large deficits. This partly holds also for thermo-chemical models in the whole flight regime considered in this book.

Solution: Remember the mass fractions of the undisturbed air and find eventually: a) $p_{N_2} = 883.448\ PA$, $p_{O_2} = 313.919\ Pa$, b) $p_{N_2} = 0.76429\ PA$, $p_{O_2} = 0.27156\ Pa$.

Problem 2.3 The Knudsen number of a probe with diameter $D = 0.02\ m$ at 70.0 km altitude is $Kn = 0.20585$. It is to be tested in a ground-simulation facility. Pressure and temperature in the test section of the facility are $p = 50.0\ Pa$ and $T = 100.0\ K$. What is the mean free path in the test section? What probe diameter is needed in the facility in order to have the same Knudsen number as in the flight situation?

Solution: Determine the density in the test section, take the simple approximation for the viscosity of air, Section 4.2, $\mu = 0.702{\cdot}10^{-7}T$, find $\lambda_{facility} = 0.00304\ m$, and $D_{facility} = 0.0147\ m$.

References

1. NN. "Guide to Reference and Standard Atmosphere Models". ANSI/AIAA G-003A-1996, 1997.
2. NN. "U. S. Standard Atmosphere". Government Printing Office, Washington D. C., 1976.
3. H. KUCZERA. "Bemannte europäische Rückkehrsysteme - ein Überblick". Space Course 1995, Vol. 1, University Stuttgart, 1995, pp. 161 - 172.
4. E. H. HIRSCHEL. "Aerothermodynamic Phenomena and the Design of Atmospheric Hypersonic Airplanes". *J. J. Bertin, J. Periaux, J. Ballmann (eds.), Advances in Hypersonics, Vol. 1, Defining the Hypersonic Environment.* Birkhäuser, Boston, 1992, pp. 1 - 39.
5. E. H. HIRSCHEL, H. G. HORNUNG, J. MERTENS, H. OERTEL, W. SCHMIDT. "Aerothermodynamik von Überschallflugzeugen (Aerothermodynamics of Supersonic Aircraft)". MBB/LKE122/HYPAC/1/A, 1987.
6. D. E. KOELLE, H. KUCZERA. "SÄNGER - An Advanced Launcher System for Europe". IAF-Paper 88-207, 1988.
7. R. M. WILLIAMS. "National Aerospace Plane: Technology for America´s Future". Aerospace America, Vol. 24, No. 11, 1986, pp. 18 - 22.
8. G. KOPPENWALLNER. "Fundamentals of Hypersonics: Aerodynamics and Heat Transfer". VKI Short Course *Hypersonic Aerothermodynamics.* Von Kármán Institute for Fluid Dynamics, Rhode-Saint-Genese, Belgium, LS 1984-01, 1984.
9. W. G. VINCENTI, C. H. KRUGER. "Introduction to Physical Gas Dynamics". John Wiley & Sons, New York/London/Sydney, 1965. Reprint edition, Krieger Publishing Comp., Melbourne, Fl., 1975
10. R. B. BIRD, W. E. STEWART, E. N. LIGHTFOOT. "Transport Phenomena". John Wiley & Sons, New York, 2nd edition, 2002 .
11. J. N. MOSS. "Computation of Flow Fields for Hypersonic Flight at High Altitudes". *J. J. Bertin, J. Periaux, J. Ballmann (eds.), Advances in Hypersonics, Vol. 3, Computing Hypersonic Flows.* Birkhäuser, Boston, 1992, pp. 371 - 427.

12. H. SCHLICHTING. "Boundary Layer Theory". 7th edition, McGraw-Hill, New York, 1979.

13. W. D. HAYES, R. F. PROBSTEIN. "Hypersonic Flow Theory, Volume 1, Inviscid Flows". Academic Press, New York/London, 1966.

14. S. WEINGARTNER. "SÄNGER - The Reference Concept of the German Hypersonics Technology Programme". AIAA-Paper 93-5161, 1993.

3 The Thermal State of the Surface

The "thermal state of a surface" was introduced in Section 1.4. It was shown that a distinction is appropriate between "thermal loads" and "thermal surface effects". In this chapter we discuss in detail the thermal state of a vehicle surface, which can be an external or an internal surface, past which a high-speed flow happens. External surfaces of hypersonic flight vehicles usually are radiation cooled. Basics of surface thermal radiation cooling are treated in Section 3.2. With simple approximations we try to obtain a basic understanding of the aerothermodynamics of radiation-cooled surfaces. Results of computations with numerical methods are used to illustrate the findings. A case study, Section 3.3, gives a detailed account. Thermo-chemical and especially viscous-flow thermal-surface effects as such are treated in Chapters 5, 7, 8, and 9. Thermal loads on structures are not a central topic of this book.

3.1 Definitions

The thermal state of a surface is governed by at least one temperature (T), *and* at least one heat flux q (heat transported through an unit area per unit time). Hence both a surface temperature *and* a temperature gradient, respectively a heat flux, must be given to define it. Large heat fluxes can be present at low surface temperature levels, and vice versa. If a surface is radiation cooled, at least three different heat fluxes at the surface must be distinguished. In the slip-flow regime, which basically belongs to the continuum regime, two temperatures must be distinguished at the surface, Sections 4.3.2 and 9.4.

The heat transported per unit area and unit time (heat flux) towards a flight vehicle is:

$$\overline{q}_{\infty} = \rho_{\infty} v_{\infty} h_t, \tag{3.1}$$

with ρ_{∞} and v_{∞} the free-stream density and speed (their product is the mass flux per unit area towards the flight vehicle), and h_t the total enthalpy (per unit mass) of the free stream, i. e., of the undisturbed atmosphere. It is composed of the enthalpy of the undisturbed atmosphere h_{∞} and the kinetic energy $v_{\infty}^2/2$ of the flow relative to the flight vehicle:

$$h_t = h_\infty + \frac{v_\infty^2}{2}. \tag{3.2}$$

At hypersonic speed the kinetic energy is dominant, and hence the total enthalpy is more or less proportional to the flight velocity squared.

A considerable part of the heat transported towards the vehicle is finally transported by diffusion mechanisms towards the vehicle surface. This part is locally the heat flux, which we call the "heat flux in the gas at the wall", q_{gw} (this heat flux usually is dubbed somewhat misleading the "convective" heat flux). It is directed towards the wall. However, q_{gw} can also be directed away from the wall.

Literature is referring usually only to one heat flux $q_w \equiv q_{gw}$, neglecting the fact that radiation cooling may be present. We distinguish them and use q_w to denominate the heat flux into the wall.

A dimensionless form of the heat flux q_{gw} is the Stanton number:

$$St = \frac{q_{gw}}{\bar{q}_\infty}, \tag{3.3}$$

which simplifies for high flight speeds to:

$$St = \frac{2q_{gw}}{\rho_\infty v_\infty^3}. \tag{3.4}$$

Other forms of the Stanton number are used, for instance:

$$St = \frac{q_{gw}}{\rho_\infty v_\infty (h_r - h_w)}, \tag{3.5}$$

with h_r and h_w being the enthalpies related to the recovery temperature (r) and the actual wall (w) temperature.

For the following discussion first a wall with finite thickness and finite heat capacity is assumed, which is completely insulated from the surroundings, except at the surface, where it is exposed to the (viscous) flow. Without radiation cooling, the wall material will be heated up by the flow, depending on the heat amount penetrating the surface and the heat capacity of the material. The surface temperature will always be that of the gas at the wall: $T_w \equiv T_{gw}$, apart from a possible temperature jump, which can be present in the slip-flow regime. If enough heat has entered the wall material (function of time), the temperature in the entire wall and at the surface will reach an upper limit, the recovery temperature $T_w = T_r$ (the heat flux goes to zero). The surface is then called an adiabatic surface: no exchange of heat takes place between gas and wall material. With steady flow conditions the recovery (adiabatic) temperature T_r is somewhat smaller than the total temperature T_t, but always of the same order of magnitude. It serves as a conservative temperature estimate for the consideration of thermal-surface effects, and of thermal loads.

It was mentioned above that the total enthalpy at hypersonic flight is proportional to the flight velocity squared. This holds also for the total temperature, if perfect gas behaviour (Chapter 5) can be assumed, which is permitted for $v_\infty \lessapprox 1.0\ km/s$. The total temperature T_t then is only a function of the total enthalpy h_t, eq. (3.2), which can be expressed as function of the flight Mach number $M_\infty = v_\infty/a_\infty$, a_∞ being the speed of sound in the undisturbed atmosphere:

$$T_t = T_\infty \left(1 + \frac{\gamma - 1}{2} M_\infty^2\right) . \tag{3.6}$$

The recovery temperature T_r can be estimated with the flat-plate relation [1]:

$$T_r = T_\infty \left(1 + r \frac{\gamma - 1}{2} M_\infty^2\right) , \tag{3.7}$$

with the recovery factor r, and, like in eq. (3.6), the ratio of specific heats γ, and the free-stream Mach number M_∞. For laminar boundary layers we have $r = \sqrt{Pr}$, Pr being the Prandtl number. For turbulent boundary layers $r = \sqrt[3]{Pr}$ can be taken [1].

Eq. (3.7), like eq. (3.6), can also be formulated in terms of local parameters from the edge $'e'$ of the boundary layer at a body:

$$T_r = T_e \left(1 + r \frac{\gamma - 1}{2} M_e^2\right) . \tag{3.8}$$

Eqs. (3.7) and (3.8) suggest a constant recovery temperature on the surface of a configuration. Actually in general the recovery temperature is weakly varying and these equations can serve only to establish its order of magnitude.

At flight velocities larger than approximately 1.0 km/s they lose their validity, since high-temperature real-gas effects appear, Chapter 5. The temperature in thermal and chemical equilibrium becomes a function of two variables, for instance the enthalpy and the density. At velocities larger than approximately 5.0 km/s, non-equilibrium effects can play a role, complicating even more these relations. Since the above relations are of approximate character, actual local recovery temperatures will have to be obtained by numerical solutions of the governing equations of the respective flow fields.

If surface-radiation cooling is employed, the situation changes completely in so far, as a, usually large, fraction of the heat flux (q_{gw}) coming to the surface is radiated away from it (q_{rad}). For the case considered above, but with radiation cooling, the "radiation-adiabatic temperature" T_{ra} will result: no heat is exchanged between gas and material, but the surface radiates heat away[1]. With steady flow conditions and a steady heat flux q_w into the wall, T_{ra} also is a conservative estimate of the surface temperature. Depending on

[1] This situation is also called radiation equilibrium.

the employed structure and materials concept (either a cold primary structure with a thermal protection system (TPS), or a hot primary structure), and on the flight trajectory segment, the actual wall temperature during flight may be somewhat lower, but will be in any case near to the radiation-adiabatic temperature, Sub-Section 3.2.2.

Interesting is the low-speed case after high-speed flight, where T_w in general will be larger than the momentary recovery temperature T_r (thermal reversal), due to the thermal inertia of the TPS or the hot structure. In [2] it is reported that the lift to drag ratio of the Space Shuttle in the supersonic and the subsonic regime was underestimated during design by up to 10 per cent. This may be at least partly attributable to the thermal reversal, which is supported by preliminary numerical investigations of Longo and Radespiel [3]. Turbulent viscous drag on the one hand depends strongly on the wall temperature. The higher the wall temperature, the lower is the turbulent skin friction, Sub-Section 7.2.3. On the other hand, the thickening of the boundary layer due to the presence of the hot surface may reduce the tile-gap induced drag, which is a surface roughness effect, Section 8.2.3.

Besides radiation cooling, which is a passive cooling means, active cooling, for instance of internal surfaces, but of course also of external ones, can be employed. In such cases a prescribed wall temperature is to be supported by a finite heat flux into the wall and towards the heat exchanger. Other cooling means are possible, see, e.g., [4].

In the following we summarize the above discussion. Neglecting tangential heat fluxes, a possible temperature jump[2], and assuming, as above, that the radiative transport of heat is directed away from the surface, we arrive at the situation shown in Fig. 3.1.

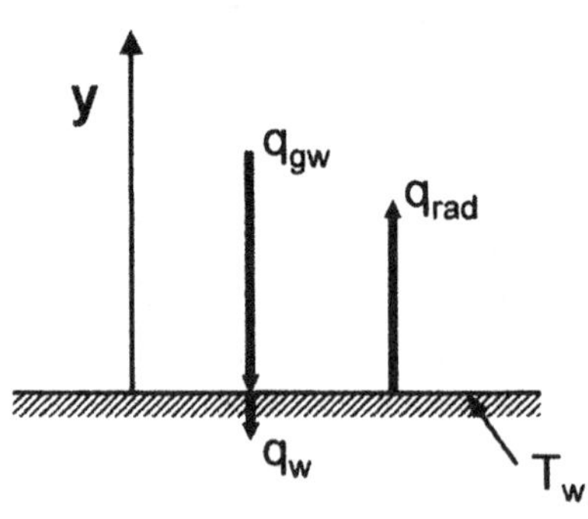

Fig. 3.1. Schematic description of the thermal state of the surface in the continuum regime, hence $T_{gw} \equiv T_w$. Tangential fluxes and non-convex radiation cooling effects are neglected, y is the surface-normal coordinate. q_{gw}: heat flux in the gas at the wall, q_w: heat flux into the wall, q_{rad}: surface radiation heat flux.

[2] We neglect here also non-convex effects, Sub-Section 3.2.4, and other possible external heat radiation sources (for instance the gas itself), see next section.

Five cases can be distinguished:

1. Radiation-adiabatic wall: $q_w = 0$, $q_{rad} = -q_{gw}$. The wall temperature is the radiation-adiabatic temperature: $T_w = T_{ra}$, which is a consequence of the flux balance $q_{gw} = -q_{rad}$.
2. The wall temperature T_w without radiation cooling ($q_{rad} = 0$) is prescribed (e. g. because of a material constraint), or it is simply given, like with a (cold) wind-tunnel model surface: the wall-heat flux is equal to the heat flux in the gas at the wall $q_w = q_{gw}$.
3. Adiabatic wall: $q_{gw} = q_w = 0$, $q_{rad} = 0$. The wall temperature is the recovery temperature: $T_w = T_r$.
4. The wall temperature T_w in presence of radiation cooling ($q_{rad} > 0$) is prescribed (e. g. because of a material constraint): the wall-heat flux q_w is the consequence of the balance of q_{gw} and q_{rad} at the prescribed T_w.
5. The wall-heat flux q_w is prescribed (e. g. in order to get available a certain amount of heat in a heat exchanger): the wall temperature T_w is a consequence of the balance of all three heat fluxes.

3.2 The Radiation-Adiabatic Surface

3.2.1 Introduction and Local Analysis

Surface-(thermal[3]-)radiation cooling is the basic cooling mode of high-speed vehicles operating in the earth atmosphere at speeds below 8.0 km/s. At this condition, typical of re-entry flight from low earth orbit, radiation emission and absorption processes of heat in the air-stream past the vehicle can be neglected.

Radiative heating by heat emission from the gas in the shock layer due to the high compression and the resulting rise of the gas temperature in that region potentially is the most important source of radiation towards the surface [5]. The effect is strongest in the stagnation point region of a body, which can have a rather large extension, if a RV-type flight vehicle at high angle of attack is considered. Here a large sub-sonic flow portion exists between the bow shock and the body surface.

In order to show that shock-layer radiation indeed can be neglected in the flight situations considered here, we compare the ordinary stagnation-point (s) heat flux $q_{gw,s}$ with the heat flux due to shock-layer radiation $q_{rad,s}$. We employ simple engineering formulas.

---

[3] The reader may note that we usually omit this word in this book, writing simply "radiation cooling".

For the stagnation-point heat flux a good approximation to the relation of Fay and Riddell, [6], for a sphere with radius R is given by Detra and Hidalgo, [7], (see also [8]):

$$q_{gw,s} = 11.03 \cdot 10^3 \frac{1}{\sqrt{R}} \left(\frac{\rho_\infty}{\rho_{SL}}\right)^{0.5} \left(\frac{v_\infty}{v_{co}}\right)^{3.15} . \tag{3.9}$$

With the nose radius R in m, the speed v_∞ in m/s, and the circular orbit velocity $v_{co} = 7.95 \cdot 10^3\ m/s$, we get $q_{gw,s}$ in W/cm^2.

For the estimation of the radiative heat flux from the shock layer to the body surface we use a formula from [9] for flight speeds $v_\infty \gtrapprox 6{,}000.0\ m/s$ with the same dimensions as in eq. (3.9):

$$q_{rad,s} = 7.9 \cdot 10^7\ R \left(\frac{\rho_\infty}{\rho_{SL}}\right)^{1.5} \left(\frac{v_\infty}{10^4}\right)^{12.5} . \tag{3.10}$$

It is important to note that this radiation heat flux, $q_{rad,s}$, to the blunt-nose surface is directly proportional to the nose radius R, while the heat flux in the gas at the wall, $q_{gw,s}$, is proportional to $1/\sqrt{R}$.

We compare now the two heat fluxes for a speed of $v_\infty = 8.0 \cdot 10^3\ m/s$ at 90.0 km altitude, where we have $\rho_\infty = 3.121 \cdot 10^{-6}\ kg/m^3$. With a nose radius of the body $R = 1.0\ m$ and the density at sea level $\rho_{SL} = 1.225\ kg/m^3$ we obtain:

$$\frac{q_{rad,s}}{q_{gw,s}} = \frac{0.0197 \frac{W}{cm^2}}{17.78 \frac{W}{cm^2}} \approx 0.001. \tag{3.11}$$

This shows that indeed shock layer radiation towards the body surface can be neglected at speeds below 8.0 km/s in the earth atmosphere below approximately 100.0 km altitude.

If, for whatever reason, radiation cooling is not sufficient, other (additional) cooling means, for instance transpiration cooling, or regenerative cooling et cetera, must be employed. Usually this results in extra weight, enlarged system complexity, or restricted re-usability of the vehicle.

Radiation cooling is very effective. Fig. 3.2 shows with data computed by Monnoyer, [10], [11], that radiation cooling reduces the wall temperature of a re-entry vehicle by such a degree that present-day TPS materials can cope with it without additional cooling. Without radiation cooling (emissivity $\varepsilon = 0$), the recovery temperature T_r near the stagnation point of the HERMES re-entry vehicle reaches around 6,000.0 K at approximately 70.0 km altitude[4]. With radiation cooling (emissivity $\varepsilon = 0.85$), the wall temperature $T_w = T_{ra}$ near the stagnation point is nearly 4,000.0 K lower than T_r, and remains below 2,000.0 K on the entire trajectory.

[4] Note that on a typical Space Shuttle trajectory the maximum thermal loads occur around 70.0 km altitude [12]. Indeed it has been shown exactly that the

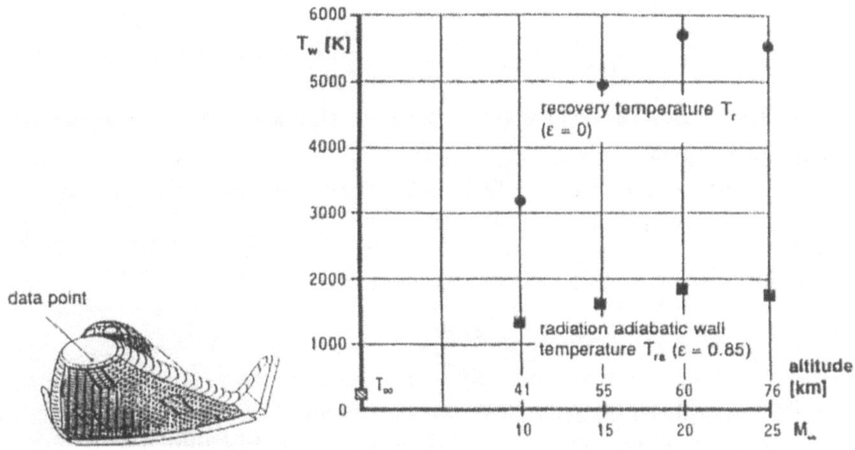

Fig. 3.2. Effect of radiation cooling at a location close to the stagnation point of HERMES ($x = 1.0$ m on the lower symmetry line, $\alpha = constant = 40°$), laminar flow, equilibrium real-gas model, at different trajectory points (M_∞, altitude H) [10] (coupled Euler/second order boundary-layer method [11]).

For the following simple analysis it is assumed that the continuum approach is valid (for a general introduction to energy transport by thermal radiation see, e. g., Eckert and Drake [14], Bird, Stewart and Lightfoot [15]). Slip effects as well as high-temperature real-gas effects are not regarded. Non-convex effects will be treated later, Sub-Section 3.2.4.

The basic assumption is that of a locally one-dimensional heat-transfer mechanism (see the situation given in Fig. 3.1). This implies the neglect of changes of the thermal state of the surface in directions tangential to the vehicle surface at the location under consideration. It implies in particular that heat radiation is directed locally normal to and away from the vehicle surface.

The general balance of the heat fluxes vectors is:

$$\underline{q}_w + \underline{q}_{gw} + \underline{q}_{rad} = 0. \tag{3.12}$$

Case 1 of the discussion at the end of the preceding sub-section is now the point of departure for our analysis. With the heat flux radiated away from the surface :

$$q_{rad} = \varepsilon \sigma T^4|_w, \tag{3.13}$$

where ε is the emissivity coefficient ($0 \leqq \varepsilon \leqq 1$), and σ the Stefan-Boltzmann constant, Section 13.1, we find:

maximum heat flux q_{gw} in the stagnation point occurs at approximately 73.0 km altitude [13].

$$q_w = q_{gw} + q_{rad} = 0 = -k_w \frac{\partial T}{\partial y}|_w + \varepsilon \sigma T^4|_w, \tag{3.14}$$

with k_w being the thermal conductivity at the wall, and y the direction normal to the surface.

A finite difference is introduced for the derivative of T, with Δ being a characteristic length normal to the surface of the boundary layer, i. e., a characteristic boundary-layer thickness, and T_r the recovery temperature of the problem. After re-arrangement we find:

$$T^4_{ra} \approx \frac{k_w}{\varepsilon \sigma} \frac{T_r}{\Delta}(1 - \frac{T_{ra}}{T_r}). \tag{3.15}$$

Since we wish to identify basic properties only in a qualitative way, we introduce for Δ simple proportionality relations for two-dimensional flow.

For laminar flow Δ is assumed to be the thickness δ_T of the thermal boundary layer, which we introduce with the lowest-order ansatz, flat plate (Blasius) boundary layer with reference-temperature extension [16] to compressible flow, eq. (7.100) in Sub-Section 7.2.1:

$$\frac{\delta_T}{x} \sim \frac{1}{Pe^{0.5}_{\infty,x}} \left(\frac{T^*}{T_\infty}\right)^{0.5(1+\omega_\mu)} = \frac{1}{(Pr_\infty Re_{\infty,x})^{0.5}} \left(\frac{T^*}{T_\infty}\right)^{0.5(1+\omega_\mu)}, \tag{3.16}$$

where $Pe_{\infty,x}, Pr_\infty, Re_{\infty,x}$ are the free-stream Peclét number, the Prandtl number and the Reynolds number, respectively. T^* is the reference temperature, eq. (7.70), and ω_μ the exponent of a viscosity power law, eq. (4.14) or eq. (4.15).

For turbulent flow the thickness of the viscous sub-layer is the relevant thickness, Sub-Section 7.2.1: $\Delta = \delta_{vs}$. We use here, however, the turbulent scaling thickness δ_{sc}, eq. (7.106), also with reference-temperature extension:

$$\frac{\delta_{sc,c}}{x} \sim \frac{1}{(Re_{\infty,x})^{0.8}} \left(\frac{T^*}{T_\infty}\right)^{0.8(1+\omega_\mu)}. \tag{3.17}$$

Eq. (3.16) introduced into eq. (3.15) yields for laminar flow:

$$T^4_{ra_{lam}} \sim \frac{k_w}{\varepsilon \sigma} Pr^{0.5}_\infty \left(\frac{T_\infty}{T^*}\right)^{0.5(1+\omega_\mu)} \frac{(Re_{\infty,L})^{0.5}}{(x/L)^{0.5}} \frac{1}{L} T_r (1 - \frac{T_{ra}}{T_r}), \tag{3.18}$$

with L the body length, and $Re_{\infty,L} = \rho_\infty v_\infty L/\mu_\infty$.

For turbulent flow eq. (3.17) is introduced into eq. (3.15), yielding:

$$T^4_{ra_{turb}} \sim \frac{k_w}{\varepsilon \sigma} \left(\frac{T_\infty}{T^*}\right)^{0.8(1+\omega_\mu)} \frac{(Re_{\infty,L})^{0.8}}{(x/L)^{0.2}} \frac{1}{L} T_r (1 - \frac{T_{ra}}{T_r}). \tag{3.19}$$

We omit now the proportionalities to constant or not strongly varying parameters, assume $T^* \approx T_{ra}$, and introduce from Section 4.2 the power-law relation for the thermal conductivity: $k \sim T_{ra}^{\omega_k}$,eq. (4.22). With $\omega_k = \omega_{k_2} = 0.75$, and $\omega_\mu = \omega_{\mu_2} = 0.65$, eq. (4.15), we see that for laminar flow k_w and $T_{ra}^{0.5(1+\omega_\mu)}$ cancel each other approximately.

The following equations are to be used to explain basic properties of radiation-cooled surfaces, but also to scale the radiation-adiabatic temperature, Sub-Section 3.2.5. Therefore we write for more generality the Reynolds number in terms of reference data: $Re_{ref,L} = \rho_{ref} u_{ref} L/\mu_{ref}$. For CAV-type flight vehicles (slender vehicles flying at low angles of attack) the free-stream data '∞' can be taken as reference data. For RV-type flight vehicles (blunt vehicles flying at large angle of attack), boundary-layer edge data would be the proper choice.

For laminar boundary-layer flow we obtain:

$$T_{ra_{lam}}^4 \sim \frac{1}{\varepsilon} \frac{(Re_{ref,L})^{0.5}}{(x/L)^{0.5}} \frac{1}{L} T_r \left(1 - \frac{T_{ra}}{T_r}\right), \tag{3.20}$$

and in terms of the unit Reynolds number $Re^u_{ref} = \rho_{ref} u_{ref}/\mu_{ref}$:

$$T_{ra_{lam}}^4 \sim \frac{1}{\varepsilon} \frac{(Re^u_{ref})^{0.5}}{(x/L)^{0.5}} \frac{1}{L^{0.5}} T_r \left(1 - \frac{T_{ra}}{T_r}\right). \tag{3.21}$$

If $T_{ra} \ll T_r$, the equations can be simplified further. We show this only for laminar flow, eq. (3.20):

$$T_{ra_{lam}}^4 \sim \frac{1}{\varepsilon} \frac{(Re_{ref,L})^{0.5}}{(x/L)^{0.5}} \frac{1}{L} T_r. \tag{3.22}$$

For turbulent boundary layers we obtain in a similar way, however by assuming for a convenient later generalization of the results in eq. (3.25), $k_w/(T^*)^{0.8(1+\omega_\mu)} \approx T_{ra}^{-0.6}$:

$$T_{ra_{turb}}^4 \sim \frac{1}{\varepsilon} T_{ra}^{-0.6} \frac{(Re_{ref,L})^{0.8}}{(x/L)^{0.2}} \frac{1}{L} T_r \left(1 - \frac{T_{ra}}{T_r}\right), \tag{3.23}$$

and

$$T_{ra_{turb}}^4 \sim \frac{1}{\varepsilon} T_{ra}^{-0.6} \frac{(Re^u_{ref})^{0.8}}{(x/L)^{0.2}} \frac{1}{L^{0.2}} T_r \left(1 - \frac{T_{ra}}{T_r}\right). \tag{3.24}$$

Of course also these equations can be further simplified, if $T_{ra} \ll T_r$.

Eqs. (3.20) and (3.23) are combined to yield:

$$T_{ra}^4 \sim \frac{1}{\varepsilon} T_{ra}^{2n-1} \frac{(Re_{ref,L})^{1-n}}{(x/L)^n} \frac{1}{L} T_r \left(1 - \frac{T_{ra}}{T_r}\right), \tag{3.25}$$

with $n = 0.5$ for laminar flow, and $n = 0.2$ for turbulent flow. This can be done for eqs. (3.21), (3.22), and (3.24), too.

It can be argued to employ in the above eqs. (3.20) to (3.25) the total temperature T_t instead of the recovery temperature T_r. However, only using the latter yields, in the frame of the chosen formulation, the expected limiting properties:

$$\begin{array}{lcl} \varepsilon \rightarrow 0 & : & T_{ra} \rightarrow T_r, \\ Re^u_{ref} \rightarrow 0 & : & T_{ra} \rightarrow 0, \\ Re^u_{ref} \rightarrow \infty & : & T_{ra} \rightarrow T_r, \\ L \rightarrow 0 & : & T_{ra} \rightarrow T_r, \\ \dfrac{x}{L} \rightarrow 0 & : & T_{ra} \rightarrow T_r, \\ \dfrac{x}{L} \rightarrow \infty & : & T_{ra} \rightarrow 0, \\ T_r \rightarrow 0 & : & T_{ra} \rightarrow 0. \end{array} \qquad (3.26)$$

With our simple approximations and proportionalities, eqs. (3.15) to (3.25), we find now the qualitative result that the radiation-adiabatic wall temperature T_{ra} on a flat surface approximately:

- is inversely proportional to the surface emissivity: $\varepsilon^{-0.25}$,
- is inversely proportional to the characteristic boundary-layer length $\Delta^{-0.25}$, with Δ being either the thickness of the thermal boundary-layer ($\delta_T \approx \delta_{flow}$), if the flow is laminar, or the turbulent scaling thickness (δ_{sc}), if the flow is turbulent,
- is proportional to the recovery temperature $T_r^{0.25}$, and with that in perfect-gas flow proportional to $M_\infty^{0.5}$,
- falls with laminar flow with increasing running length: $(x/L)^{-0.125}$, which is a result, that Lighthill gave in 1950 [17],
- falls with turbulent flow much less strong with increasing running length: $(x/L)^{-0.05}$,
- with laminar flow is proportional to $(Re^u_{ref})^{0.125}$ (note that the recovery temperature does not depend on the Reynolds number),
- with turbulent flow is proportional to $(Re^u_{ref})^{0.2}$, and hence is appreciably larger for turbulent flow than for laminar flow, because the turbulent scaling thickness δ_{sc} is smaller than the thermal boundary layer thickness δ_T, if the wall would be laminar at the same location, Sub-Section 7.2.1.

In contrast to the adiabatic wall the radiation-adiabatic wall is also a "cold wall", however with variable surface temperature, which depends on several flow parameters. It is interesting to compare the above qualitative results with those for the true cold-wall situation of a surface, for instance that of a cold-surface flight vehicle model tested in a ground-simulation facility. Experimental data show that the heat flux there falls also with some power of the boundary-layer running length, and the boundary-layer state, laminar or turbulent, also has a large influence [1].

Indeed the above discussed dependencies and trends are comparable for both. A relation similar to eq. (3.25) describes the situation at the cold wall with $n = 0.5$ for laminar and $n = 0.2$ for turbulent flow:

$$q_{w,cold} \approx \frac{k_w}{\Delta} T_r \left(1 - \frac{T_w}{T_r}\right) \sim k_w \left(\frac{T^*}{T_\infty}\right)^{(n-1)(1+\omega_\mu)} \frac{Re_L^{1-n}}{(x/L)^n} \frac{1}{L} T_r \left(1 - \frac{T_w}{T_r}\right). \tag{3.27}$$

The general behaviour is the same, however, with the important difference, that T_w is given, whereas in eq. (3.25) it is $T_w = T_{ra}$, which is the unknown.

As will be seen later from actual computations with numerical and approximate methods of two-dimensional flow and of not too strongly three-dimensional flow in this chapter, and also in Chapters 5 and 7, the above qualitative results for the radiation-adiabatic temperature are indeed well reproduced.

Approximate relations for the determination of T_{ra} for both laminar and turbulent flow at stagnation points (laminar), swept leading edges and flat configuration parts are given in Sub-Sections 7.2.6 and 7.2.5.

The radiation-adiabatic temperature cannot be determined directly from measured cold-surface heat fluxes or Stanton numbers. This is due to the highly non-linear coupling of it to the boundary-layer thickness, which itself depends on the wall temperature. With reasonable valid scaling laws (Sub-Section 3.2.5) an iterative procedure for attached flow possibly can be devised [18].

3.2.2 The Radiation-Adiabatic Surface and Reality

It must be noted that all results given in this book for radiation cooled surfaces cannot be verified in ground-test facilities, as is discussed in Section 10.3. Nevertheless, we can show at least partly their plausibility with Space Shuttle flight data [19], [20], and the results of a numerical study by Wüthrich, Sawley and Perruchoud [21], which are given in Fig. 3.3. The appropriateness of the radiation-adiabatic surface as approximation of reality is also supported by these results, at least for RV-type flight vehicles with a thermal protection system like the Space Shuttle has.

The computations with a coupled Euler/boundary-layer method were made with the laminar-turbulent transition locations taken from flight data, and a surface-emissivity coefficient $\varepsilon = 0.85$. Despite the scatter of the flight data in the transition region for $M_\infty = 7.74$, we see a rather good agreement

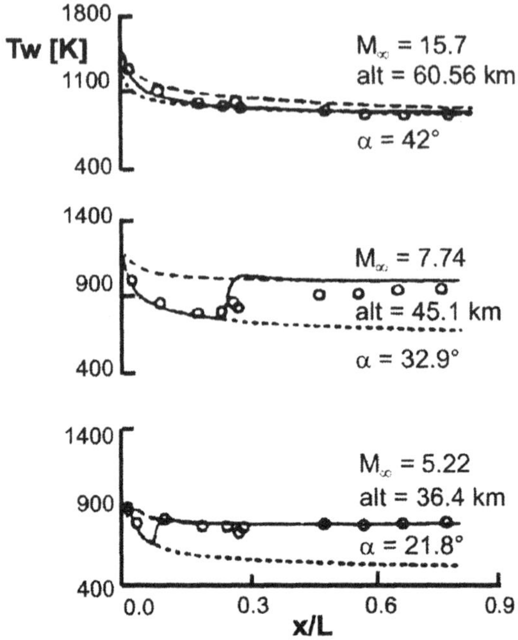

Fig. 3.3. The radiation-adiabatic temperature ($T_w = T_{ra}$) in the windward center line of the Space Shuttle [21]. Computed data: $M_\infty = 15.7$: - - - fully catalytic surface, ··· non-catalytic surface, — partially catalytic surface, $M_\infty = 7.74$ and $M_\infty = 5.22$: - - - fully turbulent, ··· fully laminar, — with transition. Space Shuttle flight data: ∘.

of computed and flight data. At the large Mach number the flow is fully laminar, surface catalytic recombination appears to be small. At the two lower Mach numbers surface catalytic recombination does not play a role, the flow is more or less in thermochemical equilibrium, Chapter 5.

We clearly see for the two smaller flight Mach numbers, that the radiation-adiabatic temperature T_{ra} drops much faster for laminar than for turbulent flow, and that indeed behind the transition location $T_{ra_{turb}}$ is appreciably larger than $T_{ra_{lam}}$. The level of T_{ra} depends distinctly on the flight Mach number, being highest for the largest Mach number. It should be remembered, that it also depends on the boundary-layer's unit edge Reynolds number, which depends on both the flight altitude and the angle of attack.

Not shown here are the results from [21] regarding the pressure and the skin-friction coefficient. The former is rather insensitive to both surface catalytic recombination and the state (laminar or turbulent) of the boundary layer. The latter at large Mach numbers reacts weakly on surface catalytic recombination, but at lower Mach numbers, i. e. lower altitudes, strongly on the state of the boundary layer, laminar or turbulent, like the radiation-adiabatic temperature.

Fig. 3.3 tells us also, with all caution, that the radiation-adiabatic surface obviously is a not too bad approximation of reality, at least for the type of thermal protection system employed on the Space Shuttle. The measured and the computed temperatures are very close to each other, hence the heat flux into or out of the wall q_w must be small.

It has not yet been quantified, how the thermal state of the surface evolves during an actual re-entry flight. It is to be surmised that a weak thermal reversal sets in after the peak heating in approximately 70.0 km altitude has occurred, page 34. The properties of the TPS (heat conduction (depending on temperature and pressure in case of the Space Shuttle [22], and heat capacity, in sum the thermal inertia of the TPS) will play a role. In any case, the thermal reversal appears to be small on a large part of the trajectory. It will become large only at the low-speed segment.

Less is known today regarding CAV-type flight vehicles. To get a feeling for the involved heat fluxes and temperatures there consider the example in Fig. 3.4, which, however, covers only a small and low Mach number flight span.

We look first at the M_∞-T_w surface at $q_w = 0$. The recovery temperature is smaller everywhere than the total temperature. Radiation cooling reduces the temperature at $M_\infty = 5.6$ by approximately 350.0 K compared to the recovery temperature. At smaller flight Mach numbers, and hence flight altitudes, the radiation cooling loses fast its effectiveness. This is due to the high unit Reynolds numbers, Fig. 2.3, which reduce the boundary layer thickness.

Look now at the q_w-T_w surface at $M_\infty = 5.6$. To sustain a wall temperature of, for instance, $T_w = 1{,}000.0\ K$, a heat flux into the wall of $q_w \approx 14.0\ kW/m^2$ would be necessary without radiation cooling. With radiation cooling the temperature would anyway only be $T_{ra} \approx 870.0\ K$. Assume now an actual heat flux into the wall of $q_w \approx 10.0\ kW/m^2$. This is approximately 27 per cent of the radiation heat flux q_{rad} in the case of the radiation-adiabatic surface. This in per cent not so small actual heat flux reduces the wall temperature compared to the radiation adiabatic temperature by $\Delta T \approx 70.0\ K$.

We have noted in that the radiation-adiabatic temperature under certain conditions is a conservative estimation of the wall temperature, and that the actual temperature in any case will lie close to it, $T_w \approx T_{ra}$. Coupled flow-structure analyses, for instance show, that this in general is true, see, e. g. [24], [25]. We have mentioned above the likelihood of a thermal reversal, and add that also with internal heat-transfer mechanisms, for instance transverse radiation cooling through the structure of a control surface to the lee side of it, [4], $T_w > T_{ra}$ can result. If for the quantification of a given design-critical thermal-surface effect T_{ra} is too crude an approximation, the true T_w must be determined with a multidisciplinary ansatz.

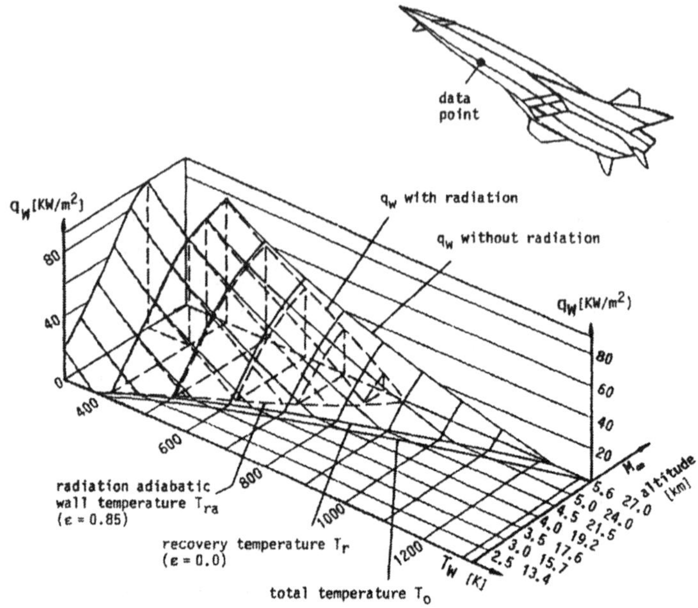

Fig. 3.4. Heat fluxes and temperatures at the lower symmetry line of a hypersonic flight vehicle at different trajectory points (flight Mach number M_∞ and altitude) [23]. The data point lies 5.0 m downstream of the vehicle nose. $\varepsilon = 0.85$, turbulent flow, approximate method. q_w is the heat flux into the wall, $T_0 \equiv T_t$.

3.2.3 Qualitative Behaviour of the Radiation-Adiabatic Temperature on Real Configurations

In the design of hypersonic flight vehicles the qualitative behaviour of the radiation-adiabatic temperature on more or less flat surface portions is not the only interesting aspect. The behaviour at the vehicle nose and at the leading edges is of particular interest, because on the one hand in the stagnation-point region and along attachment lines the boundary layer is very thin, and on the other hand nose and leading-edge radii, and leading-edge sweep govern the wave drag of a vehicle[5].

Considering only the proportionality $T_{ra}^4 \sim 1/\delta$, and that the boundary-layer thickness δ is inversely proportional to the square root of the flow acceleration ($\delta \sim 1/\sqrt{du_e/dx}$, x along the surface), Sub-Section 7.2.2, and this in

[5] This leads to the dilemma in the design of CAV-type flight vehicles that on the one hand nose and leading edge radii must be small in order to minimize the wave drag, while on the other hand they must be large enough to permit an effective surface radiation cooling. This dilemma does not exist for RV-type flight vehicles. They have on purpose large surface radii and fly at large angle of attack for an effective deceleration. The large radii at the same time also support very effectively surface radiation cooling.

turn is inversely proportional to the nose or leading-edge radius, Sub-Section 6.7.2, gives approximately the general trends shown in Table 3.1.

Table 3.1. Qualitative dependency of T_{ra}^4 on nose radius R and leading-edge sweep φ ($\varphi = 0$: non-swept case (cylinder)).

Item	δ	T_{ra}^4
Spherical nose (spherical stagnation point)	$\sim \sqrt{R}$	$\sim 1/\sqrt{R}$
Cylinder (two-dimensional stagnation point)	$\sim \sqrt{R}$	$\sim 1/\sqrt{R}$
Swept leading edge (laminar), swept cylinder	$\sim \sqrt{R}/\sqrt{\cos\varphi}$	$\sim \sqrt{\cos\varphi}/\sqrt{R}$

Not surprisingly the fourth power of the radiation-adiabatic temperature is, like the cold-wall heat flux, inversely proportional to the square root of the nose radius [6], and decreases at a leading edge with increasing sweep. Thick boundary layers, which occur at stagnation areas with large radii, and at leading edges with large sweep (and large radius), lead to large efficiency of surface radiation cooling. This is, with regard to thermal loads, the phenomenon, which leads to the blunt shapes of RV-type flight vehicles.

The result for swept leading edges needs a closer inspection. Actually it reflects the behaviour of the attachment-line boundary layer. Accordingly it must be generalized, because primary attachment lines move at slender configurations with increasing angle of attack away from the leading edge towards the lower (windward) side of the configuration[6]. Secondary and even tertiary attachment lines occur at the leeward side of a configuration at large angle of attack, together with separation lines.

What happens at such lines? At an attachment line, due to the diverging flow pattern, the boundary layer is thinner than that in the vicinity [26], Fig. 3.5. At a separation line, the flow has a converging pattern and hence the tendency is the other way around, Fig. 3.6.

Consequently the characteristic boundary-layer thickness Δ is reduced at attachment lines, and one has to expect there a rise of the radiation-adiabatic temperature compared to that in the vicinity. Indeed, as will be shown in Section 3.3, a hot-spot situation arises at attachment lines (attachment-line heating), whereas at separation lines the opposite happens, i. e. a cold-spot situation ensues. Attachment-line heating was observed during the first flights of the Space Shuttle at the orbital maneuvering system (OMS) pod, and at that time was dubbed "vortex scrubbing", probably because the respec-

[6] Due to the different characteristic angles of attack, Fig. 1.3, the primary (wing) attachment lines at a CAV-type flight vehicle (small α) lie indeed at the leading edge (see Figs. 7.7 and 7.8 in Section 7.3), but in contrast to that lie on RV-type vehicles (large α) on the windward side (see Figs. 3.16 and 3.17 in Section 3.3, and Figs. 9.4 and 9.5 in Section 9.1).

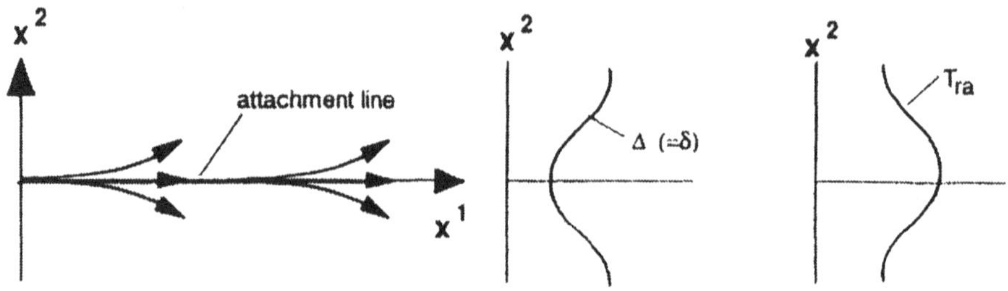

Fig. 3.5. Pattern of skin-friction lines at an attachment line, Δ-reduction, T_{ra}-increase, x^1 and x^2 are the surface-tangential coordinates (all schematically).

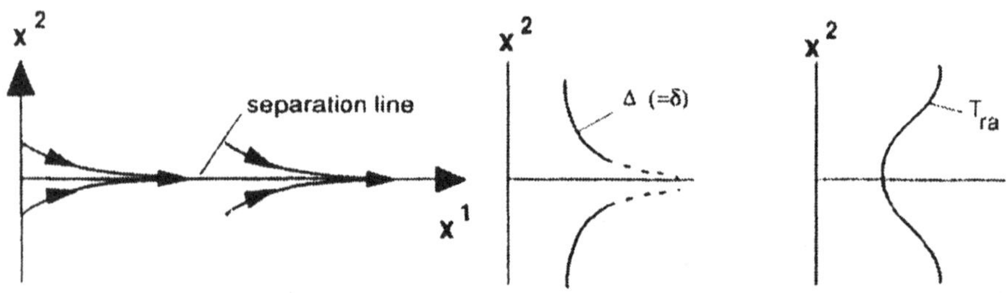

Fig. 3.6. Pattern of skin-friction lines at a separation line, Δ-increase, T_{ra}-decrease, x^1 and x^2 are the surface-tangential coordinates (all schematically).

tive attachment line was changing its location with flight attitude and speed [2]. The Δ-behaviour can be observed in three-dimensional boundary-layer calculations, [26], see also Sub-Section 7.2.1. The mathematical proof, however, is lengthy and is not given here, nor in [26]. Finally it is emphasized, that attachment-line heating, like attachment-line laminar-turbulent transition (see Section 8), is connected not only to leading edges, but to the topology of the skin-friction field of the entire flow-field under consideration.

3.2.4 Non-Convex Effects

Up to now radiation cooling on completely convex surfaces was considered. On real configurations surfaces may at least partly face at each other, which reduces the radiation-cooling effect: the surfaces receive radiation from each other and therefore the cooling effect is partly cancelled. We call this a "non-convex effect". Such effect happens at wing roots, fin roots, between fuselage and winglet et cetera. In the extreme, in an inlet, in a combustor, but also in a gap of a control surface no radiation cooling is possible at all. However, radiative energy transport may nevertheless play a role in such cases [14].

In the following the reduction of radiation cooling due to non-convex effects is treated after Höld and Fornasier [27], [28]. It is important to understand that the situation at radiation-cooled aerothermodynamic surfaces is different from that at surfaces of, for instance, satellites. Here the (characteristic) thicknesses of the boundary layers of the involved surfaces, which themselves depend to a large extent on their wall temperatures, introduce a strong non-linear coupling. This coupling does not exist for satellites. There the heat is simply exchanged without any couplings to flow properties. Nevertheless, the descriptive problems are the same (heat balances, sight lines between surface elements).

Consider the situation in Fig. 3.7.

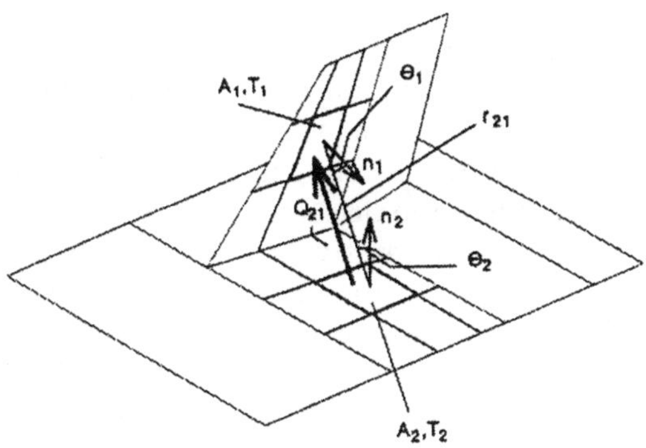

Fig. 3.7. Schematic of non-convex effects in radiation cooling at a generic fin configuration [27].

The rate of energy Q_{21} radiated from A_2 and acting on A_1 is [27]:

$$Q_{21} = \frac{\sigma\varepsilon}{\pi}\int\int\frac{T_2^4\cos\Theta_1\cos\Theta_2}{r_{21}^2}dA_1dA_2. \tag{3.28}$$

The surface element A_1 absorbs the heat flux:

$$q_{1,ab} = -\varepsilon\frac{Q_{21}}{A_1}. \tag{3.29}$$

Note that the absorption coefficient is equal to the emissivity coefficient, Kirchhoff's law, when radiation is in equilibrium with the surface [15]. This is the case with the radiation-adiabatic surface. The actual flight situation, with a finite, but generally small, heat flux q_w into the wall, can be considered as a quasi-equilibrium situation.

With the heat flux emitted from A_1

$$q_{1,em} = \varepsilon\sigma T_1^4, \tag{3.30}$$

a balance can be made for A_1:

$$\Delta q_1 = q_{1,em} + q_{1,ab} = \varepsilon\left(\sigma T_1^4 - \frac{Q_{21}}{A_1}\right). \tag{3.31}$$

If we treat this equation in analogy to the Stefan-Boltzmann law, a "fictitious" emissivity coefficient ε_f can be defined[7], which in a computation method simply replaces the surface emissivity coefficient ε in the radiation wall-boundary condition [27]:

$$\varepsilon_f = \frac{\Delta q_1}{\sigma T_1^4}. \tag{3.32}$$

This is the basic formulation of the problem. In [28] it is shown, that actually diffuse reflection must be regarded too in order to get results with high accuracy.

To apply the fictitious emissivity coefficient method with or without diffuse reflection, an influence matrix for the whole discretised configuration is to be computed, taking into account the sight lines between the individual surface elements. Because of the coupling of the radiation cooling effect to the boundary-layer thickness (see the discussion above), the fictitious emissivity coefficient must be determined iteratively.

This can easily be done during time-dependent integrations of the Navier-Stokes equations. With approximate methods or boundary-layer methods, special iteration approaches are necessary. The authors of [27] have devised a General Thermal Radiation (GETHRA) module, which can be incorporated into any computation scheme. It regards the mentioned diffuse reflection and also "third party" reflections, which may be important in half-cavity situations, like, for instance, the front part of an inlet.

The fictitious emissivity approach is valid for external flow-field applications and high surface-material emissivity coefficients, but can also be employed in computation methods for rocket nozzle flows, TPS gap flows, et cetera. It has the advantage that it gives a vivid picture of non-convex effects and not only the plain result. For internal radiation problems, for instance in combustors, less simplified methods, like the Poljak method et cetera (see e. g. [29]) should be applied.

In the following figures we show the computed (Navier-Stokes/RANS equations) distribution of the fictitious emissivity coefficient at a generic stabilizer configuration, Fig. 3.7, [28], and the temperature distribution at the afterbody of the X-38 with and without non-convex effects regarded in the computation method, [30]. The flight parameters are given in Table 3.2, the GETHRA module was employed for both cases.

The generic stabilizer configuration has a negative angle of attack in order to generate a "bow" shock, which does not interfere with the fins. The flow field was computed with a Navier-Stokes method with perfect-gas assumption. Fig. 3.8 shows that due to non-convex effects a reduction of the effective emissivity occurs especially in the fin-root regions down to $\varepsilon_f \approx 0.1$. Consequently the radiation-adiabatic temperature (not shown) rises in such

[7] Note that the original emissivity coefficient ε is a property only of the vehicle surface material.

Table 3.2. Flight parameters for the illustration of non-convex effects.

Case	M	H [km]	Re_L	L [m]	α [°]	ε	Boundary layer	Ref.
Generic config.	7	27.0	$4.3 \cdot 10^7$	7.0	-5.0	0.85	laminar	[28]
X-38	15	54.1	$1.52 \cdot 10^6$	8.41	40.0	0.80	turbulent	[30]

regions up to 1,500.0 K, compared to about 800.0 K ahead of the fins, where the non-convex effects diminish rapidly.

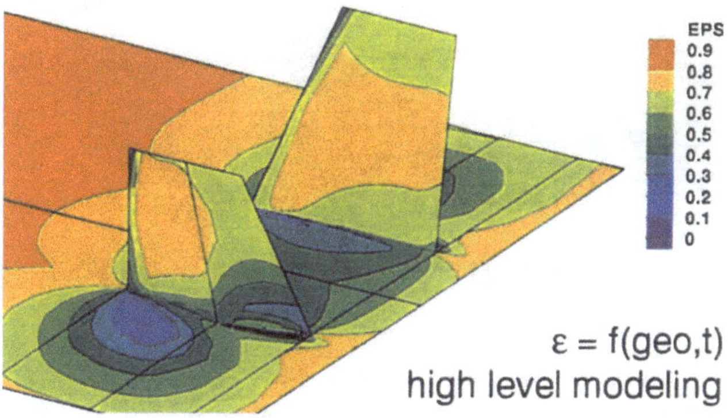

Fig. 3.8. Non-convex effects: distribution of the fictitious emissivity coefficient ε_f with high-level modeling at a generic stabilizer configuration [28].

At the X-38 we see a dramatic influence of non-convex effects in the half-cavity which is formed by the upper side of the body flap and the lower side of the fuselage, but also between the winglets and the fuselage, Fig. 3.9. In the half-cavity the fictitious emission coefficient goes down to $\varepsilon_f \approx 0.05$ (not shown). Similar low values are reached, although only in small patches, also at the winglets roots. The maximum temperatures in the half-cavity are around 1,200.0 K, if non-convex effects are regarded and only around 500.0 K, if they are not regarded. Between the winglets and the fuselage an average rise of approximately 100.0 K is observed, if non-convex effects are taken into account.

The forces and moments are not affected, which was to be expected because the X-38 as RV-type vehicle is compressibility-effects dominated, Table 1.1. The situation is quite different in this regard at a CAV-type vehicle, see Fig. 7.10 in Section 7.3.

These examples show that it is very important in hypersonic vehicle design to monitor, and to quantify if necessary, non-convex effects, at least in order to arrive at accurate predictions of thermal loads. However, because thermal-surface effects influence, for instance, the viscous drag, Section 7.1, and other

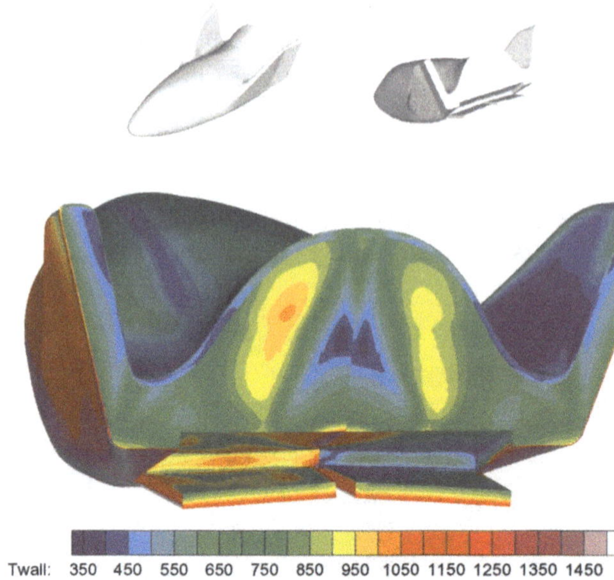

Fig. 3.9. Non-convex effects: distribution of the radiation-adiabatic temperature T_{ra} with high-level modeling at the afterbody of the X-38 [30]. Left side with, right side without non-convex effects taken into account.

wall-near flow phenomena, they must also be considered in view of vehicle aerodynamics, especially regarding CAV-type flight vehicles.

3.2.5 Scaling of the Radiation-Adiabatic Temperature

The radiation-adiabatic temperature is, in contrast to the recovery temperature, Reynolds number and scale dependent, Sub-Section 3.2.1. Therefore, data from different cases, even with the same total enthalpy, cannot directly be compared. The radiation-adiabatic wall temperature on a small body (wind-tunnel model) would be much larger than that on a large one (real configuration), if all flow parameters are the same. This makes a full simulation in an hypothetical ideal wind tunnel impossible, Section 10.3. This holds also for an experimental or demonstrator vehicle, which could be a scaled-down version of the reference concept.

If in design work an input about the distribution of the radiation-adiabatic temperature is quickly needed, a scaling of it from another, already computed case, can be made, provided that the flow topology qualitatively and quantitatively is the same in both cases (geometrical affinity, the same angles of attack and yaw), that it is not too strongly three-dimensional, and that (high-temperature) real-gas effects are similar. In addition, the assumption must be made that the recovery temperature T_r is constant over the entire vehicle surface (flat-plate behaviour), although different for laminar and turbulent flow portions. This assumption, however, may be violated to a certain extent in reality, but without posing a serious restriction for the use of the following scaling laws.

If the two different cases 1 and 2 are considered, we obtain with eq. (3.25):

$$\frac{T_{ra1}}{T_{ra2}} = \left[\frac{\varepsilon_2}{\varepsilon_1} \frac{T_{ra1}^{2n-1}}{T_{ra2}^{2n-1}} \frac{(Re_{ref,L})_1^{1-n}}{(Re_{ref,L})_2^{1-n}} \frac{L_2}{L_1} \frac{T_{r1}}{T_{r2}} \frac{(1 - T_{ra1}/T_{r1})}{(1 - T_{ra2}/T_{r2})} \right]^{\frac{1}{4}} , \tag{3.33}$$

which does not depend on x/L, i. e., it holds for every point on the surface, if T_r if sufficiently constant. Of course the respective portions on the vehicle surface with laminar flow (scaling with $n = 0.5$) and with turbulent flow (scaling with $n = 0.2$) must be the same in case 1 and 2. If the flow is fully laminar (RV-type flight vehicles above 40.0 to 60.0 km altitude) this is no problem. At flight of a vehicle of either vehicle class below that altitude, one must be aware that this should hold at least approximately. If that is not the case, one has to make sure that is does not curtail seriously the result.

If T_{ra} is small compared to T_r in both cases, eq. (3.33) can be reduced to

$$\frac{T_{ra1}}{T_{ra2}} = \left[\frac{\varepsilon_2}{\varepsilon_1} \frac{T_{ra1}^{2n-1}}{T_{ra2}^{2n-1}} \frac{(Re_{ref,L})_1^{1-n}}{(Re_{ref,L})_2^{1-n}} \frac{L_2}{L_1} \frac{T_{r1}}{T_{r2}} \right]^{\frac{1}{4}} . \tag{3.34}$$

Eq. (3.33) can be written in terms of the unit Reynolds number $Re^u = \rho_{ref} v_{ref} / \mu_{ref}$, for instance, for the employment with the RHPM-flyer, Chapter 11:

$$\frac{T_{ra1}}{T_{ra2}} = \left[\frac{\varepsilon_2}{\varepsilon_1} \frac{T_{ra1}^{2n-1}}{T_{ra2}^{2n-1}} \frac{(Re^u_{ref})_1^{1-n}}{(Re^u_{ref})_2^{1-n}} \frac{L_2^n}{L_1^n} \frac{T_{r1}}{T_{r2}} \frac{(1 - T_{ra1}/T_{r1})}{(1 - T_{ra2}/T_{r2})} \right]^{\frac{1}{4}} . \tag{3.35}$$

With the formulations given in Sub-Section 7.2.6 scaling relations can be derived also for stagnation point regions and attachment lines. All relations permit a wide range of parameters to be covered in the scaling process. All can be further reduced like above for unit Reynolds numbers et cetera.

The parameter variation range of a scaling law of this kind unfortunately cannot be firmly established from first principles. It should be established for each given case before application in design work.

Figs. 3.10 and 3.11 demonstrate scaling with a slightly modified equation (3.34) for the windward side of the HERMES configuration with laminar flow. Three cases were studied [31]. The numerical results were obtained with the coupled Euler/second-order boundary-layer method described in [11]. The flight conditions are given in Table 3.3.

Case 1 scales the radiation-adiabatic temperature of HERMES with the original length $L_a = 13.0$ m with that of a HERMES configuration enlarged

Table 3.3. Flight parameters for the scaling of the radiation-adiabatic temperature at HERMES [31].

M_∞	H $[km]$	T_∞ $[K]$	Re^u_∞ $[m^{-1}]$	α [°]	Boundary layer
25	75.0	205.3	$0.2265 \cdot 10^5$	30.0	laminar

linearly to $L_b = 34.6\ m$, the length of the Space Shuttle. In case 2 a scaling is made for two different emissivity coefficients $\varepsilon_a = 0.85$ and $\varepsilon_b = 1$ on the original HERMES configuration. In case 3 finally case 1 and case 2 are combined.

Fig. 3.10 shows at the windward side a rather good scaling with somewhat larger deviations in the nose region. There, of course, the simple flat-plate relation (3.25) is only approximately valid. The expected very good scaling for case 2 shows slight deviations in the nose region, which are probably due to the non-linear coupling between boundary-layer thickness and radiation-adiabatic temperature.

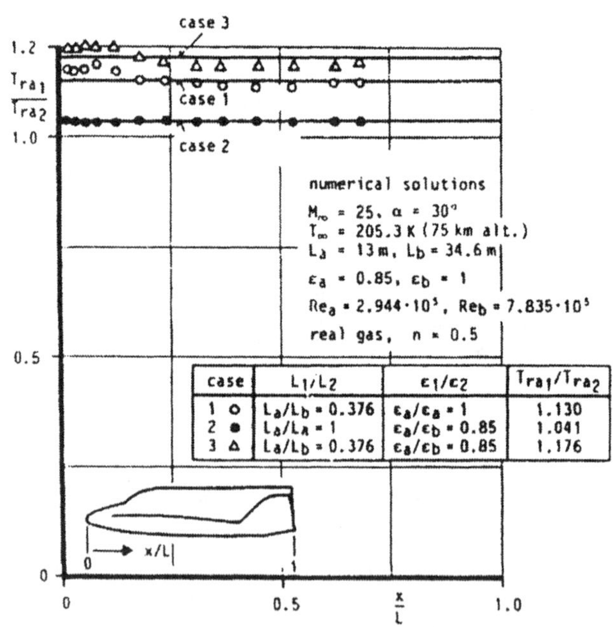

Fig. 3.10. Comparisons of T_{ra1}/T_{ra2} at the lower symmetry line of HERMES [31]. Symbols: numerical results, full lines: scaled results.

The scalings in Fig. 3.11 for the lower side cross-section at $x/L = 0.468$ show larger deviations in the vicinity of the leading edge. Here, at the attachment line, which lies well at the lower side of the configuration, the original assumption in the scaling laws of a flat-plate boundary layer introduces larger errors. At an attachment line the flow is highly three-dimensional, which will be shown in the case study in Section 3.3.

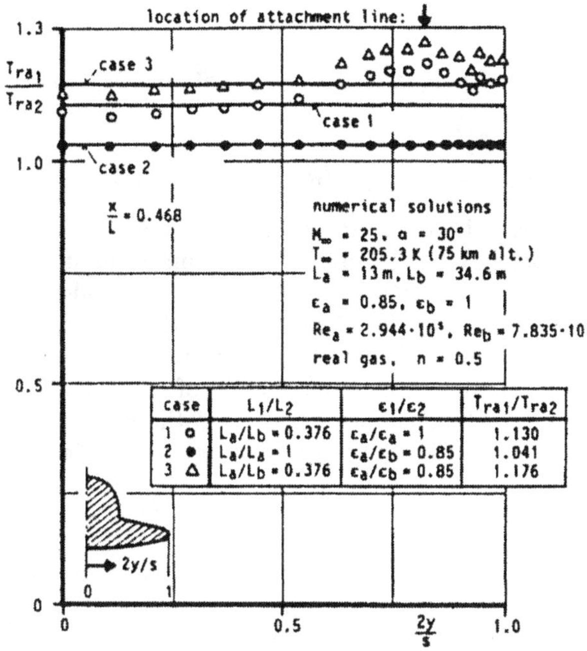

Fig. 3.11. Comparisons of T_{ra1}/T_{ra2} at the lower side cross-section $x/L = 0.468$ (half span) of HERMES [31]. Symbols: numerical results, full lines: scaled results.

3.2.6 Some Parametric Considerations of the Radiation-Adiabatic Temperature

We consider finally some general properties of the radiation-adiabatic temperature. This done in a very generic way. Neither configurational nor high-temperature real-gas effects are taken into account.

We take eq. (3.25), set for convenience $T_{ra}^{2n-1} = 1$, assume $T_{ra} \ll T_r$, and write it for T_{ra} in terms only of the emissivity coefficient ε, the unit Reynolds number Re^u, and the recovery temperature T_r:

$$T_{ra} = C' \, \varepsilon^{-0.25} (Re^u)^{0.25(1-n)} \, T_r^{0.25}, \tag{3.36}$$

with $n = 0.5$ for laminar flow, and $n = 0.2$ for turbulent flow.

Assuming further large flight Mach numbers M, we arrive at:

$$T_{ra} = C\,\varepsilon^{-0.25}\,(Re^u)^{0.25(1-n)}\,M^{0.5}. \tag{3.37}$$

The total differential of T_{ra} has the following terms:

$$\frac{\partial T_{ra}}{\partial \varepsilon} = -0.25\,C\,\varepsilon^{-1.25}\,(Re^u)^{0.25(1-n)}\,M^{0.5}, \tag{3.38}$$

$$\frac{\partial T_{ra}}{\partial Re^u} = C\,\varepsilon^{-0.25}\,0.25\,(1-n)\,(Re^u)^{-(0.75+0.25n)}\,M^{0.5}, \tag{3.39}$$

$$\frac{\partial T_{ra}}{\partial M} = C\,\varepsilon^{-0.25}\,(Re^u)^{0.25(1-n)}\,0.5\,M^{-0.5}. \tag{3.40}$$

For the discussion of these terms we generate a few quantitative data. We chose the trajectory points $H = 60.56\ km$ (point 1, laminar flow) and $H = 36.4\ km$ (point 2, turbulent flow) from Fig. 3.3 and consider the location $x/L = 0.5$ on the lower symmetry line, Table 3.4. For the two points the respective constants C are determined (note that they have a secondary influence on the slopes of the curves) and the functions T_{ra} of ε, Fig. 3.12, Re^u_∞, Fig. 3.13, and M_∞, Fig. 3.14 are found. They are valid in the vicinity of the two points (full lines in the figures). Away from the points they indicate only the trend (broken lines).

Table 3.4. Selected flight parameters (Fig. 3.3) of the Space Shuttle [19], [20].

Point	M_∞	$H\ [km]$	$Re^u_\infty\ [m^{-1}]$	$T_\infty\ [K]$	$T_{ra}\ [K]$ at $x/L = 0.5$	Boundary layer
1	15.7	60.56	$7.64 \cdot 10^4$	253.0	960.0	laminar
2	5.22	36.4	$7.99 \cdot 10^5$	222.0	660.0	turbulent

We get the following results:

- T_{ra} depends inversely on $\varepsilon^{0.25}$ and hence decreases with increasing ε, Fig. 3.12, for both laminar (high altitude) and turbulent (low altitude) flow. The gradient $\partial T_{ra}/\partial \varepsilon$ also decreases with increasing ε, eq. (3.38). At large ε we find the important result (be careful with the generalization of this result!) that T_{ra} becomes insensitive to changes of ε. At $\varepsilon = 0.85 \mp 0.05$, for instance, we find only $\Delta T_{ra} \approx \pm\ 10.0\ K$ at both trajectory points.
- The dependence of T_{ra} on Re^u_∞ is more complex. T_{ra} increases in any case strongly with increasing Re^u_∞, Fig. 3.13, and slightly stronger for turbulent than for laminar flow, reflecting the dependence of the respective characteristic boundary-layer thicknesses on the Reynolds number. The gradient $\partial T_{ra}/\partial Re^u$ decreases with increasing Re^u, eq. (3.39), stronger for laminar than for turbulent flow (note that in Fig. 3.13 not the true dependence on Re^u but on $Re^u(H)$ is given).

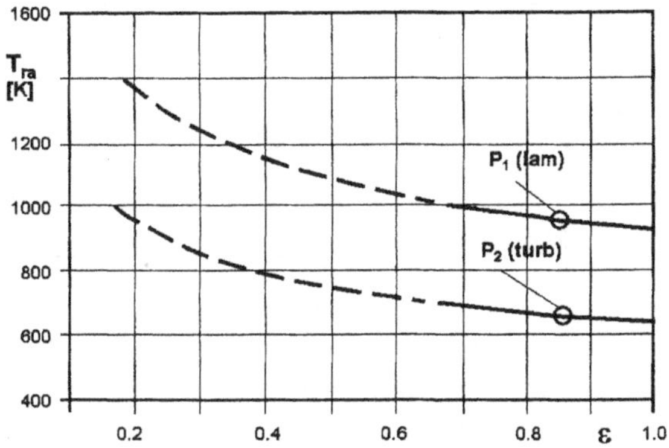

Fig. 3.12. Parametric considerations: T_{ra} at $x/L = 0.5$ on the lower symmetry line of the Space Shuttle as function of the emissivity coefficient ε around point 1 (laminar flow) and point 2 (turbulent flow) (see Table 3.4). Full lines: range of validity of the approximation, broken lines: trend extrapolation.

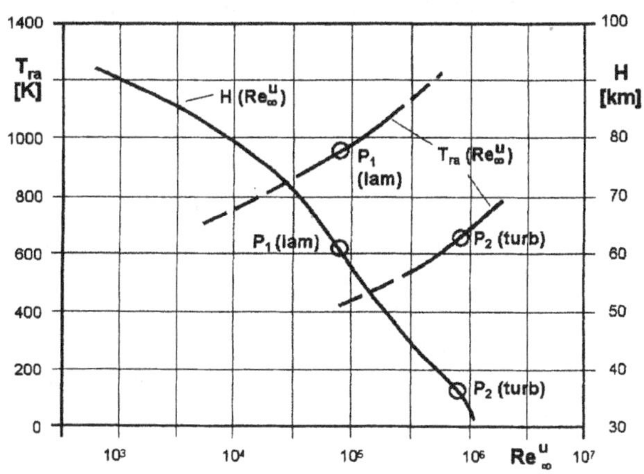

Fig. 3.13. Parametric considerations: T_{ra} at $x/L = 0.5$ on the lower symmetry line of the Space Shuttle as function of the unit Reynolds number Re^u_∞ around point 1 (laminar flow) and point 2 (turbulent flow) (see Table 3.4). Full lines: range of validity of the approximation, broken lines: trend extrapolation, $H(Re^u_\infty)$ is exact.

– With increasing M_∞ we get the expected growth of T_{ra}, Fig. 3.13. The gradient $\partial T_{ra}/\partial M$ decreases with increasing M, eq. (3.40). The different slopes of the curves are due to the different magnitudes of the constant C in point 1 and 2 (note that in Fig. 3.14 T_{ra} is given as function of $M(H)$ and not of M).

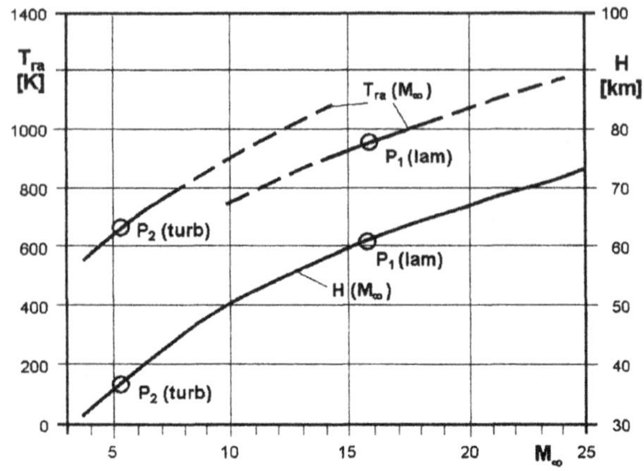

Fig. 3.14. Parametric considerations: T_{ra} at $x/L = 0.5$ on the lower symmetry line of the Space Shuttle as function of the Mach number M_∞ around point 1 (laminar flow) and point 2 (turbulent flow) (see Table 3.4). Full lines: range of validity of the approximation, broken lines: trend extrapolation, $H(M_\infty)$ is exact.

3.3 Case Study: Thermal State of the Surface of a Blunt Delta Wing

In this case study we check the qualitative results about the radiation-adiabatic temperature, which we found so far in this chapter, and also the scaling law for laminar flow, by investigating the Navier-Stokes data computed for a generic configuration, the Blunt Delta Wing. This configuration is a very strongly simplified RV-type vehicle configuration flying at moderate angle of attack with fully laminar flow. A similar check is made in Sub-Section 7.3 with data computed for a CAV-type vehicle flying at low angle of attack, with laminar and turbulent flow portions.

3.3.1 Configuration and Computation Cases

Riedelbauch [32] (see also [33]) studied the aerothermodynamic properties of hypersonic flow past the radiation-cooled surface of a generic configuration (Blunt Delta Wing (BDW) [34]). The configuration is a simple slender delta wing with blunt nose, Fig. 3.15. Navier-Stokes solutions for perfect gas were performed for flight with the conditions given in Table 3.5.

Laminar flow was assumed. A surface emissivity coefficient $\varepsilon = 0.85$ was chosen. The computations were made with an angle of attack $\alpha = 15°$ for two wing lengths $L = 14.0\ m$ and $L = 4.67\ m$.

Table 3.5. Flight parameters of the Blunt Delta Wing [32].

Case	M_∞	H [km]	T_∞ [K]	Re^u_∞ [m^{-1}]	L [m]	ε	α [°]
1	7.15	30.0	226.506	$2.69 \cdot 10^6$	14.0	0.85	15.0
2	7.15	30.0	226.506	$2.69 \cdot 10^6$	4.67	0.85	15.0

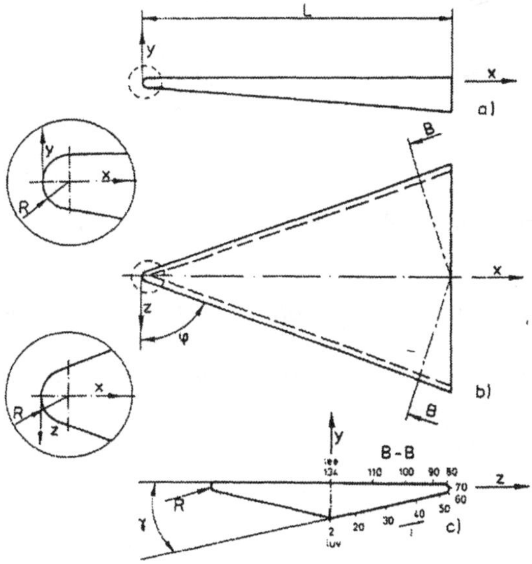

Fig. 3.15. Configuration of the BDW and the employed notation.

3.3.2 Topology of the Computed Skin-Friction and Velocity Fields

Before we discuss the computed radiation-adiabatic temperature fields, we need to have a look at the topology of the computed skin-friction field. We do this in order to identify especially attachment and separation lines, where we expect hot-spot, and cold-spot situations, respectively, Sub-Section 3.2.3. Discussed are the results for the large wing with $L = 14.0\ m$. The skin-friction line topology at the smaller wing is very similar.

Take first a look at the windward side of the configuration. In Fig. 3.16 we see the classical skin-friction line pattern on the lower side of a delta wing. Because it is not flat, the flow exhibits a slight three-dimensionality between the two primary attachment lines. These are marked by strongly divergent skin-friction lines. The stagnation point, which topologically is a singular point, in this case a nodal point, see, e. g., Peake and Tobak [35], [36], lies also on the lower side, at about 3 per cent body length. The primary attachment lines are almost from the beginning parallel to the leading edges, i. e., they don't show a conical pattern.

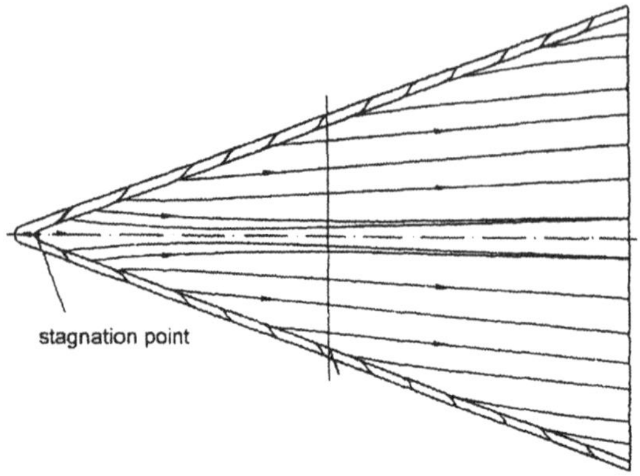

Fig. 3.16. Selected computed skin-friction lines at the lower side of the BDW [32].

The situation is quite different on the leeward side of the wing, Fig. 3.17. Here we see on the left-hand side (from the leading edge towards the symmetry line) a succession of separation and attachment lines: the primary separation line S_1, the secondary attachment line A_2, a secondary separation line S_2, and a tertiary attachment line A_3. All is mirrored on the right-hand side of the wing. Again a conical pattern is not discernible, except for a small portion near to the nose. However, the secondary separation lines are almost parallel to the single tertiary attachment line along the upper symmetry line of the wing.

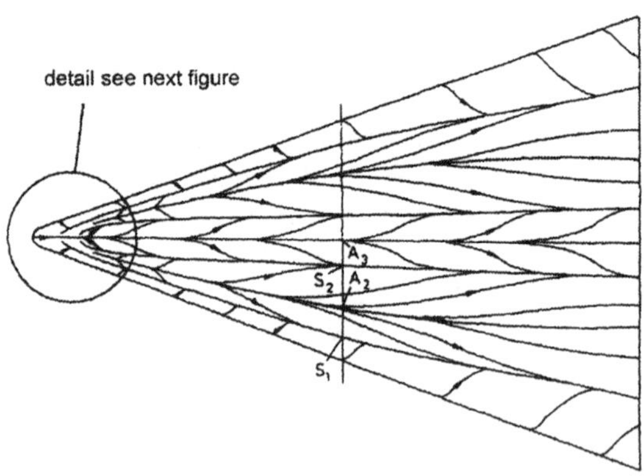

Fig. 3.17. Selected computed skin-friction lines at the upper side of the BDW [32].

Both the primary and the secondary separation are of the type "open separation", i. e. the separation line does not begin in a singular point on the surface, as was shown first by Wang [37]. Fig. 3.18 shows this for the primary separation line.

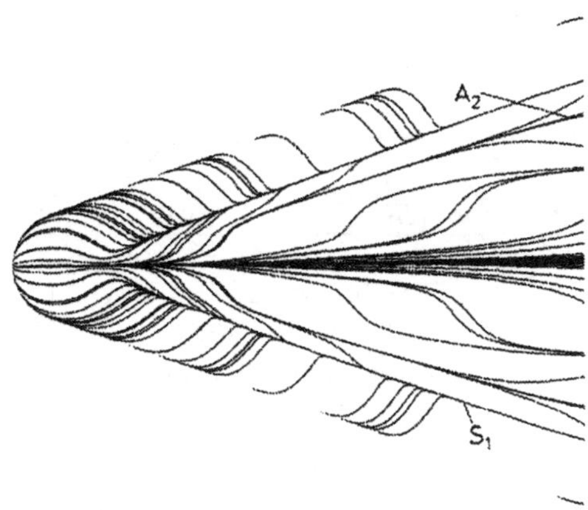

Fig. 3.18. Selected computed skin-friction lines at the upper side of the BDW near the nose (detail of Fig. 3.17) [32].

With these surface patterns we can attempt to construct qualitatively the structure of the leeward-side flow, Fig. 3.19. By marking the points, where the streamlines of the vortex-feeding layers penetrate a surface normal to the x-axis, one finds the Poincaré surface [38], Fig. 3.20. The computed cross-flow shocks are indicated.

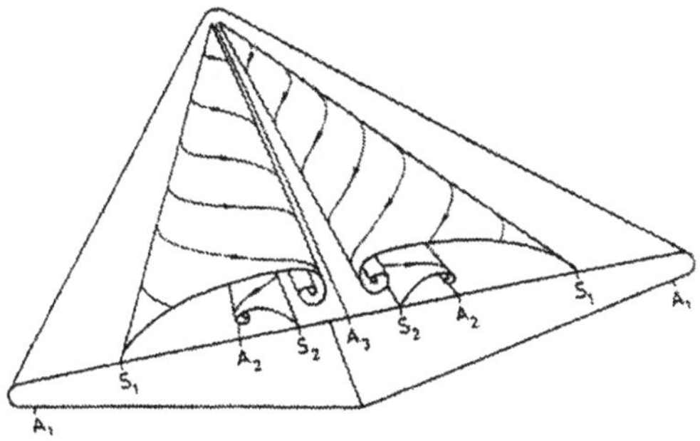

Fig. 3.19. Sketch of the leeward-side flow topology of the BDW [32]. A_1: primary attachment lines, A_2: secondary attachment lines, A_3: tertiary attachment line, S_1: primary separation lines, S_2: secondary separation lines.

In Fig. 3.20 the attachment and separation lines are marked as "half-saddles (S')" (note that the primary attachment lines are "quarter-saddles (S'')", because the flow between them is (more or less) two-dimensional. The axes of the primary and the secondary vortices are marked as "focal (F)" points, which are counted as "nodal (N)" points. Finally a "saddle (S)" point is indicated above the wing. This pattern obeys the topological rule ([35]):

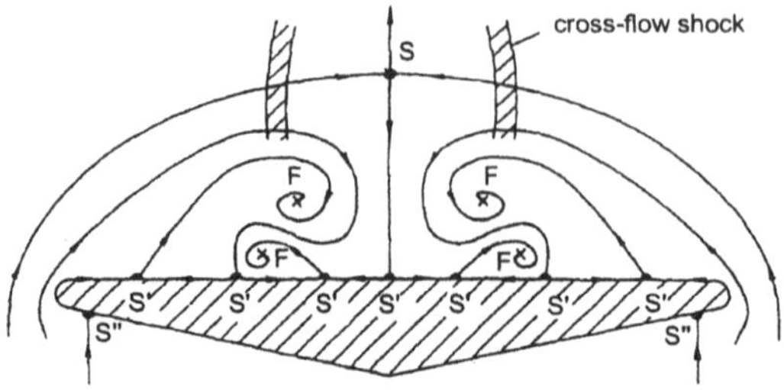

Fig. 3.20. Sketch of the topology of the BDW velocity field in the Poincaré surface at $x/L = 0.99$ [32].

$$\left(\Sigma N + \frac{1}{2}\Sigma N'\right) - \left(\sum S + \frac{1}{2}\sum S' + \frac{1}{4}\sum S''\right) = -1, \qquad (3.41)$$

and therefore is a valid topology.

Note especially that at all attachment lines not only is the characteristic length Δ small, but that also the attaching streamlines come from the outer inviscid flow and hence carry the original free-stream total enthalpy. This is in contrast to, for instance, (steady) two-dimensional separation, where the attaching separation layer does not carry the original total enthalpy, if the boundary layer ahead of the separation point is radiation or otherwise cooled.

3.3.3 The Computed Radiation-Adiabatic Temperature Field

In the following we discuss only the Navier-Stokes solution for the larger wing with $L = 14.0\ m$. Fig. 3.21 gives an overview of the results on the lower and the upper surface. On the left-hand sides of the figure the radiation heat flux (q_{rad}) distributions are shown, and on the right-hand sides the radiation-adiabatic temperature (T_{ra}) distributions. Unfortunately the colour scales are not the same in the two parts a) and b) of the picture. For quantitative data at the two locations $x/L = 0.14$ and $x/L = 0.99$ see Fig. 3.24.

We concentrate on the distributions of the radiation-adiabatic temperatures on the right-hand sides. Part a) of Fig. 3.21 shows on the lower side of the wing the almost parallel flow between the primary attachment lines. The radiation adiabatic temperature reduces in downstream direction as expected. On the larger portion of the lower side it lies around 800.0 K (see also Fig. 3.22). Along the attachment line attachment-line heating ensues with a nearly constant temperature of approximately 1,100.0 K.

On the upper side, part b), along the leading edge we also see a nearly constant temperature of about 1,050.0 K. This high temperature is due to the small boundary-layer thickness, which is a result of the strong expansion of the flow around the leading edge. At the primary separation line the temperature drops fast and a real cold-spot situation develops. The secondary

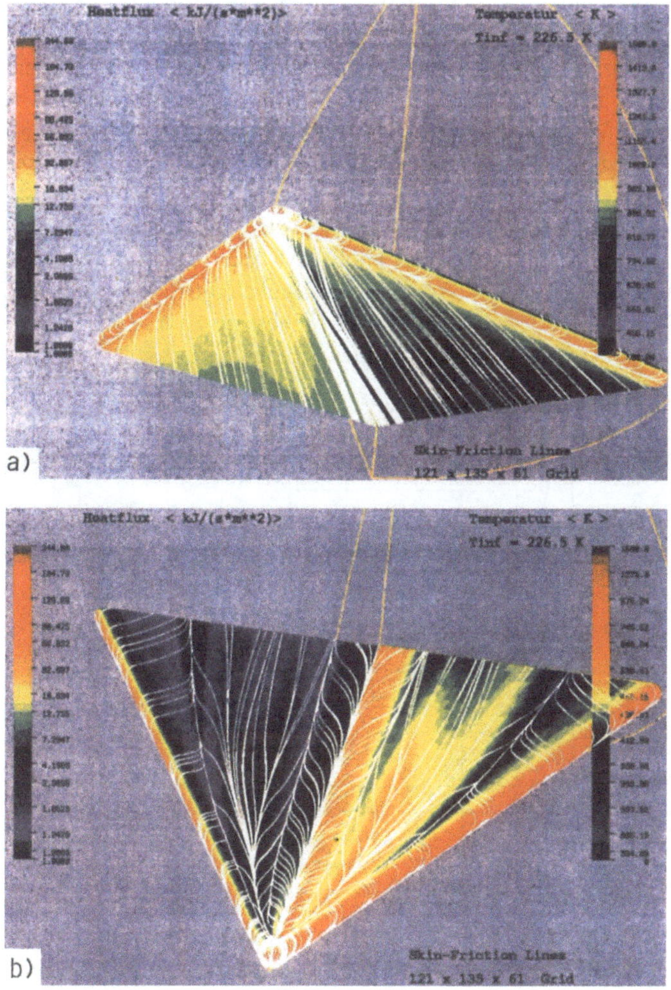

Fig. 3.21. Computed skin-friction lines, and distributions of the surface radiation heat flux q_{rad} (left) and the radiation-adiabatic surface temperature T_{ra} (right) at **a**) the lower (windward) side, and **b**) the upper (leeward) side of the BDW [32].

attachment line seems to taper off at about 40 per cent body length. Possibly a tertiary vortex would develop, if the wing length would be increased (non-conical behaviour). At the secondary separation line again a cold-spot situation, however weaker than that at the primary separation line, develops. The tertiary attachment line shows the expected attachment-line heating with an almost constant temperature of approximately 650.0 K along the upper symmetry line.

Fig. 3.22 displays the temperature distribution on the upper and the lower symmetry line of the wing. Note that the abscissa is given with the computation-grid parameter i. The inset shows its correspondence to x/L. The temperature on the windward side is typically higher than that on the leeward side. We will come back to this temperature differential. The kink in the windward-side curve at $i \approx 26$ seems to be due to the curvature jump of the surface there (see Fig. 3.15). In the leeward-side curve the effect is not visible. The temperature at the leeward side symmetry line is almost constant with approximately 650.0 K. At the windward side we find only very approximately the flat-plate behaviour suggested by eq. (3.20), i. e. the drop of $T_{ra}^4 \sim (x/L)^{-0.5}$. This is due to the up to $x/L \approx 0.34$ first slightly convergent and then divergent skin-friction line pattern at the windward symmetry line, Fig. 3.16.

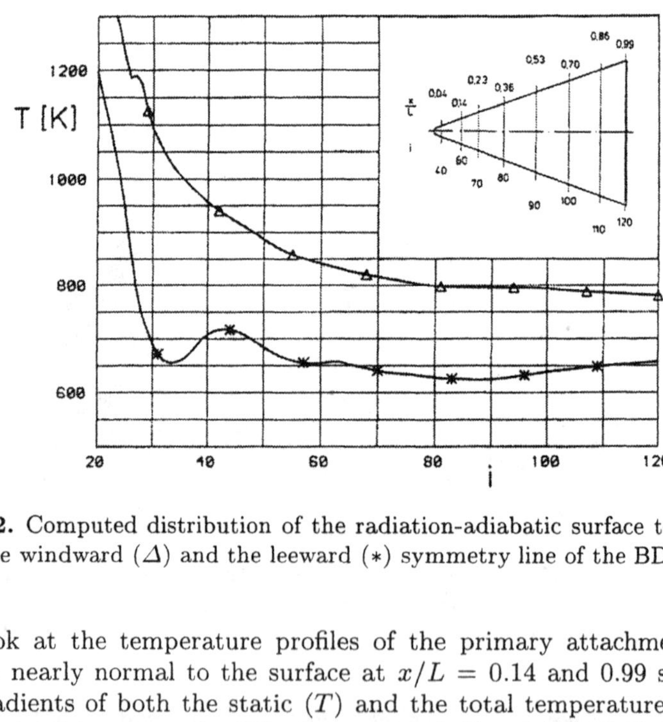

Fig. 3.22. Computed distribution of the radiation-adiabatic surface temperature T_{ra} on the windward (Δ) and the leeward ($*$) symmetry line of the BDW [32].

A look at the temperature profiles of the primary attachment line in direction nearly normal to the surface at $x/L = 0.14$ and 0.99 shows very steep gradients of both the static (T) and the total temperature (T_t), Fig. 3.23. The total temperature of the free-stream is $T_t = 2{,}542.35\ K$. The curves for the total temperature indicate a thicker boundary layer at the aft location. The curves for the static temperature seem to indicate a thinner boundary layer there. Fig. 3.16 shows more or less an infinite-swept wing situation at the primary attachment lines (the leading-edge radius is constant).

The contradictory data may be due to the fact, that the data are taken from a coordinate line, which is not a locally monoclinic line. In any case the behaviour of the static temperature near to the wall supports the assumption made in eq. (3.15) regarding the appropriate characteristic length of the laminar boundary layer ($\Delta = \delta_T$) for the analysis of the radiation-adiabatic temperature. However, since no computation was made with an adiabatic wall, no information is available about the recovery temperature at the attachment line. Eq. (3.7) for the recovery temperature is approximately valid only for flat surfaces. At attachment and separation lines strong departures can be observed from the data found with that equation, so that it is only of restricted value here.

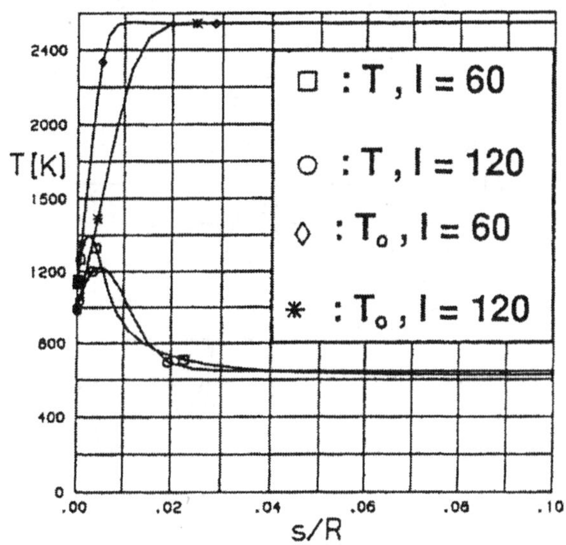

Fig. 3.23. Profiles of the static temperature T and the total temperature $T_0 \equiv T_t$ nearly normal to the surface on the primary attachment line of the BDW at x/L = 0.14 (I = 60) and x/L = 0.99 (I = 120) [32].

In Fig. 3.24 we present results of a scaling of the distribution of the radiation-adiabatic temperature at the two locations x/L = 0.14 and 0.99. For the purpose of scaling a Navier-Stokes solution was made also for the wing with a length $L = 4.67\ m$. All other parameters were the same, so that the scaling relation (3.33) can be reduced, with $n = 0.5$, and $T_{r1} = T_{r2}$, to:

$$\frac{T_{ra1}}{T_{ra2}} = \left[\frac{L_2}{L_1}\frac{(1 - T_{ra1}/T_{r1})}{(1 - T_{ra2}/T_{r2})}\right]^{\frac{1}{4}} . \qquad (3.42)$$

The results in Fig. 3.24 are given in the computation-grid parameter space j, Fig. 3.15. At the left side of the abscissa ("luv") we are on the lower symmetry line (windward side) (j = 2), at j = 70 at the leading edge, and at the right hand side ("lee") on the upper symmetry line (leeward side) (j = 134).

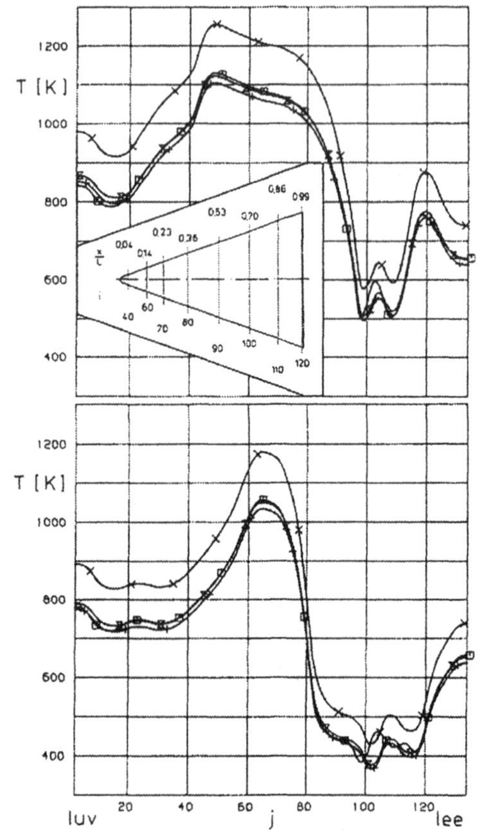

Fig. 3.24. Scaling of radiation-adiabatic surface temperatures at $x/L = 0.14$ (**upper part** of the figure) and $x/L = 0.99$ (**lower part** of the figure) of the BDW [32]. $\times$: numerical solution for $L = 4.67\ m$, $\square$: numerical solution for $L = 14.0\ m$, ∇: result of scaling with eq. (3.42), $+$: results of scaling with modified eq. (3.42). Lower symmetry line: $j = 2$, leading edge: $j = 70$, upper symmetry line: $j = 134$.

The general pattern of the temperature distributions is discussed in the following, while taking into account also the patterns of the skin-friction lines in Figs. 3.16 to 3.18. On the lower symmetry line lies a weak relative maximum, followed by a plateau, which has a large width at $x/L = 0.99$, and a small one at $x/L = 0.14$. The smallness of the latter is due to the fact, that the primary attachment line lies at a constant distance from the leading edge. This is also the reason, why the temperature maximum (attachment-line heating) in it lies at approximately $j = 48$ and in the aft location $x/L = 0.99$ at $j = 65$.

On the leeward side again we have a small plateau at the first location and a wide one at the second. Around $j = 100$ we see the cold-spot situation at the primary separation line, followed by the hot-spot situation at the secondary attachment line. This attachment line lies in the forward location at $j \approx 105$ and in the aft location at $j \approx 110$. The cold-spot situation in the secondary separation line follows at $j \approx 110$ at the forward location, and in the aft location, despite its off-tapering, at $j \approx 117$. At the second location the hot-spot situation at the tertiary attachment line in the leeward-side symmetry line is well discernible. In the first location a hot-spot situation lies at $j = 120$ with a temperature distinctly higher than that in the symmetry line.

The reason for that is not fully clear. In Fig. 3.21 (and also in Fig. 3.18) we see that the skin-friction lines in that region bend very strongly out of the tertiary attachment line. Also a local maximum is discernible in the radiation heat flux parallel and close to the symmetry line. A cross-flow shock [32] in that area, if it is sufficiently oblique, see Section 6.6, would lead to a thinning of the boundary layer downstream of it. Most probably this is the cause for the phenomenon, since the temperature level in the symmetry line there is already that found at $x/L = 0.99$.

The scaling with eq. (3.42) would give for design purposes very good results. All phenomena are at least qualitatively well reproduced. At the leeward side small differences can be seen regarding the location and the magnitude of the cold and hot-spot situations.

From the qualitative results given in Section 3.2.1 it can be deduced that at a radiation-cooled flight vehicle the radiation-adiabatic surface temperature is higher on the windward side than on the leeward side, which is supported by the computed data shown in Figs. 3.22 and 3.24. This temperature differential is due to the fact that the boundary layer thickness is smaller on the lower and larger on the upper side of the vehicle.

We demonstrate this with data for the RHPM-flyer from Chapter 11. For $M_\infty = 7.15$ and $\alpha = 15°$ the ratio of the (constant) unit Reynolds numbers of the lee (l) side and the windward (w) side is $Re^u_l / Re^u_w \approx 0.09$. This amounts to a ratio of the boundary-layer thicknesses, without taking into account the surface temperature, of $\delta_l / \delta_w \approx 3.33$. Using the scaling relation (3.35) we find a ratio of the temperatures of the lee side and the windward side $T_{ra,l}/T_{ra,w} \approx 0.74$. Of course we must be careful with a comparison because of the hot- and cold-spot situation on the upper side of the BDW. Nevertheless, when comparing this ratio with the ratio of the computed data at the two stations of the BDW, $x/L = 0.14$ and $x/L = 0.99$, or with the ratio of the computed data in Fig. 3.22, we see that this result illustrates fairly the temperature differentials, which we find on hypersonic vehicles flying at angle of attack.

3.4 Results of Analysis of the Thermal State of the Surface in View of Flight-Vehicle Design

Regarding the design of hypersonic vehicles, we note that the approximations and proportionalities derived in the preceding sub-sections imply with regard to the thermal state of the surface in case of radiation cooling:

- The thermal state of the surface, i. e., surface temperature $T_w \approx T_{ra}$ and temperature gradient, respectively heat flux in the gas at the wall q_{gw}, of radiation-cooled surfaces depend on the flight speed v_∞, respectively the flight Mach number M_∞, *and* the flight altitude H.
- It depends on the flight trajectory of the vehicle and on the structure and materials concept, whether the thermal state of the surface can be considered as quasi-steady phenomenon or must be treated as unsteady phenomenon.
- The efficiency of radiation cooling increases with decreasing flight unit Reynolds number, and hence increasing boundary layer thickness, because q_{gw} decreases with increasing boundary-layer thickness. In general radiation cooling becomes more efficient with increasing flight altitude.
- The radiation-adiabatic temperature, in contrast to the adiabatic temperature, changes appreciably in main-flow direction (e. g. along the fuselage). It decreases strongly with laminar, and less strongly with turbulent flow, according to the growth of the related boundary-layer thicknesses.
- The radiation-adiabatic temperature is significantly lower than the adiabatic temperature. The actual wall temperature in general is near the radiation-adiabatic temperature, depending somewhat on the structure and materials concept, and the actual trajectory part. Therefore the radiation-adiabatic temperature should be taken as wall temperature estimate in vehicle design rather than the adiabatic or any other mean temperature. On low trajectory segments possibly a thermal must be taken in account.
- The state of the boundary layer, laminar or turbulent, affects much stronger the radiation-adiabatic temperature than the adiabatic temperature.
- Turbulent skin-friction is much higher on radiation-cooled surfaces than on adiabatic surfaces (in general it is the higher the colder the surface is), the laminar skin friction is not as strongly affected.
- On flight vehicles at angle of attack significant differentials of the wall temperature occur between windward side and lee side, according to the different boundary-layer thicknesses there.
- At attachment lines radiation cooling leads to hot-spot situations (attachment-line heating), at separation lines the opposite is observed.

3.5 Problems

Problem 3.1 Find from eq. (3.2) with the assumption of perfect gas the total temperature T_t as function of T_∞, γ, M_∞^2.

Solution: eq. (3.6).

Problem 3.2 The RHPM-flyer flies with zero angle of attack with $v_\infty = 1.0$ km/s at 30.0 km altitude. Determine the total temperature T_t, and the recovery temperatures T_r for laminar and turbulent boundary-layer flow. Take $\gamma = 1.4$ and the Prandtl number Pr at 600.0 K, Sub-Section 4.2.3.

Solution: $T_t = 724.27\ K$, $T_{r,lam} = 656.37\ K$, $T_{r,turb} = 677.9\ K$.

Problem 3.3 Check for laminar flow the behaviour of $T_w(x/L) = T_{ra}(x/L)$ at $M_\infty = 15.7$ in Fig. 3.3.

Solution: Simplify eq. (3.25) to $T_{ra} = c(x/L)^{-n/4}$. Measure in Fig. 3.3 the temperature at $x/L = 0.1$: $T_w \approx 1100.0\ K$ and find with $n = 0.5$ the constant $c = 824.88\ K$. Measure now the temperature at $x/L = 0.75$: $T_w \approx 880.0\ K$ and compare with the temperature from the simplified equation at that location: $T_w = 855.1\ K$. How do you rate the result?

Problem 3.4 Check for laminar flow the behaviour of $T_w(x/L) = T_{ra}(x/L)$ at $M_\infty = 5.22$ in Fig. 3.3.

Solution: Proceed like in Problem 3.3. Measure $T_W \approx 656.0\ K$ at $x/L = 0.1$. Measure the temperature at $x/L = 0.75$: $T_w \approx 522.0\ K$ and compare with the temperature from the simplified equation at that location: $T_w = 509.9\ K$. How do you rate the result?

Problem 3.5 Check for turbulent flow with $n = 0.2$ (exponent in the relation for the turbulent scaling thickness, Sub-Section 7.2.1) the behaviour of $T_w(x/L) = T_{ra}(x/L)$ at $M_\infty = 5.22$ in Fig. 3.3.

Solution: Measure $T_W \approx 874.0\ K$ at $x/L = 0.1$.Find $c = 745.1\ K$. Measure in Fig. 3.3 the temperature at $x/L = 0.75$: $T_w \approx 820.0\ K$ and compare with the temperature from the simplified equation at that location: $T_w = 790.2\ K$. How do you rate the result?

Problem 3.6 Check for turbulent flow with $n = 0.1$ (exponent in the relation for the thickness of the viscous sub-layer, Sub-Section 7.2.1) the behaviour of $T_w(x/L) = T_{ra}(x/L)$ at $M_\infty = 5.22$ in Fig. 3.3.

Solution: $T_w = 831.1\ K$. Compare with the result from Problem 3.5.

Problem 3.7 If at 60.56 km altitude, Fig. 3.3, the Space Shuttle would fly at $M_\infty = 17$, what would the wall temperature approximately be at $x/L = 0.5$?

Solution: Assume that the Reynolds number remains unchanged and that the wall temperature changes are small. Assume further $T_{ra} \ll T_r$ and simplify eq. (3.25) to $T_{ra} \sim T_r^{0.25}$. Choose $\gamma_{eff} = 1.3$ and $Pr = 1.0$ and find $T_{ra,M_\infty=17}/T_{ra,M_\infty=15.7} = 1.0396$. Hence $T_w \approx 952.2\ K$ compared to originally $T_w \approx 916.0\ K$.

Problem 3.8 If at 60.56 km altitude, Fig. 3.3, the Space Shuttle would fly at $M_\infty = 14$, what would the wall temperature approximately be at $x/L = 0.5$?

Solution: Proceed like in Problem 3.7 and find $T_{ra,M_\infty=14}/T_{ra,M_\infty=15.7} = 0.946$. Hence $T_w = 866.4\ K$.

References

1. H. SCHLICHTING. "Boundary Layer Theory". 7th edition, McGraw-Hill, New York, 1979.
2. J. P. ARRINGTON, J. J. JONES (EDS.). "Shuttle Performance: Lessons Learned". NASA CP-2283, 1983.
3. J. M. LONGO, R. RADESPIEL. "Numerical Simulation of Heat Transfer Effects on 2-D Steady Subsonic Flows". AIAA-Paper 95-0298, 1995.
4. E. H. HIRSCHEL. "Heat Loads as Key Problem of Hypersonic Flight". Zeitschrift für Flugwissenschaften und Weltraumforschung (ZFW), Vol. 16, No. 6, 1992, pp. 349 - 356.
5. W. SCHNEIDER. "Radiation Gasdynamics of Planetary Entry". Astronautica Acta, Vol. 18 (Supplement), 1974, pp. 193 - 213.
6. J. A. FAY, F. R. RIDDELL. "Theory of Stagnation Point Heat Transfer in Dissociated Gas". Journal of Aeronautical Science, Vol. 25, No. 2, 1958, pp. 73 - 85.
7. R. W. DETRA, H. HIDALGO. "Generalized Heat Transfer Formulas and Graphs for Nose Cone Re-Entry Into the Atmosphere". ARS Journal, March 1961, pp. 318 - 321.
8. J. J. BERTIN. "Hypersonic Aerothermodynamics". AIAA Education Series, Washington, 1994.
9. R. B. HILDEBRAND. "Aerodynamic Fundamentals". *H. H. Koelle (ed.), Handbook of Astronautical Engineering.* McGraw-Hill, New York/Toronto/London, 1961, pp. 5-1 to 5-42.
10. F. MONNOYER. Personal communication. 1992.
11. F. MONNOYER, CH. MUNDT, M. PFITZNER. "Calculation of the Hypersonic Viscous Flow Past Reentry Vehicles with an Euler/Boundary-Layer Coupling Method". AIAA-Paper 90-0417, 1990.
12. E. H. HIRSCHEL. "Aerothermodynamische Probleme bei Hyperschall-Fluggeräten". *Jahrestagung der DGLR, München, October 9 and 10, 1986.* Also MBB/LKE122/PUB/S/270,1986.
13. E. V. ZOBY, R. N. GUPTA, A. L. SIMMONDS. "Temperature-Dependent Reaction-Rate Expression for Oxygen Recombination at Shuttle-Entry Conditions". AIAA-Paper 84-0224, 1984.
14. E. R. G. ECKERT, R. M. DRAKE. "Heat and Mass Transfer". MacGraw-Hill, New York, 1950.
15. R. B. BIRD, W. E. STEWART, E. N. LIGHTFOOT. "Transport Phenomena". John Wiley & Sons, New York/London/Sydney, 2nd edition 2002 .
16. E. R. G. ECKERT. "Engineering Relations of Friction and Heat Transfer to Surfaces in High-Velocity Flow". J. Aeronautical Sciences, Vol. 22, No. 8, 1955, pp. 585 - 587.

17. M. J. LIGHTHILL. "Contributions to the Theory of Heat Transfer Through a Laminar Boundary Layer". *Proc. Royal Society of London, Series A, Mathematical and Physical Sciences, Vol. 202*, Cambridge University Press, 1950, pp. 359 - 377.

18. G. SIMEONIDES. "Extrapolation-to-Flight of Convective Heating Measurements and Determination of Radiation-Equilibrium Surface Temperature in Hypersonic/High Enthalpy Flow". To be submitted to Shock Waves.

19. S. D. WILLIAMS. "Columbia, the First Five Flights Entry Heating Data Series, Volume 1: an Overview". NASA CR - 171 820, 1984.

20. S. D. WILLIAMS. "Columbia, the First Five Flights Entry Heating Data Series, Volume 3: the Lower Windward Surface Center Line". NASA CR - 171 665, 1983.

21. S. WÜTHRICH, M. L. SAWLEY, G. PERRUCHOUD. "The Coupled Euler/ Boundary-Layer Method as a Design Tool for Hypersonic Re-Entry Vehicles". Zeitschrift für Flugwissenschaften und Weltraumforschung (ZFW), Vol. 20, No. 3, 1996, pp. 137 - 144.

22. S. D. WILLIAMS, D. M. CURRY. "An Analytical and Experimental Study for Surface Heat Flux Determination". J. Spacecraft, Vol. 14, No. 10, 1977, pp. 632 - 637.

23. E. H. HIRSCHEL, A. KOÇ, S. RIEDELBAUCH. "Hypersonic Flow Past Radiation-Cooled Surfaces". AIAA-Paper 91-5031, 1991.

24. G. C. RUFOLO, D. TESCIONE, S. BORRELLI. "A Multidisciplinary Approach for the Analysis of Heat Shielded Space Structures". *Proc. 4th European Workshop on Hot Structures and Thermal Protection Systems for Space Vehicles, Palermo, Italy, 2002.* ESA SP-521, 2003, pp. 231 - 238.

25. TH. EGGERS, PH. NOVELLI, M. HAUPT. "Design Studies of the JAPHAR Experimental Vehicle for Dual Mode Ramjet Demonstration". AIAA-Paper 2001-1921, 2001.

26. E. H. HIRSCHEL. "Evaluation of Results of Boundary-Layer Calculations with Regard to Design Aerodynamics". AGARD-R-741, 1986, pp. 6-1 to 6-29.

27. R. K. HÖLD, L. FORNASIER. "Investigation of Thermal Loads of Hypersonic Vehicles with Emphasis on Surface Radiation Effects". ICAS-Paper 94-4.4.1, 1994.

28. R. K. HÖLD. "Modeling of Surface Radiation Effects by a Fictitious Emissivity Coefficient". *V. V. Kudriavtsev, C. R. Kleijn (eds.), Computational Technologies for Fluid/Themal/Chemical Systems with Industrial Applicions*, ASME PVP-Vol. 397-1, 1999, pp. 201 - 208.

29. W. M. ROHSENOW (ED.). "Developments in Heat Transfer". Edward Arnold ltd., London, 1964.

30. R. BEHR. "Hot, Radiation Cooled Surfaces". Rep. TET-DASA-21-TN-2410, astrium, Munich, Germany, 2002.

31. E. H. HIRSCHEL, CH. MUNDT, F. MONNOYER, M. A. SCHMATZ. "Reynolds-Number Dependence of Radiation-Adiabatic Wall Temperature". MBB-FE122-AERO-MT-872, 1990.

32. S. RIEDELBAUCH. "Aerothermodynamische Eigenschaften von Hyperschallströmungen über strahlungsadiabate Oberflächen (Aerothermodynamic Properties of Hypersonic Flows past Radiation-Cooled Surfaces)". Doctoral Thesis, Technische Universität München, Germany, 1991. Also DLR-FB 91-42, 1991.

33. S. RIEDELBAUCH, E. H. HIRSCHEL. "Aerothermodynamic Properties of Hypersonic Flow over Radiation-Adiabatic Surfaces". Journal of Aircraft, Vol. 30, No. 6, 1993, pp. 840 - 846.

34. J.-A. DÉSIDÉRI, R. GLOWINSKI, J. PERIAUX (EDS.). "Hypersonic Flows for Reentry Problems". Volume 1 and 2, Springer, Berlin/Heidelberg/New York, 1991.

35. D. J. PEAKE, M. TOBAK. "Three-Dimensional Interaction and Vortical Flows with Emphasis on High Speeds". AGARDograph No. 252, 1980.

36. TOBAK, M., D. J. PEAKE. "Topology of Three-Dimensional Separated Flows". Annual Review of Fluid Mechnics, Vol. 14, 1982, pp. 61 - 85.

37. K. C. WANG. "Separating Patterns of Boundary Layer Over an Inclined Body of Revolution". AIAA Journal, Vol. 10, 1972, pp. 1044 - 1050.

38. U. DALLMANN, A. HILGENSTOCK, S. RIEDELBAUCH, B. SCHULTE-WERNING, H. VOLLMERS. "On the Footprints of Three-Dimensional Separated Vortex Flows Around Blunt Bodies. Attempts of Defining and Analyzing Complex Vortex Structures". AGARD-CP-494, 1991, pp. 9-1 to 9-13.

4 Transport of Momentum, Energy and Mass

Fluid flow is characterized by transport of mass, momentum and energy. In this chapter we treat the transport of these three entities and its mathematical description in a basic way. Similarity parameters and, especially, surface boundary conditions will be discussed in detail.

Similarity parameters enable us to distinguish and choose between phenomenological models and the respective mathematical models. The Knudsen number, which we met in Chapter 2.3, is an example. It is used in order to distinguish flow regimes. As we will see, for the transport of the three entities different phenomenological models exist, which we characterize with the aid of appropriate similarity parameters.

Surface boundary conditions receive special attention because they govern, together with the free-stream conditions and the flight-vehicle geometry, the flow and thermo-chemical phenomena in the flow-regime of interest. In addition, they govern aerodynamic forces and moments as well as mechanical and thermal loads on the flight vehicle.

After a general introduction, which also shows how to distinguish between steady and unsteady flows, we look briefly at the transport properties of air. Then we treat in detail the equations of motion, for convenience, although with some exceptions, in two dimensions and Cartesian coordinates (the formulations of the governing equations in general coordinates are given in Chapter 12), similarity parameters, and surface boundary conditions. We treat first momentum transport, because here the boundary-layer concept plays the major role. The transport of the two other entities is treated in an analogous way.

Far-field or external boundary conditions as well as initial conditions are in general not considered in detail in this book. In a flight situation these are the free-stream conditions. With internal flows, also wind-tunnel flows, as well as with special phenomenological models like the boundary-layer, however special considerations become necessary. They will be discussed in the respective sections.

4.1 Transport Phenomena

In order to ease the discussion we consider in general steady compressible and two-dimensional flow in Cartesian coordinates. The nomenclature is given in Fig. 4.1. The coordinate x is the stream-wise coordinate, tangential to the surface. The coordinate y is normal to the surface. The components of the velocity vector $\underline{V}$ in these two directions are u and v, the magnitude[1] of the speed is $V = |\underline{V}| = \sqrt{u^2 + v^2}$.

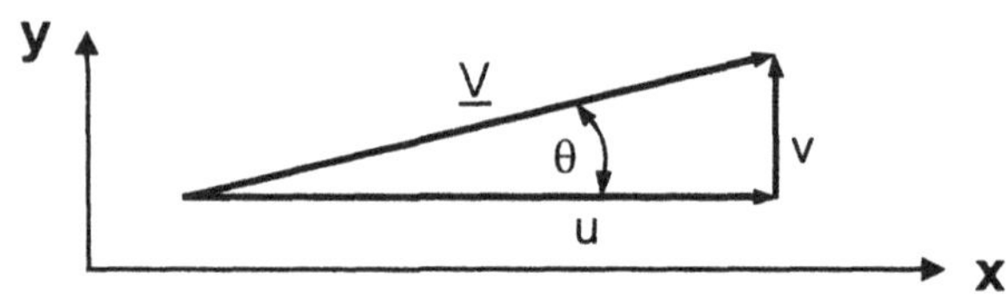

Fig. 4.1. Schematic of Cartesian coordinates and velocity components.

The governing equations of fluid flow will be discussed as differential equations in the classical formulation per unit volume, see e. g. [1], [2]. The fluid is assumed to be a mixture of thermally perfect gases, Chapter 2. However, for the discussion we usually consider it only as a binary mixture with the two species A and B.

Fluid flow transports the three entities

- momentum (vector entity),
- energy (scalar entity),
- mass (scalar entity).

The transport mechanisms of interest in this book are

1. convective transport,
2. molecular[2] transport,
3. turbulent transport,
4. radiative transport.

For a detailed discussion the reader is referred to the literature, for instance [2].

Under "convective" transport we understand the transport of the entity under consideration by the bulk motion of the fluid. In steady flow this happens along streamlines (in unsteady flow we have individual fluid particle

[1] In three dimensions the coordinate z and the velocity component w are orthogonal to the picture plane.

[2] Often also called diffusive transport.

"path lines", and also "streak lines", which represent the current locus of particles, which have passed previously through a fixed point). Under which circumstances a flow can be considered as steady will be discussed at the end of this section. If convective transport is the dominant transport mechanism in a flow, we call it "inviscid" flow.

"Molecular" transport happens by molecular motion relative to the bulk motion. It is caused by non-uniformities, i. e. gradients, in the flow field [3]. The phenomena of interest for us are viscosity, heat conduction, and mass diffusion, Table 4.1.

Table 4.1. Molecular transport phenomena.

Phenomenon	Molecular transport of	Macroscopic cause
Viscosity	momentum	non-uniform flow velocity
Heat conduction	energy	non-uniform temperature
Diffusion	mass	non-uniform concentration of species, pressure and temperature gradients

Molecular transport occurs in all directions of the flow field. Dominant directions can be present, for instance in boundary layers or, generally, in shear layers.

If molecular transport plays a role in a flow field, we call the flow summarily "viscous flow[3]". In fact, as we also will see later, all fluid flow is viscous flow. It is a matter of dominance of the different transport mechanisms, whether we speak about inviscid or viscous flow.

"Turbulent" transport is an apparent transport due to the fluctuations in turbulent flow. In fluid mechanics we treat turbulent transport of momentum, energy and mass usually in full analogy to the molecular transport of these entities, however with apparent turbulent transport properties. In general the "effective" transport of an entity Φ_{eff} is defined by adding the laminar (molecular) and the turbulent (apparent) part of it:

$$\Phi_{eff} = \Phi_{lam} + \Phi_{turb}. \tag{4.1}$$

"Radiative" transport in the frame of this book is solely the transport of heat away from the surface for the purpose of radiation cooling (Chapter 3). In principle it occurs in all directions and includes absorption and emission processes in the gas, [4], [2], which, however, are neglected in our considerations.

In Table 4.2 we summarize the above discussion.

The three entities considered here, loosely called the flow properties, change in a flow field in space and in time. This is expressed in general by the substantial time derivative, also called convective derivative [2]:

[3] To be precise, viscous flow due to molecular transport refers to "laminar flow".

Table 4.2. Schematic presentation of transport phenomena.

Item	Momentum	Energy	Mass
Convective transport	x	x	x
Molecular transport	x	x	x
Turbulent transport	x	x	x
Radiative transport		x	

$$\frac{D}{Dt} = \frac{\partial}{\partial t} + u\frac{\partial}{\partial x} + v\frac{\partial}{\partial y} + w\frac{\partial}{\partial z}. \tag{4.2}$$

This is the derivative, which follows the motion of the fluid, i. e. an observer of the flow would simply float with the fluid.

We note in this context that the aerothermodynamic flow problems we are dealing with in general are Galilean invariant [1]. Therefore we can consider a flight vehicle in our mathematical models and in ground simulation (computational simulation, ground-facility simulation) in a fixed frame with the air-stream flowing past it. In reality the vehicle flies through the - quasi-uniform - atmosphere.

The first term on the right-hand side, $\partial/\partial t$, is the derivative with respect to time, the partial time derivative, which describes the change of an entity in time at a fixed locus x, y, z in the flow field.

The substantial time derivative itself is a specialization of the total time derivative

$$\frac{d}{dt} = \frac{\partial}{\partial t} + \frac{\partial}{\partial x}\frac{dx}{dt} + \frac{\partial}{\partial y}\frac{dy}{dt} + \frac{\partial}{\partial z}\frac{dz}{dt}, \tag{4.3}$$

where now the observer moves with arbitrary velocity with the Cartesian velocity components dx/dt, dy/dt, and dz/dt.

In our applications we speak about steady, quasi-steady, and unsteady flow problems. The measure for the distinction of these three flow modes is the Strouhal number Sr. We find it by means of a normalization with proper reference values, and hence a non-dimensionalization, of the constituents q of the substantial time derivative eq. (4.2):

$$q = q_{ref} q^{\star}. \tag{4.4}$$

Here $q = t, u, v, w, x, y, z$, while q_{ref} are the respective normalization parameters. q and q_{ref} have the same dimension, whereas $q^{\star}$ are the corresponding non-dimensional parameters, which are of the order one.

Introduction of eqs. (4.4) into eq. (4.2) and rearrangement yields:

$$\frac{L_{ref}}{t_{ref} v_{ref}} \frac{\partial}{\partial t^{\star}} + u^{\star}\frac{\partial}{\partial x^{\star}} + \cdots. \tag{4.5}$$

The term in front of the partial time derivative is the Strouhal number:

$$Sr = \frac{L_{ref}}{t_{ref} v_{ref}}. \tag{4.6}$$

We introduce the residence time t_{rs} of a fluid particle in the domain under consideration:

$$t_{res} = \frac{L_{ref}}{v_{ref}}. \tag{4.7}$$

This time could be the time, which a fluid particle needs to travel with the reference speed $v_{ref} = v_\infty$ past a body with the reference length $L_{ref} = L$, Fig. 4.2.

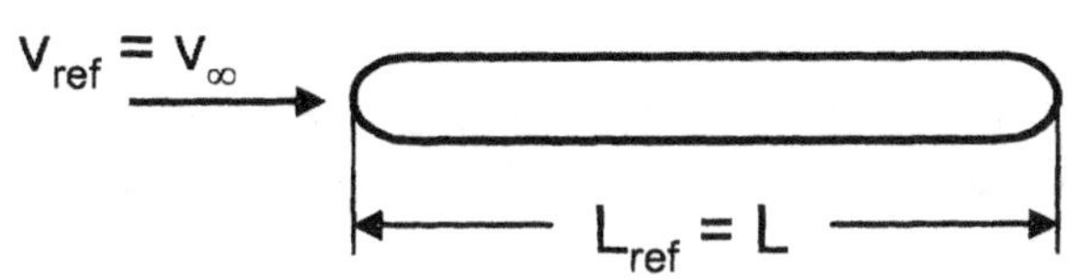

Fig. 4.2. Schematic of a reference situation: flow past a body.

Hence the Strouhal number becomes:

$$Sr = \frac{t_{res}}{t_{ref}}. \tag{4.8}$$

If the residence time is small compared to the reference time, in which a change of flow parameters happens, we consider the flow as quasi-steady, because $Sr \to 0$. Steady flow is characterized by $Sr = 0$, i. e., it takes infinitely long for the flow to change in time ($t_{ref} \to \infty$).

Unsteady flow is present if $Sr = O(1)$. For practical purposes the assumption of quasi-steady flow, and hence the treatment as steady flow, is permitted for $Sr \lessapprox 0.2$, that is, if the residence time t_{res} is at least five times shorter than the reference time t_{ref}. This means that a fluid particle "travels" in the reference time t_{ref} five times past the body with the reference length L_{ref}.

Again we must be careful with our considerations. For example, the movement of a flight vehicle may be permitted to be considered as at least quasi-steady, while at the same time truly unsteady movements of a control surface may occur. In addition there might be configuration details, where highly unsteady vortex shedding is present.

The flows treated in this book are considered to be steady flows. In the following sections, however, we present and discuss sometimes the governing equations also in the general formulation for unsteady flows.

4.2 Transport Properties

4.2.1 Introduction

The molecular transport of the three entities momentum, energy, and mass basically obeys similar laws, which combine linearly the gradients of flow velocity, temperature and species concentration with the coefficients of the respective transport properties: viscosity, thermal conductivity, diffusivity, see e. g. [2]. Fluids, which can be described in this way are called "Newtonian fluids".

The basic formulation for molecular momentum transport is Newton's law of friction:

$$\tau_{yx} = -\mu \frac{du}{dy}. \tag{4.9}$$

Here τ_{yx} is the shear stress exerted on a fluid surface in x-direction by the y-gradient of the velocity u in that direction. The coefficient μ is the fluid viscosity.

Similarly, Fourier's law of heat conduction reads:

$$q_y = -k \frac{dT}{dy}. \tag{4.10}$$

The heat flux in the y-direction, q_y, is proportional to the temperature gradient in that direction. The coefficient k is the thermal conductivity. If thermo-chemical non-equilibrium effects are present in the flow, heat is transported also by mass diffusion, Section 4.3.2. Note that always the transport is in the negative y-direction.

This also holds for the molecular transport of mass, which is described by Fick's first law:

$$j_{A_y} = -j_{B_y} = -\rho D_{AB} \frac{d\omega_A}{dy}. \tag{4.11}$$

The diffusion mass flux j_{Ay} is the flux of the species A in y-direction relative to the bulk velocity v in this direction. It is proportional to the mass-fraction gradient $d\omega_A/dy$ in that direction. $D_{AB} = D_{BA}$ is the mass diffusivity in a binary system with the species A and B. The reader is referred to e. g. [2] for equivalent forms of Fick's first law. Besides the concentration-driven diffusion also pressure-, and temperature-gradient driven diffusion can occur, Sub-Section 4.3.3. Molecular transport of mass occurs in flows with thermo-chemical non-equilibrium, but also in flows with mixing processes, for instance in propulsion devices.

The transport properties viscosity μ, thermal conductivity k, and mass diffusivity D_{AB} of a gas are basically functions of the temperature. We give in the following sub-sections relations of different degree of accuracy for the determination of transport properties of air. Emphasis is put on simple power-law approximations for the viscosity and for the thermal conductivity. They can be used for quick estimates, and are also used for the basic analytical considerations throughout the book. Of course they are valid only below approximately $T = 2{,}000.0\ K$, i. e. for not or only weakly dissociated air. Models of high-temperature transport properties are considered in Sub-Section 4.2.5.

4.2.2 Viscosity

The viscosity of pure monatomic, but also of polyatomic gases, in this case air, can be determined in the frame of the Chapman-Enskog theory with [2]:

$$\mu = 2.6693 \cdot 10^{-6} \frac{\sqrt{MT}}{\sigma^2 \Omega_\mu}. \tag{4.12}$$

The dimensions and the constants for air in the low temperature domain, [2], are: viscosity μ $[kg/ms]$, molecular weight $M = 28.97\ kg/kmole$, Temperature T $[K]$, collision diameter (first Lennard-Jones parameter) $\sigma = 3.617 \cdot 10^{-10}\ m$. The dimensionless collision integral Ω_μ, [2], is a function of the quotient of the temperature and the second Lennard-Jones parameter $\epsilon/k = 97.0\ K$, Section 13.1.

An often used relation for the determination of the viscosity of air is the Sutherland equation:

$$\mu_{Suth} = 1.458 \cdot 10^{-6} \frac{T^{1.5}}{T + 110.4}. \tag{4.13}$$

A simple power-law approximation is $\mu = c_\mu T^{\omega_\mu}$. In [5] different values of ω_μ are proposed for different temperature ranges. We find for the temperature range $T \lessapprox 200.0\ K$ the approximation, with the constant $c_{\mu 1}$ computed at $T = 97.0\ K$:

$$\mu_1 = c_{\mu 1} T^{\omega_{\mu 1}} = 0.702 \cdot 10^{-7} T, \tag{4.14}$$

and for $T \gtrapprox 400.0\ K$, with the constant $c_{\mu 2}$ computed at $T = 407.4\ K$:

$$\mu_2 = c_{\mu 2} T^{\omega_{\mu 2}} = 0.04644 \cdot 10^{-5} T^{0.65}. \tag{4.15}$$

In Table 4.3 and Fig. 4.3 we compare results of the four above relations in the temperature range up to $T = 2{,}000.0\ K$, with the understanding, that a more detailed consideration due to possible dissociation above $T \approx 1{,}500.0\ K$ might be necessary. The data computed with eq. (4.12) were obtained by linear interpolation of the tabulated collision integral Ω_μ [2], Section 13.1.

The table shows that the data from the Sutherland relation compare well with the exact data of eq. (4.12) except for the large temperatures, where they are noticeable smaller. The power-law relation for $T \lessapprox 200.0\ K$ fails above $T = 200.0\ K$. The second power-law relation gives good data for T

Table 4.3. Comparison of viscosity data μ $[kg/ms]$ of air computed with eqs. (4.12) to (4.15) for some temperatures T $[K]$ (see also Fig. 4.3).

T	$\mu \cdot 10^5$ eq.(4.12)	$\mu_{Suth} \cdot 10^5$	$\mu_1 \cdot 10^5$	$\mu_2 \cdot 10^5$
50.0	0.349	0.321	0.351	-
75.0	0.525	0.511	0.526	-
100.0	0.702	0.693	0.702	0.927
200.0	1.335	1.328	1.402	1.454
300.0	1.846	1.846	2.106	1.892
400.0	2.279	2.285	2.808	2.281
600.0	3.016	3.016	-	2.969
800.0	3.655	3.623	-	3.580
1,000.0	4.226	4.152	-	4.139
1,500.0	5.453	5.260	-	5.387
2,000.0	6.632	6.179	-	6.495

$\geqq 300.0\ K$. For $T = 200.0\ K$ the error is less than nine per cent. At large temperatures the exact data are better approximated by this relation than by the Sutherland relation.

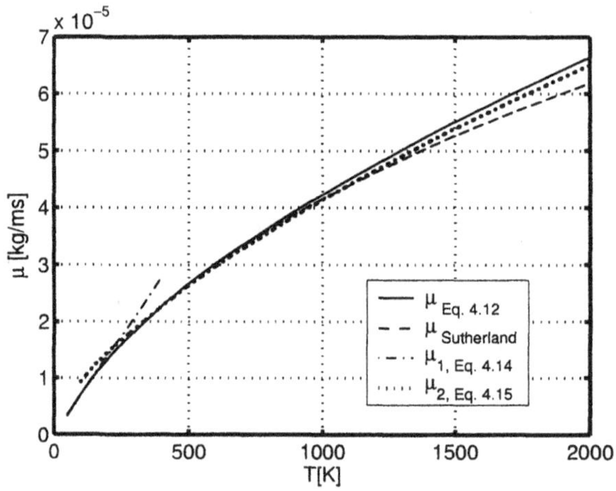

Fig. 4.3. Viscosity μ of air, different approximations, Table 4.3, as function of the temperature T.

4.2.3 Thermal Conductivity

The thermal conductivity of pure monatomic gases can be determined in the frame of the Chapman-Enskog theory [2]:

$$k = 8.3225 \cdot 10^{-2} \frac{\sqrt{T/M}}{\sigma^2 \Omega_k}. \tag{4.16}$$

The dimension of k is W/mK. The dimensionless collision integral Ω_k is identical with that for the viscosity Ω_μ. With this identity it can be shown for monatomic gases:

$$k = \frac{15}{4} \frac{R_0}{M} \mu. \tag{4.17}$$

The Chapman-Enskog theory gives no relation similar to eq. (4.16) for polyatomic gases. An approximate relation, which takes into account the exchanges of rotational as well as vibration energy of polyatomic gases, is the semi-empirical Eucken formula, where c_p is the specific heat at constant pressure :

$$k = \left(c_p + \frac{5}{4} \frac{R_0}{M} \right) \mu. \tag{4.18}$$

The monatomic case is included, if for the specific heat $c_p = 2.5\, R_0/M$ is taken.

From eq. (4.18) the relation for the Prandtl number Pr can be derived:

$$Pr = \frac{\mu c_p}{k} = \frac{c_p}{c_p + 1.25 R_0/M} = \frac{4\gamma}{9\gamma - 5}, \tag{4.19}$$

which is a good approximation for both monatomic and polyatomic gases [3].

For temperatures up to 1,500.0 K to 2,000.0 K, an approximate relation due to C. F. Hansen similar to Sutherland's equation for the viscosity of air can be used [6]:

$$k_{Han} = 1.993 \cdot 10^{-3} \frac{T^{1.5}}{T + 112.0}. \tag{4.20}$$

A simple power-law approximation can also be formulated for the thermal conductivity: $k = c_k T^{\omega_k}$. For the temperature range $T \lessapprox 200.0\ K$ the approximation reads, with the constant c_{k1} computed at $T = 100.0\ K$:

$$k_1 = c_{k1} T^{\omega_{k1}} = 9.572 \cdot 10^{-5}\, T, \tag{4.21}$$

and for $T \gtrapprox 200.0\ K$, with the constant c_{k2} computed at $T = 300.0\ K$:

$$k_2 = c_{k2} T^{\omega_{k2}} = 34.957 \cdot 10^{-5}\, T^{0.75}. \tag{4.22}$$

In Table 4.4 and Fig. 4.4 we compare the results of the four above relations in the temperature range up to T = 2,000.0 K, again with the understanding, that a more detailed consideration due to possible dissociation above $T \approx$ 1,500.0 K might be necessary. The data computed with eq. (4.18) were obtained for non-dissociated air with vibration excitation from:

$$c_p = 3.5R + \omega_{O_2} c_{v_{vibr_{O_2}}} + \omega_{N_2} c_{v_{vibr_{N_2}}} . \qquad (4.23)$$

For the determination of the specific heats at constant volume $c_{v_{vibr}}$ see Section 5.2. The mass fractions $\omega_{O_2} = 0.26216$, and $\omega_{N_2} = 0.73784$ were taken from Sub-Section 2.2 for air in the low-temperature range.

Table 4.4. Comparison of thermal-conductivity data k $[W/mK]$ computed with eqs. (4.18) to (4.22) for some temperatures T $[K]$ of air (see also Fig. 4.4). Included are the specific heat at constant pressure c_p (second column), and the Prandtl number Pr based on eq. (4.19) (last column).

T	c_p/R	$k \cdot 10^2$ eq.(4.18)	$k_{Han} \cdot 10^2$	$k_1 \cdot 10^2$	$k_2 \cdot 10^2$	Pr
50.0	3.500	0.476	0.435	0.478	-	0.7368
75.0	3.500	0.715	0.692	0.717	-	0.7368
100.0	3.500	0.957	0.940	0.957	1.100	0.7368
200.0	3.500	1.820	1.807	1.914	1.859	0.7368
300.0	3.508	2.521	2.514	2.872	2.519	0.7373
400.0	3.538	3.131	3.114	3.833	3.126	0.7389
600.0	3.667	4.256	4.114	-	4.238	0.7458
800.0	3.830	5.329	4.945	-	5.258	0.7539
1,000.0	3.974	6.335	5.668	-	6.216	0.7607
1,500.0	4.204	8.535	7.183	-	8.425	0.7708
2,000.0	4.242	10.453	8.440	-	10.454	0.7724

The table shows that the data from the Hansen relation compare well with the Eucken data except for the temperatures above approximately 600.0 K, where they are noticeably smaller. The power-law relation for $T \lessapprox 200.0\ K$ fails for $T \gtrapprox 200.0\ K$. The second power-law relation gives good data for $T \gtrapprox 200.0\ K$.

It should be noted, that non-negligible vibration excitation sets in already around $T = 400.0\ K$. This is reflected in the behaviour of the Prandtl number, Fig. 4.5, where also c_p/R is given. To obtain, as it is often done, the thermal conductivity simply from eq. (4.19) with a constant Prandtl number would introduce errors above $T \approx 400.0\ K$. The Prandtl number of air generally is $Pr < 1$ in a large temperature and pressure range[4], [6].

4.2.4 Mass Diffusivity

The Chapman-Enskog theory gives for the mass diffusivity of a binary gas a relation similar to those for the viscosity and the thermal conductivity:

[4] In the literature values for the Prandtl number of air at ambient temperatures are given as low as $Pr = 0.72$, compared to $Pr = 0.7368$ in Table 4.4. A gas-kinetic theory value of $Pr = 0.74$ for $T = 273.2\ K$, compared to an observed value of $Pr = 0.73$, is quoted in [2].

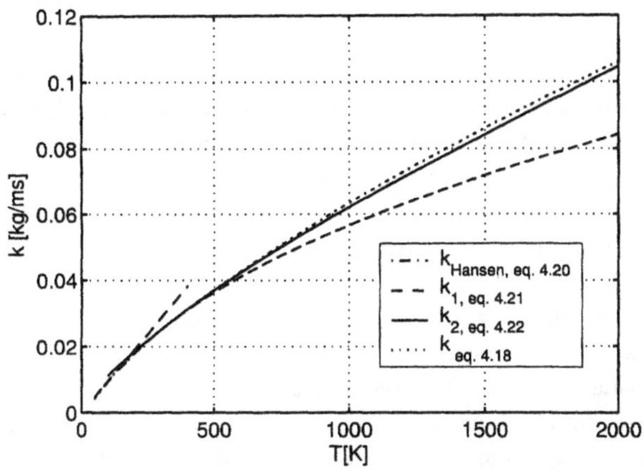

Fig. 4.4. Thermal conductivity k, different approximations, specific heat at constant pressure c_p/R, and the Prandtl number Pr of air, Table 4.4, as function of the temperature T.

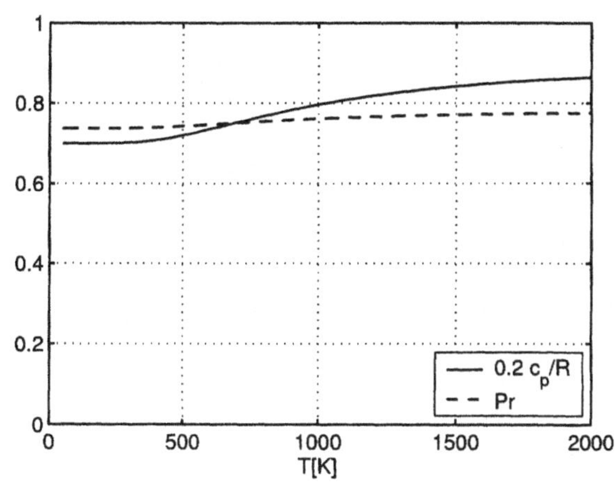

Fig. 4.5. Prandtl number Pr and specific heat at constant pressure c_p of air, Table 4.4, as function of the temperature T.

$$D_{AB} = D_{BA} = const. \frac{\sqrt{T^3 \left(\frac{1}{M_A} + \frac{1}{M_B}\right)}}{p\sigma_{AB}^2 \Omega_{D_{AB}}}. \tag{4.24}$$

The dimension of D_{AB} is m^2/s. Since we do not intend to derive and use approximate relations we renounce a detailed discussion and refer the reader to, for instance, [3], [2].

4.2.5 Computation Models

The relations for the determination of the transport properties viscosity μ and thermal conductivity k, which we have considered in the preceding subsections, are valid in the lower temperature domain, i. e., as long as dissociation doesn't occur. Air begins to dissociate at temperatures between 1,500.0 K and 2,000.0 K, depending on the density level. In general the following holds: the lower the density, the lower is the temperature at which dissociation occurs.

If appreciable dissociation is present, the transport properties are determined separately for each involved species. Mixing formulas (exact, and approximate ones such as Wilke's and other), [3], [2], are then employed in order to determine the transport properties of dissociated air.

Wilke's semi-empirical formula [2], for example, reads:

$$\mu_{mix} = \sum_{i=1}^{n} \frac{x_i \mu_i}{\sum_{j=1}^{n} x_i \Phi_{ij}}, \tag{4.25}$$

with

$$\Phi_{ij} = \frac{1}{\sqrt{8}} \left(1 + \frac{M_i}{M_j}\right)^{-0.5} \left[1 + \left(\frac{\mu_i}{\mu_j}\right)^{0.5} \left(\frac{M_j}{M_i}\right)^{0.25}\right]^2, \tag{4.26}$$

where μ_i, x_i and M_i are viscosity, mole fraction and molecular weight of species i, respectively, and j is a dummy subscript.

Wilke's formula is used in aerothermodynamics with good results, also for the determination of the thermal conductivity of gas mixtures. For multi-component diffusion coefficients the available mixing formulae, [2], are less satisfactory, [7].

The transport properties of gas mixtures are in both thermo-chemical equilibrium and non-equilibrium at temperatures above the temperatures, at which dissociation occurs, always functions of two thermodynamic variables, for example internal energy e, and density ρ.

The nowadays fully accepted theoretical base for the determination of transport properties is the Chapman-Enskog theory for gases at "low" density with extensions also for dissociated air in equilibrium or in non-equilibrium. In the methods of numerical aerothermodynamics curve-fitted state surfaces are employed, see, e. g., [8], in order to obtain in a fast and exact manner the needed data. They use extensive data bases, e. g., [9], [10], [11].

In general it is accepted that transport properties of multi-species air at the flow conditions considered here can be determined to a sufficient degree of accuracy [7], [12]. The situation is different with flow problems exhibiting very high temperatures, and with combustion and combustion-wake problems of hypersonic flight propulsion systems.

Examples of curve-fitted state surfaces of viscosity $\mu(\rho, e)$ and thermal conductivity $k(\rho, p/\rho)$ [8] for the typical pressure/density/internal energy range encountered by the flight vehicle classes in the background of this book are shown in Figs. 4.6 and 4.7.

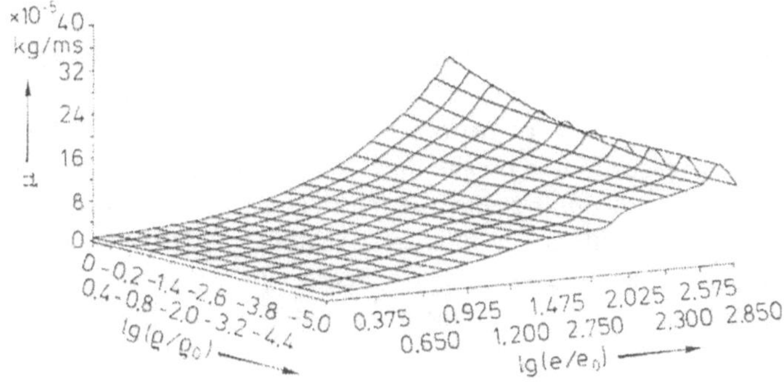

Fig. 4.6. Viscosity μ of air as function of density ρ and internal energy e [8], database from [9] ($\rho_0 = 1.243\ kg/m^3$, $e_0 = 78{,}408.4\ m^2/s^2$).

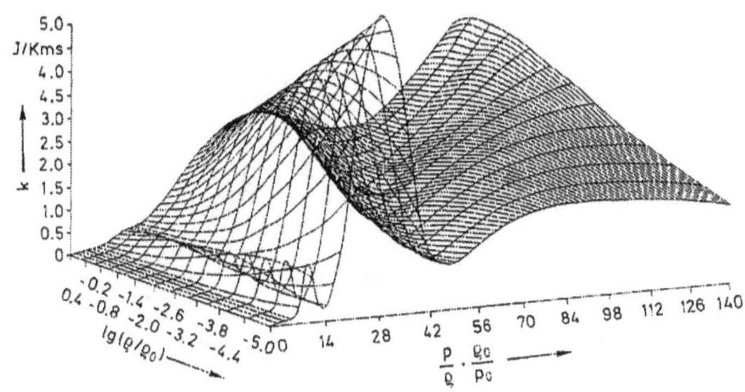

Fig. 4.7. Thermal conductivity k of air as function of density ρ and pressure/density p/ρ [8], database from [9] ($\rho_0 = 1.292\ kg/m^3$, $p_0 = 1.0133 \cdot 10^5\ Pa$).

4.3 Equations of Motion, Initial Conditions, Boundary Conditions, and Similarity Parameters

4.3.1 Transport of Momentum

Governing equations Momentum transport is described by means of the Navier-Stokes equations. Consider, for two-dimensional flow, the equation for momentum transport in x-direction:

$$\rho \frac{\partial u}{\partial t} + \rho u \frac{\partial u}{\partial x} + \rho v \frac{\partial u}{\partial y} = -\frac{\partial p}{\partial x} - \left(\frac{\partial \tau_{xx}}{\partial x} + \frac{\partial \tau_{yx}}{\partial y} \right) - \left(\frac{\partial \tau_{xx_{diff}}}{\partial x} + \frac{\partial \tau_{yx_{diff}}}{\partial y} \right). \tag{4.27}$$

The first term on the left-hand side represents the rate of increase of momentum in the unit volume with time, the second and third the gain of momentum by convective transport. On the right-hand side the first term stands for the pressure force on the unit volume, the first bracket for the gain of momentum by molecular transport, and the second bracket for the gain of momentum by mass diffusion. This term usually can be neglected.

The equation for the momentum transport in y-direction reads similarly, with the terms having the same meaning as before:

$$\rho \frac{\partial v}{\partial t} + \rho u \frac{\partial v}{\partial x} + \rho v \frac{\partial v}{\partial y} = -\frac{\partial p}{\partial y} - \left(\frac{\partial \tau_{xy}}{\partial x} + \frac{\partial \tau_{yy}}{\partial y} \right) - \left(\frac{\partial \tau_{xy_{diff}}}{\partial x} + \frac{\partial \tau_{yy_{diff}}}{\partial y} \right). \tag{4.28}$$

The components of the viscous stress tensor $\underline{\underline{\tau}}$ in eqs. (4.27) and (4.28) read for a compressible Newtonian fluid:

$$\tau_{xx} = -2\mu \frac{\partial u}{\partial x} + (\frac{2}{3}\mu - \kappa) \left(\frac{\partial u}{\partial x} + \frac{\partial v}{\partial y} \right), \tag{4.29}$$

$$\tau_{xy} = \tau_{yx} = -\mu \left(\frac{\partial u}{\partial y} + \frac{\partial v}{\partial x} \right), \tag{4.30}$$

$$\tau_{yy} = -2\mu \frac{\partial v}{\partial y} + (\frac{2}{3}\mu - \kappa) \left(\frac{\partial u}{\partial x} + \frac{\partial v}{\partial y} \right). \tag{4.31}$$

Here we meet another transport property, the bulk viscosity κ. It is connected to rotational non-equilibrium of polyatomic gases [3]. For practical purposes the formulation of this connection for air and its molecular constituents can be found in, e. g., [13]. For low-density monatomic gases $\kappa = 0$. For the flow problems considered here we can assume in general $\kappa \ll \mu$, and therefore neglect it.

By adding to eq. (4.27) the "global" continuity equation multiplied by u, Sub-Section 4.3.3:

$$u \left(\frac{\partial \rho}{\partial t} + \frac{\partial \rho u}{\partial x} + \frac{\partial \rho v}{\partial y} \right) = 0, \tag{4.32}$$

we find the conservative formulation[5] for the momentum transport in x-direction, see also Chapter 12:

[5] The governing equations need to be applied in conservative formulation in discrete numerical computation methods in order to capture shock-wave and slip surfaces, Section 10.2.

$$\frac{\partial(\rho u)}{\partial t} + \frac{\partial}{\partial x}\left(\rho u^2 + p + \tau_{xx}\right) + \frac{\partial}{\partial y}\left(\rho v u + \tau_{xy}\right) = 0. \tag{4.33}$$

The conservative formulation for the momentum transport in y-direction is found likewise:

$$\frac{\partial(\rho v)}{\partial t} + \frac{\partial}{\partial x}\left(\rho u v + \tau_{yx}\right) + \frac{\partial}{\partial y}\left(\rho v^2 + p + \tau_{yy}\right) = 0. \tag{4.34}$$

The shear stress exerted on the body surface, i. e. the transport of x-momentum in (negative) y-direction towards the surface, is found from eq. (4.30):

$$\tau_w = -\tau_{yx}|_w = \mu\left(\frac{\partial u}{\partial y} + \frac{\partial v}{\partial x}\right)|_w. \tag{4.35}$$

If distributed blowing or suction with a gradient in x-direction is not present, we arrive at the classical wall-shear stress relation:

$$\tau_w = -\tau_{xy}|_w = \mu_w \frac{\partial u}{\partial y}|_w. \tag{4.36}$$

For the surface pressure, which is a force normal to the surface, we don't have a relation, since it is an implicit result of the solution of the governing equations, [14].

Mach number, Reynolds number, and flow boundary-layer thickness We begin by comparing the convective x-momentum flux term and the pressure term in eq. (4.33):

$$\frac{\rho u^2}{p} = \frac{\rho u^2}{\rho R T} = \gamma M^2. \tag{4.37}$$

The Mach number M is defined by:

$$M = \frac{u}{a}, \tag{4.38}$$

with a the speed of sound:

$$a = \left(\frac{\partial p}{\partial \rho}\right)_s = \sqrt{\gamma R T}. \tag{4.39}$$

The magnitude of the Mach number governs compressibility effects in fluid flow, Chapter 6. Here we employ it only in order to distinguish flow types:

- $M = 0$: incompressible flow.

- $M \lessapprox M_{crit,lower}$: subsonic flow[6].
- $M_{crit,lower} \lessapprox M \lessapprox M_{crit,upper}$: transonic flow[7].
- $M_{crit,upper} \lessapprox M \lessapprox 5$: supersonic flow.
- $M \gtrapprox 5$: hypersonic flow.

Hypersonic flow usually is defined as flow at speeds larger than those at which first appreciable high-temperature real-gay effects occur: $M \approx 5$. Oswatitsch, [15], defines hypersonic flow with $M \to \infty$, Section 6.8. In practice this means a Mach number large enough that the Mach number independence principle holds.

Noting that the momentum flux is a vector entity, we compare now in a very schematic way the convective and the molecular x-momentum flux in x-direction in the first large bracket of eq. (4.33) after introducing the simple proportionality $\tau_{xx} \sim \mu(u/L)$, which here does not anticipate the presence of a boundary layer:

$$\frac{\rho u^2}{\tau_{xx}} \sim \frac{\rho u^2}{\mu(u/L)} = \frac{\rho u L}{\mu} = Re, \tag{4.40}$$

and find in this way the Reynolds number Re. The Reynolds number is the principle similarity parameter governing viscous phenomena, Chapter 7.

The following limiting cases of Re can be distinguished:

- $Re \to 0$: the molecular transport of momentum is much larger than the convective transport, the flow is the "creeping" or Stokes flow (see, e. g., [2]): the convective transport can be neglected.
- $Re \to \infty$: the convective transport of momentum is much larger than the molecular transport, the flow can be considered as inviscid, i. e. molecular transport can be neglected. The governing equations are the Euler equations, i. e. in two dimensions eqs. (4.27) and (4.28) without the molecular and mass-diffusion transport terms. If the flow is also irrotational, they can be reduced to the potential equation.
- $Re = O(1)$: the molecular transport of momentum has the same order of magnitude as the convective transport, the flow is viscous, i. e. it is boundary-layer, or in general, shear-layer flow[8].

---

[6] $M_{crit,lower}$ is the lower critical Mach number, at which first supersonic flow at the body surface appears. For airfoils $M_{crit,lower} \approx 0.7$ to 0.8. Usually $M_{crit,lower}$ is simply called critical Mach number M_{crit}

[7] $M_{crit,upper}$ is the upper critical Mach number, at which the flow past the body is fully supersonic. For airfoils $M_{crit,upper} \approx 1.1$ to 1.2.

[8] Note that in boundary-layer theory the boundary-layer equations are found for $Re \to \infty$, however, only after the "boundary-layer stretching" has been introduced, Sub-Section 7.1.3.

We refrain from a discussion of the general meaning of the Reynolds number as a similarity parameter. This can be found in text books on fluid mechanics.

For $Re = O(1)$ the convective transport of x-momentum in x-direction ρu^2 is compared now with the molecular transport of x-momentum in y-direction τ_{xy}, anticipating a boundary layer with the (asymptotic) thickness δ. We do this in the differential form given with eq. (4.27), assuming steady flow, and neglecting the second term of τ_{yx} in Eq.(4.30):

$$\rho u \frac{\partial u}{\partial x} + \cdots \approx \cdots + \frac{\partial}{\partial y}(\mu \frac{\partial u}{\partial y}) + \cdots. \tag{4.41}$$

Again we introduce in a schematic way characteristic values and find:

$$\rho u \frac{u}{L} \sim \frac{\mu u}{\delta^2}. \tag{4.42}$$

After rearrangement we obtain for the boundary-layer thickness δ:

$$\frac{\delta}{L} \sim \sqrt{\frac{\mu}{\rho u L}} \sim \frac{1}{\sqrt{Re_L}}, \tag{4.43}$$

and, using the boundary-layer running length x as characteristic length:

$$\frac{\delta}{x} \sim \sqrt{\frac{\mu}{\rho u x}} \sim \frac{1}{\sqrt{Re_x}}. \tag{4.44}$$

This boundary-layer thickness is the thickness of the flow or ordinary boundary layer $\delta \equiv \delta_{flow}$, [16]. We will identify below with the same kind of consideration the thermal, as well as the diffusion boundary layer with thicknesses, which are different from the flow boundary-layer thickness. The problem of defining actual boundary-layer thicknesses is treated in Sub-Section 7.2.1.

Boundary conditions The Navier-Stokes equations (4.27) and (4.28) have derivatives of second order of the velocity components u and v in both x- and y-direction. Hence we have to prescribe two boundary conditions for each velocity component. One pair (in two dimensions) of the boundary conditions is defined at the body surface, the other for external flow problems (far-field or external boundary conditions), in principle, at infinity away from the body. For internal flows, e. g., inlet flows, diffuser-duct flows, et cetera, boundary conditions are to be formulated in an appropriate way.

We treat first the wall-boundary conditions for u and v, and consider the situation in both the continuum and the slip-flow regime, Section 2.3.

For the tangential flow component u_w at a body surface we get [17]:

$$u_w = \frac{2-\sigma}{\sigma} \lambda \frac{\partial u}{\partial y}|_w + \frac{3}{4} \frac{\mu}{\rho T} \frac{\partial T}{\partial x}|_w. \tag{4.45}$$

Here σ is the reflection coefficient, which is depending on the pairing gas/surface material : $0 \leqq \sigma \leqq 1$. Specular reflection is given with $\sigma = 0$, and diffusive reflection with $\sigma = 1$. Specular reflection indicates perfect slip, i. e. $\sigma \to 0$: $(\partial u/\partial y)|_w \to 0$. For air on any surface usually $\sigma = 1$ is chosen. In Section 9.4 we will show results of a study on varying reflection coefficients.

The second term in the above equation

$$\frac{3}{4}\frac{\mu}{\rho T}\frac{\partial T}{\partial x}|_w$$

in general is not taken into account in aerothermodynamic computation models. It induces at the wall a flow in direction of increasing temperature. On radiation-cooled surfaces with the initially steep decrease of the radiation-adiabatic temperature in main flow direction it would reduce the magnitude of slip flow. No results are available in this regard. However, the term can be of importance for measurement devices for both hypersonic hot ground-simulation facilities and flight measurements in the slip-flow regime.

After the mean free path, eq. (2.12), has been introduced into the second term on the right-hand side, and after rearrangement, eq. (4.45) can be written in terms of the reference Knudsen number $Kn_{ref} = \lambda_{ref}/L_{ref}$:

$$\begin{aligned} u_w = &\frac{2-\sigma}{\sigma} Kn_{ref}(\lambda/\lambda_{ref})\frac{\partial u}{\partial(y/L_{ref})}|_w + \\ &+ \frac{15}{64}\sqrt{2\pi R}\, Kn_{ref}(\lambda/\lambda_{ref})\frac{1}{\sqrt{T}}\frac{\partial T}{\partial(x/L_{ref})}|_w . \end{aligned} \tag{4.46}$$

We find now in accordance with Section 2.3:

- the classical no-slip boundary condition for the continuum-flow regime:

$$Kn_{ref} \lessapprox 0.01 : \ u_w = 0, \tag{4.47}$$

- and the slip-boundary condition eq. (4.45) for the slip-flow regime:

$$0.01 \lessapprox Kn_{ref} \lessapprox 0.1 : \ u_w > 0. \tag{4.48}$$

The reference Knudsen number Kn_{ref} must be chosen according to the flow under consideration. For boundary-layer flow, for example, the length scale L_{ref} would be the boundary-layer thickness δ. Further it should be remembered that there are no sharp boundaries between the continuum and the slip-flow flow regime.

The boundary condition for the normal flow component v at the body surface usually is:

$$v_w = 0. \tag{4.49}$$

If there is suction or blowing through the surface, of course we get

$$v_w \neq 0 \tag{4.50}$$

according to the case under consideration. Whether the suction or blowing orifices can be considered as continuously distributed or discrete orifices must be taken into account. For blowing through the surface also total pressure and enthalpy of the fluid, as well as its composition must be prescribed.

Far-field or external boundary conditions must, as initially mentioned, in principle be prescribed at infinity away from the body surface, also in downstream direction. In reality a sufficiently large distance from the body is chosen. Because a flight vehicle induces velocity overshoots and velocity/total pressure defects in its vicinity and especially in the wake (due to lift, induced and viscous drag), far-field boundary conditions must ensure a passage of the flow out of the computation domain without upstream damping and reflections. For this reason special formulations of the far-field boundary conditions are introduced, especially for subsonic and transonic flow computation cases.

In supersonic and hypersonic flows, depending on the employed computation scheme, the upstream external boundary conditions can be prescribed just ahead of the bow shock if bow-shock capturing is used, or at the bow shock (via the Rankine-Hugoniot conditions, Section 6.1), if bow-shock fitting is employed. Then downstream of the body appropriate far-field boundary conditions must be given, as mentioned above.

The situation is different with two-layer computation methods, like coupled Euler/boundary-layer methods. Then external boundary conditions, found with solutions of the Euler equations, are applied at the outer edge of the boundary layer.

4.3.2 Transport of Energy

Governing equations Energy is a scalar entity. Energy transport is described by means of the energy equation. From the many possible formulations, see, e. g., [2], we choose for our initial considerations the conservative flux-vector formulation[9] (we assume here also two-dimensional flow):

$$\frac{\partial}{\partial t}\left[\rho(e+\frac{1}{2}V^2)\right] = -\left(\underline{\nabla}\cdot \underline{q}_e\right). \tag{4.51}$$

The term on the left-hand side represents the rate of increase of energy in the unit volume with time, the first term on the right-hand side the gain of energy.

The internal energy of our our model air - a mixture of thermally perfect gases - is defined by:

[9] This formulation is also employed in Chapter 12, where the whole set of governing equations is collected.

$$e\,(T,\rho) = \int c_v(T,\rho)\, dT. \tag{4.52}$$

It is composed of the contributions of the molecular (i^m) and the atomic (i^a) species:

$$e = \sum_{i^m=1}^{n^m} \omega_{i^m} e_{i^m} + \sum_{i^a=1}^{n^a} \omega_{i^a} e_{i^a}, \tag{4.53}$$

which have the parts (see Section 5.2)

$$e_{i^m} = e_{trans_{i^m}} + e_{rot_{i^m}} + e_{vibr_{i^m}} + e_{el_{i^m}} + \Delta e_{i^m}, \tag{4.54}$$

and

$$e_{i^a} = e_{trans_{i^a}} + e_{el_{i^a}} + \Delta e_{i^a}. \tag{4.55}$$

e_{trans} is the translation, e_{rot} the rotation, e_{vibr} the vibration, e_{el} the electronic excitation, and Δe the zero-point or formation energy, [3].

V is the magnitude of the velocity vector:

$$V = |\underline{V}|, \tag{4.56}$$

$\underline{\nabla}$ the Nabla operator:

$$\underline{\nabla} = \underline{e}_x \frac{\partial}{\partial x} + \underline{e}_y \frac{\partial}{\partial y}, \tag{4.57}$$

and $\underline{q}_e$ the energy-flux vector:

$$\underline{q}_e = \rho(e + \frac{1}{2}V^2)\underline{V} + \underline{q} + p\underline{V} + \underline{\underline{\tau}} \cdot \underline{V}. \tag{4.58}$$

The first term on the right-hand side represents the convective transport of energy into the unit volume, the second the molecular transport of energy, the third the work on the fluid by pressure forces (compression work), and the last by viscous forces (dissipation work).

The molecular transport of energy, $\underline{q}$, in general, i. e. chemical non-equilibrium included, has two parts:

$$\underline{q} = -k\underline{\nabla}T + \sum_{i=1}^{n} \underline{j}_i h_i. \tag{4.59}$$

The first stands for the molecular transport of thermal energy (heat conduction), and the second for the transport of thermal energy by mass diffusion due to chemical non-equilibrium of air as a mixture of thermally perfect gases.

In eq. (4.59) h_i is the enthalpy of the species i:

$$h_i = \int c_{p_i} dT, \tag{4.60}$$

and $\underline{j}_i$ the diffusion mass flux.

We find the heat transported towards the body surface, q_{gw}, from the y-component of the energy-flux vector at the wall:

$$q_{gw} = \underline{q}_{e_y}|_w = \left[\rho(e + \frac{1}{2}V^2)v + q_y + pv - \mu v\left(2\frac{\partial v}{\partial y} - \frac{2}{3}(\frac{\partial u}{\partial x} + \frac{\partial v}{\partial y})\right) - \right.$$
$$\left. -\mu u(\frac{\partial u}{\partial y} + \frac{\partial v}{\partial x})\right]_w . \tag{4.61}$$

If $v_w = 0$, that is without suction or blowing at the wall, this equation reduces to

$$q_{gw} = -k_w \frac{\partial T}{\partial y}|_w + \left(\sum_i j_{i_y} h_i\right)_w - \left(\mu u \frac{\partial u}{\partial y}\right)_w . \tag{4.62}$$

The first term is the classical heat-conduction term, the second the heat transport in chemical non-equilibrium flow, and the third finally a heat flux, which in the slip-flow regime appears in addition to the two other fluxes, [18].

Peclét number, Prandtl number, Lewis number, Eckert number, and thermal boundary-layer thickness Noting that the energy flux is a scalar entity we compare now, in the same way as we did for the momentum flux, the convective and the molecular flux in x-direction.

To ease the discussion we introduce a simpler, and more familiar form of the energy equation in terms of the enthalpy h [2]:

$$\rho\frac{\partial h}{\partial t} + \rho u\frac{\partial h}{\partial x} + \rho v\frac{\partial h}{\partial y} = -\frac{\partial q_x}{\partial x} - \frac{\partial q_y}{\partial y} + \frac{\partial p}{\partial t} + u\frac{\partial p}{\partial x} + v\frac{\partial p}{\partial y} -$$
$$- \left[\tau_{xx}\frac{\partial u}{\partial x} + \tau_{yy}\frac{\partial v}{\partial y} + \tau_{xy}(\frac{\partial u}{\partial y} + \frac{\partial v}{\partial x})\right] . \tag{4.63}$$

By adding to the left-hand side of this equation the global continuity equation times the enthalpy we find, while expressing the mass-diffusion energy-transport term in eq. (4.59) for a binary gas, the conservative form of convective and molecular transport:

$$\frac{\partial \rho h}{\partial t} + \frac{\partial}{\partial x}\left(\rho u h + k\frac{\partial T}{\partial x} + \rho D_{AB}\sum_{\alpha=A,B} h_\alpha \frac{\partial \omega_\alpha}{\partial x}\right) +$$
$$+\frac{\partial}{\partial y}\left(\rho v h + k\frac{\partial T}{\partial y} + \rho D_{AB}\sum_{\alpha=A,B} h_\alpha \frac{\partial \omega_\alpha}{\partial y}\right) = \cdots . \tag{4.64}$$

Assuming perfect gas with $h = c_p T$, and not anticipating a thermal boundary layer, we compare the convective and the conductive transport in x-direction (the first two terms in the first bracket on the left-hand side) after introduction of the simple proportionality $\partial T/\partial x \sim T/L$:

$$\frac{\rho u c_p T}{k(\partial T/\partial x)} \sim \frac{\rho u c_p T}{k(T/L)} = \frac{\rho u c_p L}{k} = Pe = \frac{\mu c_p}{k} \frac{\rho u L}{\mu} = Pr\ Re, \tag{4.65}$$

and find in this way the Peclét number:

$$Pe = \frac{\rho u c_p L}{k}, \tag{4.66}$$

and the Prandtl number:

$$Pr = \frac{Pe}{Re} = \frac{\mu c_p}{k}. \tag{4.67}$$

The Prandtl number Pr can be written:

$$Pr = \frac{\mu/\rho}{k/\rho c_p} = \frac{\mu/\rho}{\alpha}, \tag{4.68}$$

where

$$\alpha = \frac{k}{\rho c_p} \tag{4.69}$$

is the thermal diffusivity [4], which is a property of the conducting material. The Prandtl number Pr hence can be interpreted as the ratio of kinematic viscosity $\nu = \mu/\rho$ to thermal diffusivity α.

Of interest are the limiting cases of of the Péclet number Pe (compare with the limiting cases of the Reynolds number Re):

- $Pe \to 0$: the molecular transport of heat is much larger than the convective transport.
- $Pe \to \infty$: the convective transport of heat is much larger than the molecular transport.
- $Pe = O(1)$: the molecular transport of heat has the same order of magnitude as the convective transport.

Comparing now the convective transport and the transport by mass diffusion (the first and the third term in the first bracket of eq. (4.64)), also after introduction of a simple proportionality, $\partial \omega_\alpha/\partial x \sim \omega_\alpha/L$, and assuming $\omega_\alpha = 1$, we obtain:

$$\frac{\rho u c_p T}{\rho D_{AB} \sum_{\alpha=A,B} h_\alpha (\partial \omega_\alpha / \partial x)} \sim \frac{\rho u c_p T}{\rho D_{AB} c_p T (\omega_\alpha / L)} = \\ = \frac{\rho u L}{\mu} \frac{\mu c_p}{k} \frac{k}{\rho D_{AB} c_p} = \frac{Re Pr}{Le}. \tag{4.70}$$

We find in this way the Lewis number:

$$Le = \frac{\rho D_{AB} c_p}{k}, \tag{4.71}$$

which is interpreted as the ratio 'heat transport by mass diffusion' to 'heat transport by conduction' in a flow with chemical non-equilibrium. In the temperature and density/pressure range of interest in this book we have $0.5 \lessapprox Le \lessapprox 1.5$ [6].

If we non-dimensionalize eq. (4.63) without the time derivative, and with proper reference data (p is non-dimensionalized with ρu^2), we find:

$$\rho u c_p \frac{\partial T}{\partial x} + \cdots = \frac{1}{RePr} \left(\frac{\partial}{\partial x} (k \frac{\partial T}{\partial x}) + \cdots \right) + E \left(u \frac{\partial p}{\partial x} + \cdots \right) + \\ + \frac{E}{Re} \left[2\mu \frac{\partial^2 u}{(\partial x)^2} + \cdots \right]. \tag{4.72}$$

All entities in this equation are dimensionless. The new parameter is the Eckert number:

$$E = (\gamma - 1) M^2, \tag{4.73}$$

with the Mach number defined by eq. (4.38). The Eckert number can be interpreted as ratio of kinetic energy to thermal energy of the flow.

For $E \to 0$, respectively $M \to 0$, we find the incompressible case, in which of course a finite energy transport by both convection and conduction can happen, but where no compression work is done on the fluid, and also no dissipation work occurs. For $E = 0$ actually fluid mechanics and thermodynamics are decoupled.

We compare now for $Pe = O(1)$ the convective transport of heat in x-direction $\rho u c_p T$ with the molecular transport of heat in y-direction q_y, anticipating a thermal boundary layer with the thickness δ_T. We do this in the differential form given with eq. (4.64):

$$\rho u c_p \frac{\partial T}{\partial x} + \cdots \approx \cdots + \frac{\partial}{\partial y} (k \frac{\partial T}{\partial y}) + \cdots \tag{4.74}$$

Again we introduce in a schematic way characteristic data and find after rearrangement:

$$\rho u c_p \frac{T}{L} \sim \frac{kT}{\delta_T^2}. \tag{4.75}$$

From this we find the thickness δ_T of the thermal boundary layer:

$$\frac{\delta_T}{L} \sim \sqrt{\frac{k}{c_p \rho u L}} \sim \frac{1}{\sqrt{Pe_L}} \sim \frac{1}{\sqrt{Re_L Pr}}, \tag{4.76}$$

and, using again the boundary-layer running length x as characteristic length:

$$\frac{\delta_T}{x} \sim \sqrt{\frac{k}{c_p \rho u x}} \sim \frac{1}{\sqrt{Pe_x}} \sim \frac{1}{\sqrt{Re_x Pr}}. \tag{4.77}$$

The thickness of the thermal boundary layer δ_T is related to the thickness of the flow boundary layer $\delta \equiv \delta_{flow}$ by

$$\frac{\delta_T}{\delta} \sim \frac{1}{\sqrt{Pr}}. \tag{4.78}$$

We find now regarding the limiting cases of Pr [19]:

- $Pr \to 0$: the thermal boundary layer is much thicker than the flow boundary layer, which is typical for the flow of liquid metals.
- $Pr \to \infty$: the flow boundary layer is much thicker than the thermal boundary layer, which is typical for liquids.
- $Pr = O(1)$: the thermal boundary layer has a thickness of the order of that of the flow boundary layer. This is typical for gases, in our case air. However, since in the interesting temperature and density/pressure domain $Pr < 1$ [6], the thermal boundary layer is somewhat thicker than the flow boundary layer.

Boundary conditions The energy equation, either in the form of eq. (4.51), or in the form of eq. (4.63), has terms of second order of the temperature T in both the x- and the y-direction. Hence we have to prescribe two boundary conditions. Like for momentum transport, one is defined at the body surface, the other for external flow problems (far-field or external boundary conditions) in principle at infinity away from the body. In addition we have to prescribe boundary conditions for the heat transport by mass diffusion in chemical non-equilibrium flow, and by velocity slip, eq. (4.62).

First we treat the ordinary heat-flux term. We have seen in Section 3.1 that five different situations regarding the thermal state at the body surface are of practical interest. Before we look at the corresponding boundary conditions, we consider the general wall-boundary condition for T in both the continuum and the slip-flow regime, Section 2.3.

The general boundary condition for T at a body surface reads [17]:

$$T_{gw} = T_w + \frac{2-\alpha}{\alpha} \frac{2\gamma}{\gamma+1} \frac{\lambda}{Pr} \frac{\partial T}{\partial y}|_w. \tag{4.79}$$

Here α is the thermal accommodation coefficient, which depends on the pairing gas/surface material : $0 \leqq \alpha \leqq 1$. Specular reflection, which means vanishing energy exchange, is given with $\alpha = 0$, and diffusive reflection, indicating reflection accommodated to the surface temperature T_w with $\alpha = 1$. Specular reflection hence indicates perfect decoupling of the temperature of the gas at the wall T_{gw} from the wall temperature T_w, i. e. $\alpha \to 0$: $(\partial T/\partial y)|_w \to 0$. For air on any surface usually $\alpha = 1$ is chosen.

Eq. (4.79) can be written in terms of the reference Knudsen number $Kn_{ref} = \lambda_{ref}/L_{ref}$:

$$T_{gw} = T_w + \frac{2-\alpha}{\alpha}\frac{2\gamma}{\gamma+1}Kn_{ref}\frac{(\lambda/\lambda_{ref})}{Pr}\frac{\partial T}{\partial(y/L_{ref})}|_w. \tag{4.80}$$

We find now, again in accordance with Section 2.3:

– the classical wall boundary condition in the continuum-flow regime:

$$Kn_{ref} \lessapprox 0.01 : \; T_{gw} = T_w, \tag{4.81}$$

– and the temperature-jump condition eq. (4.45) in the slip-flow regime:

$$0.01 \lessapprox Kn_{ref} \lessapprox 0.1 : \; T_{gw} \neq T_w. \tag{4.82}$$

Again the reference Knudsen number must be chosen according to the flow under consideration, e. g., for boundary-layer flow it would be based on the boundary-layer thickness. And also here it should be remembered that there are no sharp boundaries between the two flow regimes.

With regard to the boundary conditions, which represent the five different situations of the thermal state of the surface (Section 3.1), we consider only the continuum-flow regime case with $T_w \equiv T_{gw}$ and find:

1. Radiation-adiabatic wall: $q_{gw} = q_{rad} \to k\frac{\partial T}{\partial y}|_w = \epsilon\sigma T_w^4$.
2. Wall temperature at wall without radiation cooling: T_w.
3. Adiabatic wall: $q_{gw} = 0 \to \frac{\partial T}{\partial y}_w = 0$.
4. Wall temperature at wall with radiation cooling: T_w, and $q_{rad} = \epsilon\sigma T_w^4$.
5. Wall-heat flux (into the wall material) is prescribed: $q_w = q_{gw}$ - q_{rad}.

There remains to consider the energy transport by mass diffusion in chemical non-equilibrium flow. The wall-boundary conditions of the diffusion flux j_{i_y} will be treated in the following Sub-Section 4.3.3. Important are the cases of finite and full catalytic recombination of atoms at the surface, Section 5.6. Catalytic recombination enhances strongly the heat transport towards the surface. On the one hand this is due to the release of dissociation energy in the recombination process. On the other hand, since the atomic species disappear partly or fully at the surface, the mass-diffusion flux as such is enlarged.

Finally the velocity-slip term in eq. (4.62) is recalled. The corresponding boundary conditions are found in Sub-Section 4.3.1.

Regarding external boundary conditions the same holds for the energy equation as for the Navier-Stokes equations. Special situations exist for internal flows, e. g., in inlets, diffuser ducts et cetera, which we do not discuss here.

4.3.3 Transport of Mass

Governing equations Mass is a scalar entity. Mass transport is described by means of the (global) continuity equation (two-dimensional):

$$\frac{\partial \rho}{\partial t} + \frac{\partial \rho u}{\partial x} + \frac{\partial \rho v}{\partial y} = 0. \tag{4.83}$$

The first term on the left-hand side represents the rate of increase of mass in the unit volume with time, the second and the third term the gain of mass by convective transport.

In the case of chemical non-equilibrium we have a "species" continuity equation for every involved species[10] [2]:

$$\frac{\partial \rho_i}{\partial t} + \frac{\partial \rho_i u}{\partial x} + \frac{\partial \rho_i v}{\partial y} = -\left(\frac{\partial j_{i_x}}{\partial x} + \frac{\partial j_{i_y}}{\partial y}\right) + Sm_i. \tag{4.84}$$

The terms in the bracket on the right-hand side represent the mass transport by diffusion, and Sm_i the source term of the species, Sections 5 and 12. The terms j_{i_x} and j_{i_y} are the components of the mass-flux vector $\underline{j}_i$ of the species i. This diffusive mass-flux is a flux relative to the convective (bulk-flow) mass flux $\rho\underline{V}$ [2]. Summation over all species continuity equations (4.84) results in the global continuity equation (4.83).

The diffusion mass-flux vector, which we have mentioned already in the preceding Section 4.3.2, has four parts [2]:

$$\underline{j}_i = \underline{j}_i^{(\omega)} + \underline{j}_i^{(p)} + \underline{j}_i^{(T)} + \underline{j}_i^{(g)}. \tag{4.85}$$

For convenience we write these four parts in terms of a binary mixture with the species A and B:

– Mass diffusion due to a concentration gradient $\underline{\nabla}\omega_A$ (Fick's first law):

$$\underline{j}_A^{(\omega)} = -\underline{j}_B^{(\omega)} = -\rho D_{AB}\underline{\nabla}\omega_A. \tag{4.86}$$

[10] Hence we have the global and n species continuity equations, one more than needed. In praxis one can use the superfluous equation for an accuracy check.

– Mass diffusion due to a pressure gradient $\underline{\nabla} p_A$:

$$\underline{j}_A^{(p)} = -\underline{j}_B^{(p)} = -\frac{M_B - M_A}{R_0 T} D_{AB}\, \omega_A (1-\omega_A) \underline{\nabla} p_A. \tag{4.87}$$

– Mass diffusion due to a temperature gradient $\underline{\nabla} T_A$ (D_A^T is the thermo-diffusion coefficient):

$$\underline{j}_A^{(T)} = -\underline{j}_B^{(T)} = -\frac{1}{T} D_A^T\, \underline{\nabla} T_A. \tag{4.88}$$

– Mass diffusion due to body forces:

$$\underline{j}_A^{(g)} = -\underline{j}_B^{(g)}. \tag{4.89}$$

Of these four diffusion mechanisms the last two usually can be neglected in aerothermodynamics. Pressure-gradient diffusion may play a role in oblique shocks [18], curved nozzles, et cetera. In aerothermodynamic diffusion problems we deal with multi-component mixtures, hence we have to use generalized relations, given for instance in [2].

Finally we write eq. (4.84) in conservative form, and in terms of the mass fraction ω_i:

$$\frac{(\partial \rho \omega_i)}{\partial t} + \frac{\partial}{\partial x}\left(\rho \omega_i u + j_{i_x}\right) + \frac{\partial}{\partial y}\left(\rho \omega_i v + j_{i_y}\right) = S m_i. \tag{4.90}$$

Schmidt number, Lewis number and mass-concentration boundary-layer thickness Noting that mass is a scalar entity we compare now, in the same way as we did for the energy flux, the convective and the molecular flux in x-direction, however only in terms of binary concentration-driven diffusion, eq. (4.86):

$$\frac{\rho \omega_A u}{\rho D_{AB} (\partial \omega_A / \partial x)} \sim \frac{\rho \omega_A u}{\rho D_{AB} (\omega_A / L)} = \frac{\rho u c_p L}{k} = \frac{\mu}{\rho D_{AB}} \frac{\rho u L}{\mu} = Sc\, Re, \tag{4.91}$$

and find the Schmidt number:

$$Sc = \frac{\mu}{\rho D_{AB}}, \tag{4.92}$$

which is related to the Lewis number Le via the Prandtl number Pr:

$$Le = \frac{Pr}{Sc} = \frac{\rho c_p D_{AB}}{k} = \frac{D_{AB}}{\alpha}. \tag{4.93}$$

The following limiting cases of Sc can be distinguished:

- $Sc \to 0$: the molecular transport of mass is much larger than the convective transport.
- $Sc \to \infty$: the convective transport of mass is much larger than the molecular transport.
- $Sc = O(1)$: the molecular transport of mass has the same order of magnitude as the convective transport.

We compare now for $Sc = O(1)$ the convective transport of mass in x-direction $\rho\omega_i u$ with the molecular transport of mass in y-direction j_y, anticipating a mass-concentration boundary layer with the thickness δ_M. We do this in the form given with eq. (4.84) after introducing the mass fractions ω_A and ω_B:

$$\frac{\partial \rho\omega_A u}{\partial x} + \cdots \approx \cdots + \frac{\partial}{\partial y}(\rho D_{AB}\frac{\partial \omega_A}{\partial y}) + \cdots. \quad (4.94)$$

Again we introduce in a schematic way characteristic data and find:

$$\frac{\rho\omega_A u}{L} \sim \frac{\rho D_{AB}\omega_A}{\delta_M^2}. \quad (4.95)$$

From this relation we obtain the thickness δ_M of the mass-concentration boundary layer :

$$\frac{\delta_M}{L} \sim \sqrt{\frac{D_{AB}}{uL}} \sim \frac{1}{\sqrt{Re_L Sc}}, \quad (4.96)$$

and in terms of the boundary-layer running length x:

$$\frac{\delta_M}{x} \sim \sqrt{\frac{D_{AB}}{ux}} \sim \frac{1}{\sqrt{Re_x Sc}}. \quad (4.97)$$

The thickness of the mass-concentration boundary layer δ_M is related to the thickness of the flow boundary layer $\delta \equiv \delta_{flow}$ by

$$\frac{\delta_M}{\delta} \sim \frac{1}{\sqrt{Sc}}. \quad (4.98)$$

We can distinguish again three cases:

- $Sc \to 0$: the mass-concentration boundary layer is much thicker than the flow boundary layer.
- $Sc \to \infty$: the flow boundary layer is much thicker than the mass-concentration boundary layer.
- $Sc = O(1)$: the mass-concentration boundary layer has a thickness of the order of that of the flow boundary layer.

The above discussion has shown that mass transport can be described analogously to momentum and energy transport. However, the discussion is somewhat academic regarding thermo-chemical non-equilibrium flows, because mass transport is not necessarily dominated by the surface boundary conditions, like in the case of the other entities. An exception possibly is the flow of a strongly dissociated gas past a fully catalytic surface.

Boundary conditions The global continuity equation, eq. (4.83), has first-order derivatives of the velocity components u, v, and the density ρ only. At the body surface the boundary conditions for u and v are those discussed for momentum transport in Sub-Section 4.3.1. For the density ρ, like for the pressure p, only a wall-compatibility equation, i. e. no boundary condition, can be prescribed at the body surface [14]. Hence, if an explicit boundary condition is necessary for ρ, it must be described at the far-field or external boundary.

The situation is different with the species-continuity equations (4.84), (4.90). Here we have second-order derivatives with each of the diffusion-driving mechanisms. Regarding the boundary conditions at the body surface we note without detailed discussion four possible general cases (see also Section 5.6):

– equilibrium conditions for the species:

$$\omega_i|_w = \omega_i(\rho, T)|_w,$$

– vanishing mass-diffusion fluxes:

$$j_{i_y}|_w = 0,$$

– fully catalytic surface recombination:

$$\omega_{atom_i}|_w = 0,$$

– finite catalytic surface recombination (k_w is the catalytic recombination rate at the wall):

$$\omega_i|_w = \omega_i(k_w, \rho, T).$$

Regarding far-field or external boundary conditions the same holds for the continuity equations (global and species) as for the Navier-Stokes equations and the energy equation.

4.4 Remarks on Similarity Parameters

In the preceding sub-sections we have studied the governing equations of fluid flow. We considered especially wall boundary conditions and similarity parameters. The latter we derived in an intuitive way by comparing flow entities of the same kind, e. g., convective and molecular transport of momentum in order to define the Reynolds number.

The Π or *Pi* theorem, see, e. g., [20], permits to perform dimensional analysis in a rigorous way. For us it is of interest that it yields parameters additional to the basic similarity parameters, which we derived above.

For the problems of viscous aerothermodynamics the ratio of wall temperature to free-stream temperature

$$\frac{T_w}{T_\infty}$$

is a similarity parameter [21].

A more general form is given in [16]:

$$\frac{T_w - T_{ref}}{T_{ref}}.$$

Other similarity parameters are the Damköhler numbers, Section 5.4, concerning reacting fluid flow, but also the binary scaling parameter ρL, Section 10.3.

There are two aspects to deal with similarity parameters. The first is that they permit, as we did above, to identify, distinguish, and model mathematically flow phenomena ("phenomena modeling"), for example, subsonic, transonic, supersonic and hypersonic flow.

The other aspect ("ground-facility simulation") is that in aerothermodynamics, like in aerodynamics, experimental simulation in ground-simulation facilities is performed with sub-scale models of the real flight vehicle.

That this is possible in principle is first of all due to the fact that our simulation problems are Galilean invariant, Section 4.1. Secondly it is necessary that the relevant flight similarity parameters are fulfilled. This is a basic problem in aerodynamic and aerothermodynamic ground-facility simulation, because in general only a few of these parameters can be duplicated. A special problem is the thermal state of the surface. Nowadays ground-facility simulation models have cold and thermally uncontrolled surfaces. We will come back to the ground-facility simulation problem in Section 10.3.

For both aspects, however, it is important to use proper reference data for the determination of similarity parameters. For phenomena modeling purposes to a certain degree fuzzy data can be used. For ground-facility simulation the data should be as correct as possible, even when the resulting similarity parameters cannot be duplicated. This is necessary in order to estimate kind and magnitude of simulation uncertainties and errors, Section

10.1. In design work margins are governed by these uncertainties and errors in concert with the design sensitivities, [22].

4.5 Problems

Problem 4.1 A flight vehicle model in a ground-simulation facility has a length of 0.10 m. The free-stream velocity in the test section is 3,000.0 m/s. How long is the residence time t_{res}, how long should the measurement time $t_{meas} = t_{ref}$ be, if a Strouhal number $Sr \leqq 0.2$ is demanded?

Solution: $t_{meas} \geqq 1.67 \cdot 10^{-4}\ s$.

Problem 4.2 Compute μ_{Suth}, μ_1, μ_2 for air at $T = 500.0\ K$. What are the differences $\Delta\mu_1$ and $\Delta\mu_2$ of μ_1 and μ_2 compared to μ_{Suth}?

Solution: Check results with the help of Fig. 4.3.

Problem 4.3 Compute k_{Han}, k_1, k_2 for air at $T = 500.0\ K$. What are the differences $\Delta\ k_1$ and $\Delta\ k_2$ of k_1 and k_2 compared to k_{Han}?

Solution: Check results with the help of Fig. 4.4.

Problem 4.4 Compute c_p for air at $T = 500.0\ K$, and find γ and Pr. Compare Pr with the result found from eq. (4.19).

Solution: $c_p = 1{,}031.28\ m^2/s^2\ K$, $\gamma = 1.386$, $Pr = 0.743$.

Problem 4.5 We speak about incompressible flow, if $M = 0$. What does $M = 0$ mean?

Solution: On page 91 it was mentioned, that in that case ($E = 0$) fluid mechanics and thermodynamics are decoupled. Discuss.

References

1. A. H. Shapiro. "Basic Equations of Fluid Flow". *V. L. Streeter (ed.), Handbook of Fluid Dynamics.* McGraw-Hill, New York/Toronto/London, 1961, pp. 2-1 to 2-19.
2. R. B. Bird, W. E. Stewart, E. N. Lightfoot. "Transport Phenomena". John Wiley & Sons, New York/London/Sydney, 2nd edition, 2002.
3. W. G. Vincenti, C. H. Kruger. "Introduction to Physical Gas Dynamics". John Wiley & Sons, New York/London/Sydney, 1965. Reprint edition, Krieger Publishing Comp., Melbourne, Fl., 1975
4. E. R. G. Eckert, R. M. Drake. "Heat and Mass Transfer". MacGraw-Hill, New York, 1950.
5. G. Koppenwallner. "Hot Model Testing Requirements and Ludwieg Tubes". Hyperschall Technologie Göttingen, HTG-Report 94-6, 1994.

6. C. F. Hansen. "Approximations for the Thermodynamic and Transport Properties of High-Temperature Air". NACA TR R-50, 1959.
7. G. S. R. Sarma. "Physico-Chemical Modelling in Hypersonic Flow Simulation". Progress in Aerospace Sciences, Vol. 36, No. 3-4, 2000, pp. 281 - 349.
8. Ch. Mundt, R. Keraus, J. Fischer. "New, Accurate, Vectorized Approximations of State Surfaces for the Thermodynamic and Transport Properties of Equilibrium Air". Zeitschrift für Flugwissenschaften und Weltraumforschung (ZFW), Vol. 15, No. 3, 1991, pp. 179 - 184.
9. S. Srinivasan, J. C. Tannehill, K. J. Weilmuenster. "Simplified Curve Fits for the Transport Properties of Equilibrium Air". ISU-ERI-Ames-88405, Iowa State University, 1987.
10. L. Biolsi, D. Biolsi. "Transport Properties for the Nitrogen System: N_2, N, N^+, and e". AIAA-Paper 83-1474, 1983.
11. L. Biolsi. "Transport Properties for the Oxygen System: O_2, O, O^+, and e". AIAA-Paper 88-2657, 1988.
12. M. Fertig, A. Dohr, H.-H. Frühauf. "Transport Coefficients for High-Temperature Nonequilibrium Air Flows". AIAA Journal of Thermophysics and Heat Transfer, Vol. 15, No. 2, 2001, pp. 148 - 156.
13. F. P. Bertolotti. "The Influence of Rotational and Vibrational Energy Relaxation on Boundary-Layer Stability". J. Fluid Mechanics, Vol. 372, 1998, pp. 93 - 118.
14. E. H. Hirschel, A. Groh. "Wall-Compatibilty Condition for the Solution of the Navier-Stokes Equations". Journal of Computational Physics, Vol. 53, No. 2, 1984, pp. 346 - 350.
15. K. Oswatitsch. "Ähnlichkeitsgesetze für Hyperschallströmung". ZAMP, Vol. II, 1951, pp. 249 - 264. Also: "Similarity Laws for Hypersonic Flow". Royal Institute of Technology, Stockholm, Sweden, KTH-AERO TN 16, 1950.
16. H. Schlichting. "Boundary Layer Theory". 7th edition, McGraw-Hill, New York, 1979.
17. S. A. Schaaf, P. L. Chambré. "Flow of Rarefied Gases". Princeton University Press, Princeton, N.J., 1961.
18. E. H. Hirschel. "Hypersonic Flow of a Dissociated Gas over a Flat Plate". *L. G. Napolitano (ed.), Astronautical Research 1970.* North-Holland Publication Co., Amsterdam, 1971, pp. 158 - 171.
19. S. I. Pai. "Laminar Flow". *V. L. Streeter (ed.), Handbook of Fluid Dynamics.* McGraw-Hill, New York/Toronto/London, 1961, pp. 5-1 to 5-34.
20. M. Holt. "Dimensional Analysis". *V. L. Streeter (ed.), Handbook of Fluid Dynamics.* McGraw-Hill, New York/Toronto/London, 1961, pp. 15-1 to 15-25.
21. B. Oskam. "Navier-Stokes Similitude". NLR Memorandum AT-91, 1991.
22. E. H. Hirschel. "Thermal Surface Effects in Aerothermodynamics". *Proc. Third European Symposium on Aerothermodynamics for Space Vehicles, Noordwijk, The Netherlands, November 24 - 26, 1998.* ESA SP-426, 1999, pp. 17 - 31.

5 Real-Gas Aerothermodynamic Phenomena

Aerothermodynamic phenomena in the context of this book are the so-called real-gas effects and flow phenomena related to hypersonic flight. Hypersonic flight usually is defined as flight at Mach numbers $M \gtrapprox 5$. Here appreciable real-gas effects begin to appear. In this chapter we discuss the important real-gas phenomena with the goal to understand them and their implications in vehicle design[1].

The basic distinction regarding real-gas phenomena is that between thermally and calorically perfect or imperfect gases. The thermally perfect gas obeys the equation of state $p = \rho RT$. A calorically perfect gas has constant specific heats c_p and c_v. We speak about a "perfect" or "ideal gas", if both is given.

A gas can be thermally perfect, but calorically imperfect. An example is air as we treat it usually in aerothermodynamics. If a gas is thermally imperfect, it will also be calorically imperfect, and hence is a "real gas". We note, however, that in the aerothermodynamic literature, and also in this book, the term "real gas" is used in a broader, but in a strict sense incorrect, way to describe gases, which are thermally perfect and calorically imperfect. We will see in the following that real-gas effects in aerothermodynamics usually are high-temperature real-gas effects.

We first have a look at the "classical" real gas, the van der Waals gas. After that the high-temperature real-gas effects are treated, which are of major interest in aerothermodynamics. Essentially, we consider air as mixture of the "thermally perfect gases" N_2, N, O_2, O, NO, implying that these are calorically imperfect. Rate effects are explained, and also catalytic surface recombination. Finally computation models are presented.

We give here only a few illustrating examples of real-gas effects, because real-gas phenomena, like flow phenomena in general, usually cannot

[1] We recall the discussion in Section 1.2. We noted there that (high-temperature) real-gas effects are strong on the windward side of RV-type flight vehicles and almost absent on their lee side due to the large angles of attack, at which these vehicles fly, Fig. 1.3. On CAV-type flight vehicles real-gas effects are small to moderate, but are present, to a different degree, on both the windward and the lee side due to the low angles of attack, Fig. 1.2.

be treated in an isolated way. More examples can be found in the following Chapters.

5.1 Van der Waals Effects

Thermally perfect gases are gases, for which it can be assumed that its constituents have no spatial extension, and no intermolecular forces acting between them, except during actual collisions [1]. This situation is given, if the gas density is small. For such gases the equation of state holds:

$$p = \rho RT. \tag{5.1}$$

If at the same time the specific heats at constant pressure c_p and constant volume c_v are independent of the temperature, we speak about thermally *and* calorically perfect gases.

If the molecular spacing is comparable to the range of the intermolecular forces, "van der Waals" forces are present. This happens at rather low temperatures and sufficiently high densities/pressures. The equation of state then is written:

$$p = \rho RTZ(\rho, T), \tag{5.2}$$

with Z being the real-gas factor, which is a function of the "virial coefficients" B, C, D, ...[2]:

$$Z(\rho, T) = 1 + \rho B(T) + \rho^2 C(T) + \rho^3 D(T) + \cdots. \tag{5.3}$$

We show for the pressure range: $0.0\ atm \leqq p \leqq 100.0\ atm$ ($1.0\ atm = 101{,}325.0\ Pa$), and the temperature range: $0.0\ K < T \leqq 1{,}500.0\ K$, the real-gas factor $Z(\rho, T)$ in Fig. 5.1, and the ratio of specific heats $\gamma(\rho, T)$ in Fig. 5.2. The data were taken from [3], where they are presented in the cited temperature and pressure range for non-dissociated air with equilibrium vibrational excitation.

Fig. 5.1 exhibits that especially at temperatures below approximately $300.0\ K$, and relatively large pressures, and hence densities, van der Waals effects play a role. However, the real-gas factor is rather close to the value one. The same is true for larger temperatures, where even at $10.0\ atm$ the factor Z is smaller than 1.004.

The ratio of specific heats, Fig. 5.2, is similarly insensitive, if the pressure is not too high. The critical temperature range, where van der Waals effects appear, even at low pressures, again is below approximately $300.0\ K$.

In order to get a feeling about the importance of van der Waals effects, we consider the parameter ranges in hypersonic flight, Sections 1.2 and 2.1, and find the qualitative results given in Table 5.1.

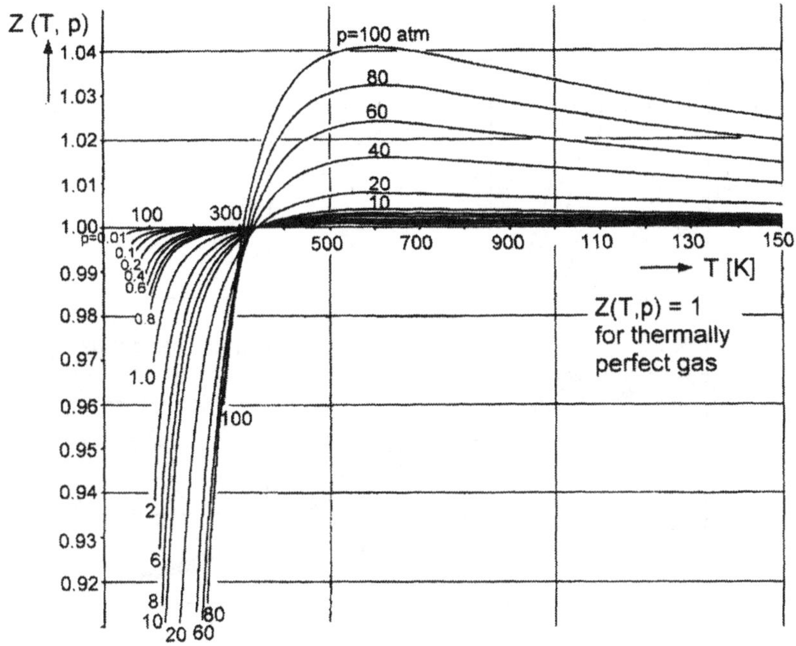

Fig. 5.1. Real-gas factor $Z(T, p)$ of air [3].

Table 5.1. Qualitative consideration of aerothermodynamic parameters in the flight free-stream, and in the stagnation region of a flight vehicle.

Item	Speed v	Pressure p	Density ρ	Temperature T	v. d. Waals effects
Free stream	large	small	small	small	small
Stagnation area	small	large	large	large	small

We observe from Table 5.1 that in hypersonic flight obviously the tendencies of pressure/density and temperature are against van der Waals effects. In the free stream at small temperatures also pressure/density are small, whereas in the stagnation area at large pressures/density also the temperature is large. Therefore in general we can neglect van der Waals effects in the flight regime covered by hypersonic vehicles in the earth atmosphere.

The situation can be different with aerothermodynamic ground-test facilities with high reservoir pressures and enthalpies. Here van der Waals effects can play a role, and hence must be quantified, and, if necessary, be taken into account, see, e. g., [4].

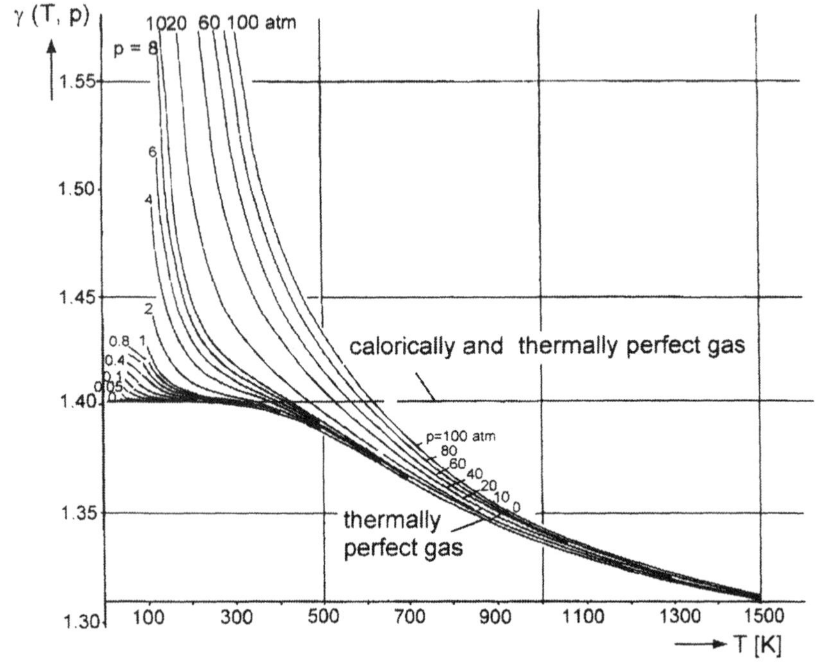

Fig. 5.2. Ratio of specific heats $\gamma(T, p)$ of air [3].

5.2 High-Temperature Real-Gas Effects

High-temperature real-gas effects are called those effects, which make a gas calorically imperfect. We note that a molecule has four parts of internal energy [1]:

$$e = e_{trans} + e_{rot} + e_{vibr} + e_{el}. \tag{5.4}$$

Here e_{trans} is the translational energy, which also an atom has. Rotational energy e_{rot} is fully present already at very low temperatures, and in aerothermodynamics in general is considered as fully excited[2]. Vibrational energy e_{vibr} is being excited in air at temperatures[3] above 300.0 K. Electronic excitation energy e_{el}, i. e. energy due to electronic excitation, is energy, which, like ionization, usually can be neglected in the flight-speed/altitude domain considered in this book.

The high-temperature real-gas effects of interest thus are vibrational excitation and dissociation/recombination. Dissociated gases can be considered

[2] The characteristic temperatures $\Theta_{rot_{N_2}}$, $\Theta_{rot_{O_2}}$, and $\Theta_{rot_{NO}}$ can be found in Section 13.1.

[3] See Table 4.4.

as mixtures of thermally perfect gases, whose molecular species are calorically imperfect.

We illustrate the parts of the internal energy by considering the degrees of freedom f, which atoms and molecules have. We do this by means of the simple dumb-bell molecule shown in Figure 5.3.

We see that molecules (and atoms) have:

- three translational degrees of freedom,
- two rotational degrees of freedom (the energy related to the rotation around the third axis a - a can be neglected),
- two vibrational degrees of freedom, i. e. one connected to the internal translation movement, and one connected to the spring energy.

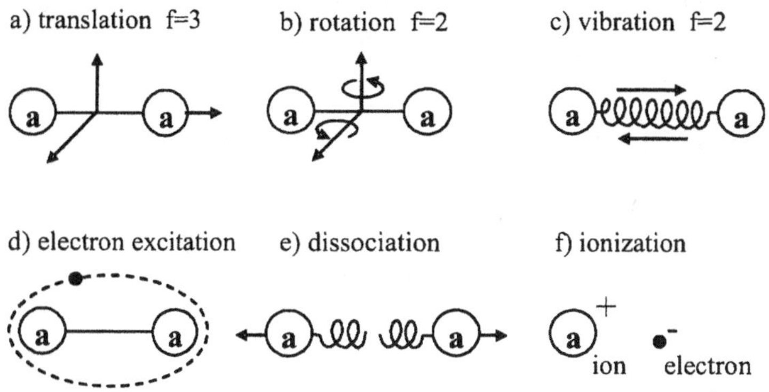

Fig. 5.3. Schematic of degrees of freedom f of a dumb-bell molecule, and illustration of other high-temperature phenomena.

The internal energy of a mixture of thermally perfect gases in equilibrium is, with ω_i being the mass fraction of species i:

$$e = \sum_{i=1}^{n} \omega_i e_i. \tag{5.5}$$

The internal energy of a species i is :

$$e_i = e_{trans_i} + e_{rot_i} + e_{vibr_i} + e_{el_i}. \tag{5.6}$$

The terms e_{trans_i} and e_{el_i} apply to both atoms and molecules, the terms e_{rot_i} and e_{vibr_i} only to molecules.

The enthalpy of a gas is defined by [1]:

$$h = e + \frac{p}{\rho}. \tag{5.7}$$

For thermally perfect gases we have, in general, with the specific heats at constant pressure c_p and constant volume c_v, and the gas constant $R = c_p - c_v$:

$$dh = c_p dT = (c_v + R)\, dT. \tag{5.8}$$

Likewise it holds for the internal energy:

$$de = c_v dT. \tag{5.9}$$

The principle of equipartition of energy [1] permits us to formulate the internal energy e, the specific heats c_v, c_p, and their ratio $\gamma = c_p/c_v$ of atoms and molecules in terms of the degree of freedom f, which gives us insight into some basic high-temperature phenomena.

We assume excitation of all degrees of freedom (translational, rotational, vibrational) of atoms and molecules. We neglect e_{el}, and obtain the general relations, [1], for a species with molecular weight M, R_0 being the universal gas constant ($R = R_0/M$):

$$e = \frac{f}{2}\frac{R_0}{M}T;\ c_v = \frac{f}{2}\frac{R_0}{M};\ c_p = \frac{f+2}{2}\frac{R_0}{M};\ \gamma = \frac{f+2}{f}, \tag{5.10}$$

which we now apply to the air species.

Atoms (N, O) For atoms we obtain with three translational degrees of freedom, $f = 3$:

$$e_{atom_i} = \frac{3}{2}\frac{R_0}{M_i}T;\ c_{v_{atom_i}} = \frac{3}{2}\frac{R_0}{M_i};\ c_{p_{atom_i}} = \frac{5}{2}\frac{R_0}{M_i};\ \gamma_{atom_i} = 1.66\bar{6}. \tag{5.11}$$

Molecules (N_2, O_2, NO)

- Molecules with translational and rotational excitation only[4], $f = 5$.

$$e_{molec_i} = \frac{5}{2}\frac{R_0}{M_i}T;\ c_{v_{molec_i}} = \frac{5}{2}\frac{R_0}{M_i};\ c_{p_{molec_i}} = \frac{7}{2}\frac{R_0}{M_i};\ \gamma_{molec_i} = 1.4. \tag{5.12}$$

- Molecules with translational, rotational, and partial vibrational excitation[5].

The vibrational energy of diatomic molecules[6] [1]:

[4] This is the case of air at low temperatures ($T \lessapprox 400.0$ K).

[5] In the cases with vibrational excitation, the translational and rotational degrees of freedom are considered as fully excited.

[6] The characteristic temperatures Θ_{vibr} of N_2 ,O_2 and NO can be found in Section 13.

$$e_{vibr} = \frac{R\Theta_{vibr}}{e^{\Theta_{vibr}/T} - 1}, \tag{5.13}$$

and the specific heat

$$c_{v_{vibr}} = R\left(\frac{\Theta_{vibr}}{T}\right)^2 \frac{e^{\Theta_{vibr}/T}}{(e^{\Theta_{vibr}/T} - 1)^2}. \tag{5.14}$$

We find now:

$$\begin{aligned} e_{molec_i} &= \frac{5}{2}\frac{R_0}{M_i}T + e_{v_{vibr_i}};\ c_{v_{molec_i}} = \frac{5}{2}\frac{R_0}{M_i} + c_{v_{vibr_i}}; \\ c_{p_{molec_i}} &= \frac{7}{2}\frac{R_0}{M_i} + c_{v_{vibr_i}};\ \gamma_{molec_i} = \gamma_{molec_i}(T) \leqq 1.4. \end{aligned} \tag{5.15}$$

– Molecules with full vibrational excitation, $f = 7$.

The case of full vibrational excitation of a molecule is hypothetical because the molecule will dissociate before it reaches this state. Nevertheless, eqs. 5.13 and 5.14 exhibit for large temperatures the limiting cases:

$$e_{vibr} = R\,T;\ c_{v_{vibr}} = R. \tag{5.16}$$

Hence we obtain:

$$e_{molec_i} = \frac{7}{2}\frac{R_0}{M_i}T;\ c_{v_{molec_i}} = \frac{7}{2}\frac{R_0}{M_i};\ c_{p_{molec_i}} = \frac{9}{2}\frac{R_0}{M_i};\ \gamma_{molec_i} = 1.285^{\cdots}. \tag{5.17}$$

– Molecules with half vibrational excitation (Lighthill gas), $f = 6$.

This case with a heat capacity twice as large as that of atoms was proposed by Lighthill in his study of the dynamics of dissociated gases [5]. It yields a good approximation for applications in a large temperature and pressure/density range[7]:

$$e_{molec_i} = \frac{6}{2}\frac{R_0}{M_i}T;\ c_{v_{molec_i}} = \frac{6}{2}\frac{R_0}{M_i};\ c_{p_{molec_i}} = \frac{8}{2}\frac{R_0}{M_i};\ \gamma_{molec_i} = 1.33\bar{3}. \tag{5.18}$$

[7] This case can be seen in the frame of the "effective ratio of specific heats" approach. $\gamma_{eff} \lessapprox 1.4$ can be used to estimate the influence of high-temperature real-gas effects.

– Molecules with an infinitely large number of degrees of freedom, $f = \infty$.

  This is a limiting case, which means:

$$e_{molec_i} = \infty; \; c_{v_{molec_i}} = \infty; \; c_{p_{molec_i}} = \infty; \; \gamma_{molec_i} = 1. \tag{5.19}$$

5.3 Dissociation and Recombination

The excitation of internal degrees of freedom of molecules happens through collisions. If a molecule receives "too much" energy by collisions, it will dissociate. Recombination is the reverse phenomenon in which two atoms (re)combine to a molecule. However, a "third body" is needed to carry away the excess energy, i. e. the dissociation energy, which is released again during the recombination process. The third body can be an atom or a molecule, but also the vehicle surface, if it is finitely or fully catalytic (catalytic surface recombination). At very high temperatures also ionization occurs in the flow field. It is the species NO, which is ionized first, because it needs the lowest ionization energy of all species.

Dissociation and recombination are chemical reactions, which alter the composition of the gas, and which bind or release heat [1], [6]. In equilibrium, Section 5.4, the five species N_2, N, O_2, O, and NO, are basically products of three reactions:

$$\begin{aligned} N_2 &\rightleftarrows N + N, \\ O_2 &\rightleftarrows O + O, \\ NO &\rightleftarrows N + O. \end{aligned} \tag{5.20}$$

If ionization occurs, we get for the first appearing product

$$NO \rightleftarrows NO^+ + e^-. \tag{5.21}$$

We distinguish between homogeneous and heterogeneous reactions. The former are given, if only the gas constituents are involved. The latter, if also the vehicle surface plays a role (catalytic surface recombination). In the following Section 5.4 we discuss basics of homogeneous reactions, and in Section 5.6 of heterogeneous reactions.

5.4 Thermal and Chemical Rate Processes

We assume the flow cases under consideration to be in the continuum regime, and at most at its border, in the slip-flow regime, Section 2.3. Nevertheless, we used in our above discussion the concept "excitation by collisions". We apply this concept now in order to get a basic understanding of rate processes, i. e. excitation or reaction processes, which in principle are time dependent.

The number of collisions per unit time depends on the density and the temperature of the gas. The number of collisions at which molecules react is much less than the number of collisions they undergo, because only a fraction of the collisions involves sufficient energy (activation factor), and only a fraction of collisions with sufficient energy leads actually to a reaction (steric factor) [1]. The orders of magnitude of the number of collisions needed to reach equilibrium of degrees of freedom or to dissociate molecules [1] are given in the following Table 5.2.

Table 5.2. Number of collisions to reach equilibrium of degrees of freedom or dissociation.

Phenomenon	Number of collisions
Translation	O(10)
Rotation	O(10)
Vibration	$O(10^4)$
Dissociation	$> O(10^4)$

The table shows that only a few collisions are needed to obtain full excitation of translation and rotation[8]. Knowing the very low characteristic rotational temperatures of the diatomic species N_2, O_2, NO, Section 13.1, we understand why for our flow cases in general rotational degrees of freedom can be considered as fully excited. Vibrational excitation needs significantly more collisions, and dissociation still more. We note that this consideration is valid also for the reverse processes, i. e. de-excitation of internal degrees of freedom, and recombination of atomic species.

Since excitation is a process in time, we speak about the "characteristic excitation or reaction time" τ, which is needed to have the necessary collisions to excite degrees of freedom or to induce reactions. Comparing this time with a characteristic flow time leads us to the consideration of thermal and chemical rate processes.

We introduce the first Damköhler number $DAM1$ [7]:

$$DAM1 = \frac{t_{res}}{\tau}. \tag{5.22}$$

[8] The reader should note that translational non-equilibrium is the cause of molecular transport (Chapter 4) of momentum (viscous stress), heat (heat conduction) and mass (mass diffusion), and rotational non-equilibrium that of bulk viscosity, [1].

Here t_{res} is the residence time of a fluid particle with its atoms and molecules in the considered flow region, which we met in another context already in Section 4.1:

$$t_{res} = \frac{L_{ref}}{v_{ref}}. \tag{5.23}$$

L_{ref} is a characteristic length of the flow region under consideration, and v_{ref} a reference speed, Fig. 4.2.

The characteristic time

$$\tau = \tau(\rho, T, state) \tag{5.24}$$

is the excitation time of a degree of freedom or the reaction time of a species, and depends on the density, the temperature and the initial state regarding the excitation or reaction [1].

We distinguish the following limiting cases:

- $DAM1 \to \infty$: the residence time is much larger than the characteristic time: $t_{res} \gg \tau$, the considered thermo-chemical process is in "equilibrium", which means that each of the internal degrees of freedom and the chemical reactions is in equilibrium. Then the actual vibrational energy or the mass fraction of a species is a function of local density and temperature only: $e_{vibr_i} = e_{vibr_i}(\rho, T)$, $\omega_i = \omega_i(\rho, T)$. In this case we speak about "equilibrium flow" or "equilibrium real gas".
- $DAM1 \to 0$: the residence time is much shorter than the characteristic time: $t_{res} \ll \tau$, the considered thermo-chemical process is "frozen". In the flow region under consideration practically no changes of the excitation state or the mass fraction of a species happen: e. g. $e_{vibr_i} = e_{vibr_{i\infty}}$, $\omega_i = \omega_{i\infty}$. We speak about "frozen flow" or "frozen real gas". In Sub-Section 5.5.1 an example is discussed.
- $DAM1 = O(1)$: the residence time is of the order of the characteristic time, the considered process is in "non-equilibrium". In the flow region under consideration the thermo-chemical processes lag behind the local flow changes, i. e. the process is a relaxation process. We call this "non-equilibrium flow" or "non-equilibrium real gas". We discuss an example in Sub-Section 5.5.

Of course for the aerothermodynamic problems at hand it is an important question whether the thermo-chemical rate process is relevant energetically. The parameter telling this is the second Damköhler number $DAM2$:

$$DAM2 = \frac{q_{ne}}{H_0}, \tag{5.25}$$

which compares the energy involved in the non-equilibrium process q_{ne} with the total energy of the flow H_0. If $DAM2 \to 0$, the respective rate process in general can be neglected.

We note that the Damköhler numbers must be applied, like any similarity parameter, with caution. A global Damköhler number $DAM1 \to \infty$ can indicate, that the flow past a flight vehicle *globally* is in equilibrium . However, locally a non-equilibrium region can be embedded. This can happen, for example, behind the bow shock in the nose region, but also in regions with strong flow expansion. At a reentry vehicle at large angle of attack, for instance, we have the situation that the fluid in the stagnation region will be heated and be in some thermo-chemical equilibrium or non-equilibrium state. Part of this fluid will enter the windward side of the flight vehicle, where only a rather weak expansion happens. Hence we will have equilibrium or non-equilibrium flow effects also here. The part entering the lee-ward side on the other hand undergoes a strong expansion. Hence it is to be expected that frozen flow exists there, with a frozen equilibrium or non-equilibrium state more or less similar to that in the stagnation region.

We note finally that in the flow of a gas mixture the Damköhler numbers of each of the thermo-chemical rate processes must be considered, since they can be vastly different. This would imply that a multi-temperature model, with, for instance, the translational temperature T_{trans}, and vibrational temperatures $T_{vibr_i}(i = N_2, O_2, NO)$, and appropriate transport equations, Chapter 12, must be employed, see, e. g., [6]. As a consequence of multi-temperature models one has to be careful in defining the local speed of sound of a flow, which is important for the stability of discrete computation methods, but also for the presentation of experimental data with the Mach number as parameter [1], [6].

In the following we consider shortly the basics of rate processes of vibrational excitation, and chemical reactions. For a thorough presentation see, e. g., [1], [6].

For the vibrational rate process of a diatomic species we can write in general[1]:

$$\frac{de_{vibr}}{dt} = \frac{e^*_{vibr} - e_{vibr}}{\tau}. \tag{5.26}$$

Here e_{vibr} is the actual vibrational energy, e^*_{vibr} the equilibrium energy, and τ the relaxation time. The equilibrium energy e^*_{vibr} is a function of the temperature T, eq. (5.13). The relaxation time τ is a function of temperature and pressure/density, as well as of characteristic data k_1, k_2, Θ_{vibr} of the species (Landau-Teller theory, [1]):

$$\tau = \frac{k_1 T^{5/6} e^{(k_2/T)^{1/3}}}{p(1 - e^{-\Theta_{vibr}/T})}, \tag{5.27}$$

and for sufficiently low temperatures:

$$\tau = c\,\frac{e^{(k_2/T)^{1/3}}}{p}. \tag{5.28}$$

Eq. (5.26) shows, since $\tau > 0$, that the vibrational energy always tends towards its equilibrium value. The relation in principle holds for large non-equilibrium, but due to the underlying assumption of a harmonic oscillator, it actually is limited to small, not accurately defined, departures from equilibrium [1].

Above it was said that full vibrational excitation cannot be reached, because the molecule will dissociate before this state is reached. The transition to dissociation, i. e. vibration-dissociation coupling, can be modeled to different degrees of complexity, see, e. g., [8], [9], [10]. Often it is neglected in computation methods, because it is not clear how important it is for the determination of flow fields, forces, thermal loads, et cetera.

Chemical rate processes without vibration-dissociation coupling can be described in the following way. For convenience we consider only the so called thermal dissociation/recombination of nitrogen [6]:

$$N_2 + M \rightleftarrows 2N + M. \tag{5.29}$$

Here M is an arbitrary other gas constituent. For the dissociation reaction (forward reaction: right arrow) it is the collision partner, which lifts the energy level of N_2 above its activation, i. e. dissociation, level. For the recombination reaction (backward reaction: left arrow) it is the already mentioned "third body" M, which carries away the dissociation energy and conserves also the momentum balance.

The rate of change equation for an arbitrary gas species i in terms of its molar concentration c_i is:

$$Sm_i = M_i \frac{dc_i}{dt} = M_i \sum_r (\nu''_{ir} - \nu'_{ir}) \left(k_{fr} \prod_i c_i^{-\nu'_{ir}} - k_{br} \prod_i c_i^{-\nu''_{ir}} \right). \tag{5.30}$$

This equation sums up all individual reactions r, which add to the overall rate of change of the species, i. e. the contributions of the reactions with different third bodies M. In case of the thermal dissociation/recombination of nitrogen, eq. (5.29), we have five reactions ($r = 5$), with M being N_2, N, O_2, O, NO.

In eq. (5.30) ν'_{ir} and ν''_{ir} are the stoichiometric coefficients of the forward (') and the backward (") reactions r of the species i with the third body M. The forward reaction rate of reaction r is k_{fr}, and the backward rate k_{br}.

If a reaction r is in equilibrium, $dc_{ir}/dt \rightarrow 0$, eq. (5.30) yields:

$$k_{fr} \prod_i c_i^{-\nu'_{ir}} = k_{br} \prod_i c_i^{-\nu''_{ir}}, \tag{5.31}$$

and the equilibrium constant:

$$K_c = \frac{k_{fr}}{k_{br}}. \tag{5.32}$$

The equilibrium constant of each reaction can be approximated by the modified Arrhenius equation [1]:

$$K_c = C_c T^{\eta_c} e^{-\Theta_{diss}/T}, \tag{5.33}$$

with Θ_{diss} the characteristic dissociation temperature, and the forward reaction rate by:

$$k_{fr} = C_f T^{\eta_f} e^{-\Theta_{diss}/T}, \tag{5.34}$$

so that the backward reaction rate is:

$$k_{br} = \frac{C_f}{C_c} T^{(\eta_f - \eta_c)}. \tag{5.35}$$

In these equations C_c, C_f, η_c, η_f are independent of the temperature T, [1].

We note that (global) equilibrium flow is only given, if all Damköhler numbers $DAM1_i \to 0$, which means that *all* reactions r of all species i are in equilibrium.

The constants and characteristic data for the above relations can be found in [1], [6]. Regarding a critical review of the state of art see, e. g., [8].

5.5 Rate Effects, Two Examples

In this section two examples from computational studies of rate effects in hypersonic flows are presented and discussed. The first example is normal shock-wave flow in presence of vibrational and chemical rate!effects ($DAM1 = O(1)$), the second nozzle flow of a ground-simulation facility with high total enthalpy ("hot" facility) with freezing phenomena ($DAM1 \to 0$).

5.5.1 Normal Shock Wave in Presence of Rate Effects

The shock wave, Section 6.1, basically is a small zone in which the velocity component normal to the wave front (v_1) drops from supersonic (M_1) to subsonic speed (v_2, M_2), Fig. 5.4. This drop is accompanied by a rise of density and temperature. Translational and rotational degrees of freedom adjust through this zone to the (new) conditions behind the shock wave, Fig. 5.4 a). Because this zone is only a few mean free paths wide, the shock wave in general is considered as discontinuity, Fig. 5.4 b).

The situations changes in high-enthalpy flow. Here the temperature rise in the shock wave leads to the excitation of vibrational degrees of freedom and to dissociation. However, because of the small extent of the zone, the adjustment to the new conditions takes longer than that of translation and rotation. The result is a "relaxation" zone with non-equilibrium flow, and hence a much increased actual thickness of the shock wave, Fig. 5.4 c).

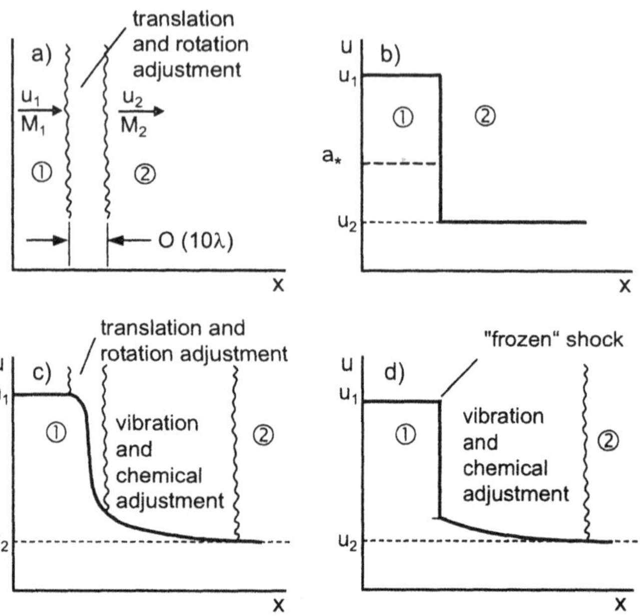

Fig. 5.4. Schematic of a normal shock wave in presence of rate effects (following [1]), '1' denotes 'ahead of shock', '2' is 'behind shock', **a**) normal shock wave, **b**) shock wave idealized as flow discontinuity (a_*: critical speed of sound (see Sub-Section 6.3.1), **c**) shock wave with all rate effects, **d**) situation **c**) idealized with "frozen" shock.

Also here we can approximate the situation. Fig. 5.4 d) assumes, that translation and rotation adjustment is hidden in the discontinuity like in Fig. 5.4 b), while vibration and chemical reactions are frozen there. Their adjustment then happens only in the relaxation zone behind the discontinuity.

We add some numerical results from a study of hypersonic flow past a cylinder [11]. The flow parameters are given in Table 5.3. The computations were made with a Navier-Stokes method in thin-layer formulation with the assumptions of perfect gas, and also of thermo-chemical non-equilibrium [11].

Fig. 5.5 shows for perfect gas a shock stand-off distance of $x/r \approx$ - 0.4, much larger than for the non-equilibrium cases with $x/r \approx$ - 0.25 (we will come back to this phenomenon in Sub-Section 6.4.1). The temperature behind the shock is nearly constant. It reaches shortly ahead of the cylinder approximately the total temperature T_0 = 8,134.0 K and drops then in the thermal boundary layer[9] to the prescribed wall temperature T_w = 600.0 K.

The translational temperature together with the rotational temperature initially reaches in the two (one with adiabatic wall) non-equilibrium cases nearly the perfect-gas temperature. It drops then fast in the zone of thermo-

[9] Note that all boundary-layer thicknesses are effectively finite at a stagnation point, and especially along attachment lines, Sub-Section 7.2.1.

Table 5.3. Parameters of the flow at the cylinder under reentry conditions at 53.5 km altitude [11].

M_∞	v_∞ $[m/s]$	T_∞ $[K]$	ρ_∞ $[kg/m^3]$	ω_{N_2}	ω_{O_2}	T_w $[K]$	R_{cyl} $[m]$
12.5	3,988.56	252.5	$3.963 \cdot 10^{-4}$	0.738	0.262	600.0/*adiabatic*	0.25

chemical adjustment due to the processes of dissociation, Fig. 5.6, and vibrational excitation.

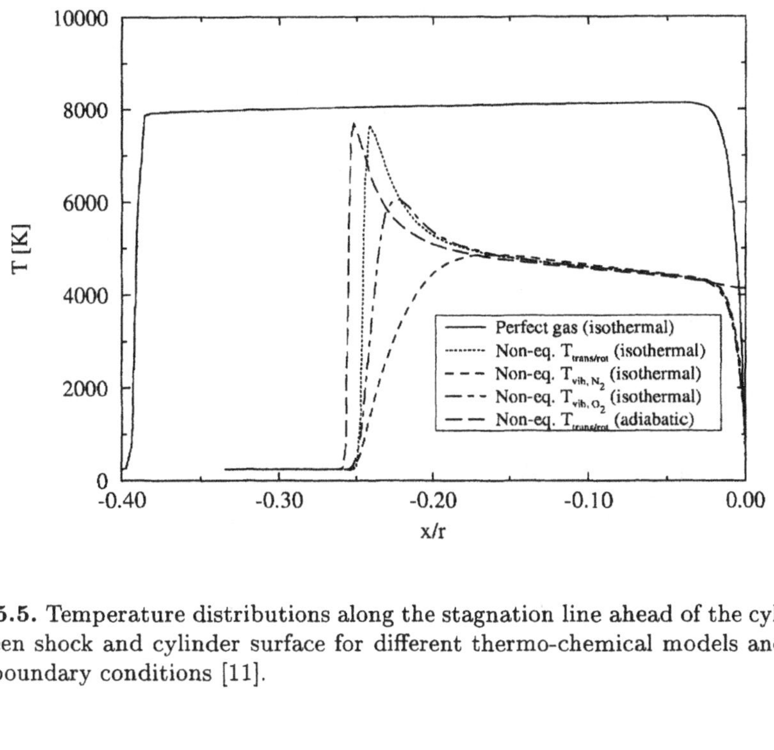

Fig. 5.5. Temperature distributions along the stagnation line ahead of the cylinder between shock and cylinder surface for different thermo-chemical models and two wall boundary conditions [11].

With the adiabatic wall case ($T_r \approx 4{,}000.0\ K$) the shock stand-off distance is slightly larger than with the isothermal wall case with $T_w = 600.0\ K$ (see also here Sub-Section 6.4.1). The vibrational temperature of oxygen $T_{vibr_{O_2}}$ adjusts clearly faster than that of nitrogen $T_{vibr_{N_2}}$. Thermal equilibrium is reached at $x/r \approx -0.17$.

The distributions of the species mass concentrations are shown in Fig. 5.6 for the fully catalytic and for the non-catalytic surface. We see that the concentration of N_2 does not change much. It reaches fast the equilibrium state. At the wall it rises slightly, and in the case of the fully catalytic wall it reaches the free-stream value, i. e. nitrogen atoms are no more present (the

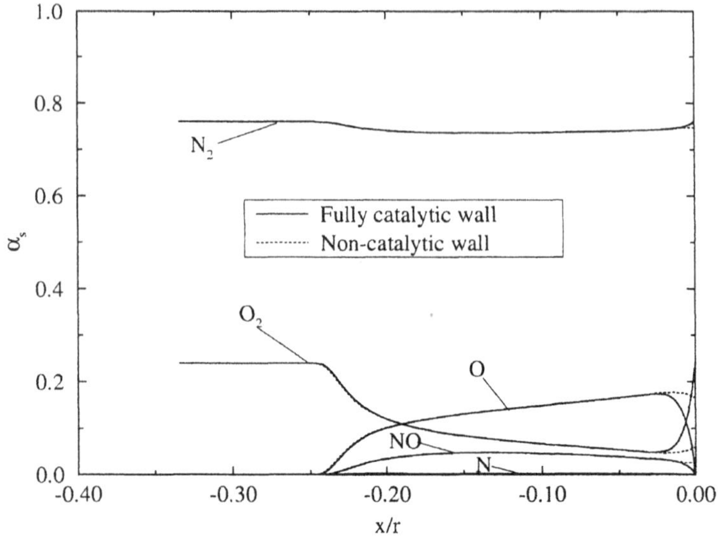

Fig. 5.6. Distributions of species mass concentrations ($\alpha_s \equiv \omega_i$) along the stagnation line ahead of the cylinder between shock and cylinder surface for different catalytic wall boundary conditions [11].

mass fraction of N anyway is very small, because the atoms combine with O to NO), and NO also disappears almost completely. The concentration of O_2 decreases up to the outer edge of the thermal boundary layer and then increases strongly in the case of the fully catalytic case, reaching, like N_2, the free-stream value. The effect of the non-catalytic wall is confined to the wall-near region.

5.5.2 Nozzle Flow in a "Hot" Ground-Simulation Facility

In [12], see also [13], results of a numerical study of flow and rate phenomena in the nozzle of the high-enthalpy free-piston shock tunnel HEG of the DLR in Göttingen, Germany, are reported. We present and discuss in the following some of these results.

The computations in [12], [13] were made with a Navier-Stokes code, employing the 5 species, 17 reactions air chemistry model of Park [6] with vibration-dissociation coupling. For the expansion flow thermo-chemical equilibrium and non-equilibrium was assumed. For the description of thermal non-equilibrium a three-temperature model was used ($T_{trans/rot}$, $T_{vibr_{N_2}}$, and $T_{vibr_{O_2}}$). The vibrational modes of NO were assumed to be in equilibrium with $T_{trans/rot}$, which applies to this nozzle-flow problem [12]. No-slip conditions were used at the nozzle wall, fully catalytic behaviour of the wall was assumed.

Geometrical data of the contoured nozzle and reservoir conditions of the case with tunnel operating condition I, for which results will be shown in the following, are given in Table 5.4. The nozzle originally was designed for a condition which is slightly different from condition I, therefore a weak recompression happens around 75 per cent of nozzle length.

Table 5.4. Geometrical nozzle data of HEG and reservoir conditions of operating condition I [12].

L_{nozzle}	D_{throat}	D_{exit}	Area ratio	Test gas	p_t	h_t	T_w
3.75 m	0.022 m	0.88 m	1,600.0	air	386.0 bar	20.19 MJ/kg	300.0 K

Fig. 5.7 shows the nozzle radius $r(x)$ and the computed pressure $p(x)$ along the nozzle axis. The expansion process is characterized by a strong drop of the pressure just downstream of the throat.

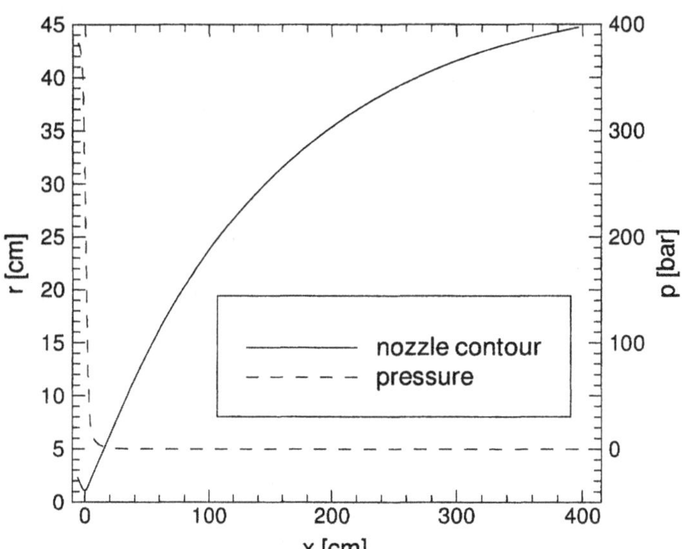

Fig. 5.7. Nozzle radius $r(x)$, and pressure $p(x)$, non-equilibrium computation, along the axis of the HEG nozzle (nozzle throat at $x = 0$) [12].

This drop goes along with a strong acceleration of the flow and a strong drop of the density, Fig. 5.8. Consequently the number of collisions between the gas species goes down. Vibrational energy will not be de-excitated, and dissociation will be frozen.

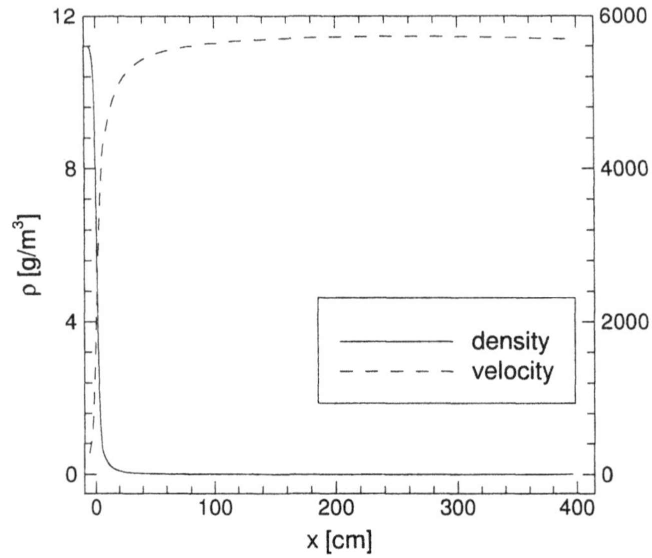

Fig. 5.8. Velocity $u(x)$ and density $\rho(x)$, non-equilibrium computation, along the axis of the HEG nozzle [12].

Comparing the flow variables and parameters at the nozzle exit for thermo-chemical equilibrium and non-equilibrium, Table 5.5, one sees, that velocity u, pressure p and density ρ are rather weakly affected by the choice of equilibrium or non-equilibrium. This is due to the fact, that the distribution of these variables along the nozzle axis is more or less a function of the ratio of 'pressure at the reservoir' to 'pressure at the nozzle exit', and, of course, of the nozzle contour, i. e. that it is largely independent of the rate processes in the nozzle, see, e. g., [14].

Table 5.5. Computed flow variables and parameters at the exit of the HEG nozzle for thermo-chemical equilibrium and non-equilibrium [12].

Case	u_{exit} $[m/s]$	p_{exit} $[Pa]$	ρ_{exit} $[kg/m^3]$	$T_{exit}[K]$	M_{exit}	Re^u_{exit} $[m^{-1}]$
Equilibrium	5,908.5	1270.7	$1.77 \cdot 10^{-03}$	2,435.3	6.49	137,993.0
Non-equilibrium	5,689.4	678.6	$2.44 \cdot 10^{-03}$	836.3	8.99	378,299.0

The different temperatures, Fig. 5.9, and the mass fractions - we present only $O(x)$ and $O_2(x)$ - Fig. 5.10, however, are strongly affected by the choice of equilibrium or non-equilibrium, hence the vastly different nozzle exit Mach numbers and unit Reynolds numbers for these cases in Table 5.5.

Fig. 5.9 shows that freezing of the vibrational temperature of N_2, $T_{vib_{N_2}}$, happens at a location on the nozzle axis shortly behind the nozzle throat at $x \approx 16.0\ cm$, while $T_{vib_{O_2}}$ freezes somewhat downstream of this location at $x \approx 23.0\ cm$. The solution for thermo-chemical equilibrium yields a temperature, T_{equil}, which drops on the nozzle axis monotonically, with a slightly steepening slope around $x \approx 320.0\ cm$, due to the weak re-compression, to $T_{equil} = 2{,}435.3\ K$ at the nozzle exit. For thermo-chemical non-equilibrium $T_{trans/rot}$ is much lower than T_{equil}. It drops to a minimum of approximately 550.0 K at $x \approx 260.0\ cm$ and rises then, also due to the weak re-compression, to $T_{trans/rot} = 836.3\ K$ at the nozzle exit. At $x = 400.0\ cm$, in the test section of the HEG facility, the computed $T_{trans/rot}$ compares well with measured data.

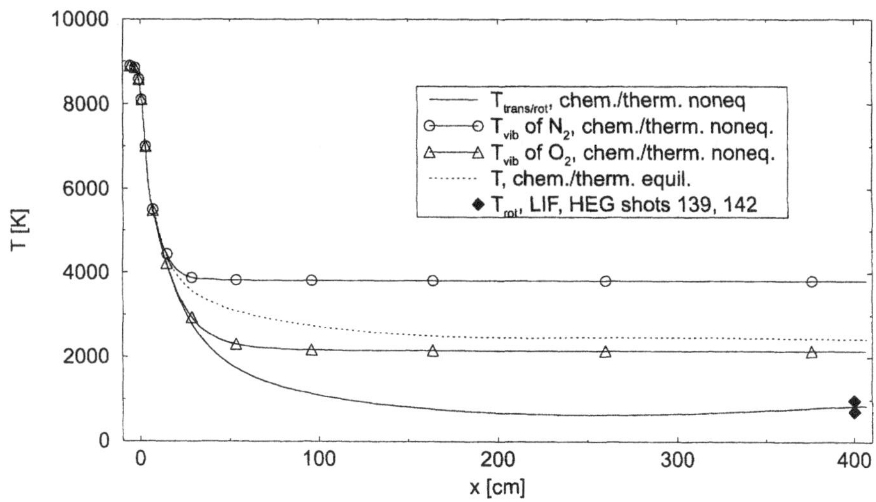

Fig. 5.9. Computed temperatures along the axis of the HEG nozzle [12].

Freezing of the composition of oxygen in the non-equilibrium case happens on the nozzle axis at $x \approx 18.0\ cm$, with $\alpha_{O,noneq} \approx$ const. ≈ 0.17 and $\alpha_{O_2,noneq} \approx$ const. ≈ 0.05 downstream of $x \approx 50.0\ cm$, Fig. 5.10. The mass fraction of atomic oxygen in the equilibrium case falls monotically from $\alpha_{O,equil} \approx 0.21$ at the nozzle throat, with a slightly steepening slope at $x \approx 320.0\ cm$, to $\alpha_{O,equil} \approx 0.02$ at the nozzle exit. The mass fraction of diatomic oxygen rises monotonically from nearly zero at the throat to $\alpha_{O_2,equil} \approx 0.21$ at the nozzle exit, with a slight increase of the slope at $x \approx 320.0\ cm$.

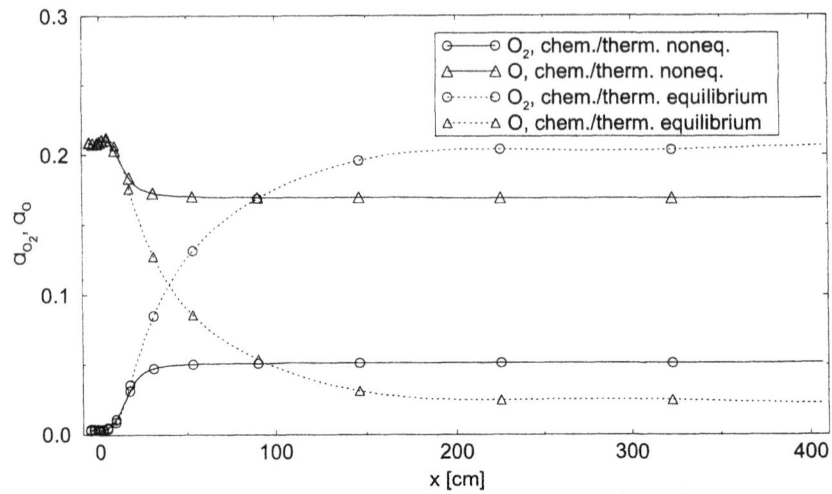

Fig. 5.10. Computed mass fractions of α_O and α_{O_2} ($\alpha_i \equiv \omega_i$) along the axis of the HEG nozzle [12].

The fact that the equilibrium temperature T_{equil} is much higher than the frozen non-equilibrium temperature $T_{trans,rot}$, can simply be explained.

We assume for both cases that the temperature in the nozzle is already so low that we can write $c_p = const.$ in the relation for the total enthalpy. We write for the equilibrium case:

$$h_t = c_p T_{equil}(x) + \frac{v^2(x)}{2}. \tag{5.36}$$

If now a part of the total enthalpy is trapped in frozen non-equilibrium vibration or dissociation, it will not participate in the expansion process. Hence we get for the "active" total enthalpy h_t':

$$h_t' < h_t. \tag{5.37}$$

With

$$h_t' = c_p T'(x) + \frac{v'^2(x)}{2}, \tag{5.38}$$

and $v' \approx v$, because, as we saw above, the velocity is only weakly affected, we obtain finally the result that the temperature in frozen non-equilibrium flow, $T_{trans,rot}$, is smaller than that in equilibrium flow:

$$T_{trans,rot}(x) = T'(x) < T^*(x). \tag{5.39}$$

As we have seen above, this effect can be very large, i. e. a significant amount of thermo-chemical energy can freeze in non-equilibrium during expansion in the nozzle of a high-enthalpy tunnel. The model then will "fly" in a frozen atmosphere with a large amount of energy hidden in non-equilibrium vibrational degrees of freedom and dissociation, compared to the actual flight of the real vehicle through a cold atmosphere consisting of N_2 and O_2, where only the degrees of freedom of translation and rotation are excited. Consequently the measured data may be not representative to a certain degree. A solution of the freezing problem is the employment of very high reservoir pressure/density, in order to maintain a density level during the expansion process, which is sufficient to have equilibrium flow. This then may have been bought with van der Waals effects in the reservoir and, at least, in the throat region of the nozzle.

The freezing phenomenon can be very complex. In a hypersonic wind tunnel, with a reservoir temperature of $T_t \approx$ 1,400.0 K for the $M = 12$ nozzle, large enough to excite vibration, de-excitation was found to happen in the nozzle near the nozzle exit [15]. The original freezing took place shortly behind the nozzle throat. The de-excitation increased static temperature, $T_\infty \equiv T_{trans/rot}$, and pressure p_∞ in the test section of the tunnel, and decreased the Mach number M_∞. Although the actual Mach-number was not much lower than that expected for the case of isentropic expansion, errors in static pressure were significant. The sudden de-excitation at the end of the nozzle was attributed to a high level of air humidity [15]: the water droplets interacted with the air molecules and released the frozen vibrational energy.

5.6 Surface Catalytic Recombination

If the surface of the flight vehicle acts as a third body in the chemical recombination reactions discussed in the previous section, we speak about catalytic surface recombination, which is a "heterogeneous" reaction. Catalytic surface recombination is connected to the thermal state of the surface. We will see that, on the one hand, it is a thermal-surface effect, depending on the surface temperature, but also influencing it. On the other hand, it has an effect on thermal loads. Because of these two effects it is of interest for flight vehicle design.

We keep in mind that of the flight vehicles considered, Chapter 1, RV-type vehicles bear the largest thermal loads. Due to material limitations currently surface temperatures of up to about 2,000.0 K can be permitted. Because we have radiation-cooled surfaces, these maximum temperatures occur only in the stagnation point region, and then drop fast to values as low as about 1,000.0 K, depending on flight speed and vehicle attitude[10]. The heat flux in the gas at the wall, q_{gw}, shows a similar qualitative behaviour.

A catalyst reduces the necessary activation energy of a reaction, and hence more collisions lead to reactions. Accordingly more reaction heat is released.

[10] Such a temperature distribution supports the tendency of surface catalytic N_2 recombination in the front part of the windward side of the vehicle, while O_2 recombination occurs only in the colder middle and aft part.

Hence the surface material (coating) should be a poor catalyst in order to reduce the release of reaction heat at the surface. Catalytic recombination helps to proceed towards equilibrium faster, but does not change equilibrium state (temperature, density) and composition of a gas.

Regarding the maximum heat flux towards the wall, q_{gw}, it is observed, that the "fully catalytic" wall gives a heat flux similar to that of the "equilibrium" wall. Therefore often the equilibrium wall is taken as the reference case for the largest heat load[11]. However, these are two concepts, which need to be distinguished, see, e. g., [16]:

- Equilibrium wall: If the flow past a flight vehicle would be in chemical equilibrium, a cold surface - we have seen above, that the surface temperature in our cases is at most around 2,000.0 K - would shift the gas composition at the wall into the respective equilibrium wall composition. This composition could be, depending on temperature and density/pressure, locally still a mixture of molecules and atoms.
- Fully catalytic wall: The fully catalytic wall, in contrast to the equilibrium wall, would lead at the wall to a recombination of all atoms, even if wall temperature and density/pressure would atoms permit to exist.

We discuss now basics of catalytic surface recombination in a phenomenological way. We introduce the recombination coefficient of atomic species γ_{i^a}, [16]:

$$\gamma_{i^a} = \frac{j_{i^a_{yr}}}{j_{i^a_y}}. \qquad (5.40)$$

Here $j_{i^a_y}$ is the mass flux of the atomic species i^a towards the surface, and $j_{i^a_{yr}}$ the mass flux of actually recombining atoms.

The recombination coefficient γ_{i^a} depends on the pairing gas/surface species, like the surface accommodation coefficients, Section 4.3, and on the wall temperature T_w.

It has been observed in experiments, [16], that the energy transferred during the recombination process is less than the dissociation energy (partial energy accommodation), so that another recombination coefficient, the energy transfer recombination coefficient can be introduced. The effect, however, seems to be of minor importance and often is neglected in computational methods [16].

[11] Close to the surface the mass-diffusion velocity is the characteristic velocity v_{ref}, Section 5.4. This means large residence time and hence $DAM1 \rightarrow \infty$. Catalytic wall behaviour, however, is again amplifying gradients, and in turn also near-wall non-equilibrium transport phenomena.

The catalytic recombination rate $k_{w_{i^a}}$ is a function of the wall temperature and of the gas-species properties. It is expressed in the form[12] (Hertz-Knudsen relation):

$$k_{w_{i^a}} = \gamma_{i^a} \sqrt{\frac{R_0 T}{2\pi M_{i^a}}}. \tag{5.41}$$

With the catalytic recombination rate the mass flux of atoms actually recombining at the surface can be written:

$$j_{i^a_{yr}} = \gamma_{i^a} k_{w_{i^a}} \rho_{i^a}. \tag{5.42}$$

The mass flux of the atoms towards the surface is

$$j_{i^a_y} = \frac{1}{\gamma_{i^a}} j_{i^a_{yr}} = k_{w_{i^a}} \rho_{i^a}. \tag{5.43}$$

We distinguish three limiting cases:

- $\gamma_{i^a} \to 0$: no recombination occurs, the surface is non-catalytic, the catalytic recombination rate goes to zero: $k_{w_{i^a}} \to 0$.
- $0 < \gamma_{i^a} < 1$: only a part of the atoms recombines, the surface is partially catalytic, the catalytic recombination rate is finite: $0 < k_{w_{i^a}} < \infty$.
- $\gamma_{i^a} \to 1$: all atoms recombine, the surface is fully catalytic. For this case one finds in literature that the recombination rate is considered to be as infinitely large: $k_{w_{i^a}} \to \infty$.

We discuss now the possible boundary conditions for the mass-transport equations, Sub-Section 4.3.3:

- Equilibrium conditions for the species:

  $$\omega_i|_w = \omega_i(\rho, T)|_w.$$

  This case would not involve the solution of species-continuity equations. Hence no boundary conditions need to be considered.
- Vanishing mass-diffusion flux of species i:

  $$j_{i_y}|_w = 0.$$

  If this holds for all species, it is the boundary condition for the case of the non-catalytic surface, in which no net flux of atoms and molecules towards the surface happens. The non-catalytic surface does not influence the gas composition at the wall.

---

[12] The heterogeneous reaction can be formulated in analogy to the homogeneous reaction, eq. (5.30), which we do not elaborate further here.

– Fully catalytic surface recombination:

$$\omega_{i^a}|_w = 0.$$

The complete vanishing of the atoms of species i is prescribed.

– Finite catalytic surface recombination:

$$j_{i^a_y}|_w = \rho_{i^a} k_{w_{i^a}}|_w .$$

The partial vanishing of atoms of species i is prescribed.

In closing this section we discuss some results from [11] and [17] in order to illustrate surface catalytic recombination effects. In Fig. 5.11, [11], distributions of the heat flux in the gas at the wall q_{gw} at an hyperbola with nose radius $R = 1.322\ m$ and opening angle $\Phi = 41.7°$ are given. The generator of the hyperbola approximates the contour of the lower symmetry line at the first seven meters of the forward part of the Space Shuttle. The flow parameters are given in Table 5.6. Although the surface temperature is still rising at that trajectory point, a constant temperature $T_w = 800.0\ K$ was chosen, like for the Direct Simulation Monte Carlo (DSMC) computations [18]. The error in T_w is about 5 per cent.

Table 5.6. Parameters of the (laminar) flow past the hyperbola under reentry conditions at 85.74 km altitude [11].

M_∞	$v_\infty\ [m/s]$	$T_\infty\ [K]$	$\rho_\infty\ [kg/m^3]$	ω_{N_2}	ω_{O_2}	$T_w\ [K]$
27.35	7,511.4	187.0	$6.824 \cdot 10^{-6}$	0.738	0.262	800.0

At the chosen trajectory point we are at the border of the continuum regime. Slip effects were not modeled in [11], the comparison with DSMC results served the validation of the Navier-Stokes method CEVCATS.

Fig. 5.11 shows everywhere a good agreement between the results of the two methods. The assumption of a non-catalytic surface yields the smallest heat flux, that of a fully catalytic surface a heat flux approximately twice as large. The finite catalytic case yields results not much higher than that of the non-catalytic case. In all cases we have to a good approximation the cold-wall laminar-flow behaviour of $q_w \sim (x/L)^{-0.5}$ as discussed in Sub-Section 3.2.1 (eq. (3.27)). Up to $x \approx 2.0\ m$ the flight data are met by the finite catalytic case, and after $x \approx 4.0\ m$ by the fully catalytic case. Here the surface is cold enough to support fully catalytic recombination. The transition from one case to the other is not predicted.

In Fig. 5.12 we show a comparison of computed, [11], and flight-measured distributions, [19], of the radiation-adiabatic temperature along the lower

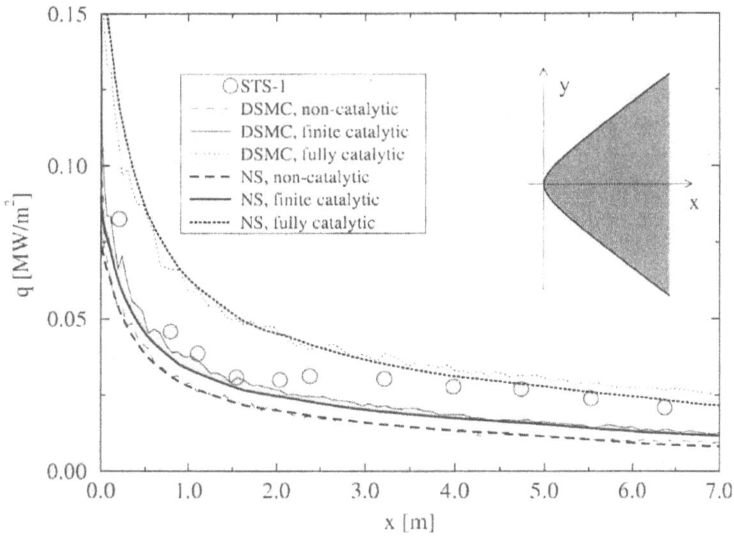

Fig. 5.11. Distribution of the heat flux in the gas at the wall ($q \equiv q_{gw}$) of the Space-Shuttle equivalent hyperbola with different surface-catalytic recombination models in comparison to in-flight measurements [11] (STS-1: first Space-Shuttle flight, DSMC: Direct Simulation Monte Carlo [18], NS: Navier-Stokes method CEVCATS).

symmetry plane of the Space Shuttle. The configuration used in the computations is the, on the lee side simplified, Space Shuttle configuration, known as HALIS configuration, which was introduced in [20].

The flow parameters are given in Table 5.7. The surface temperature is assumed to be the radiation-adiabatic temperature. The surface emissivity is assumed to be $\epsilon = 0.85$. Computations with CEVCATS were made for the non-catalytic, the finite catalytic, and the fully catalytic case. Results are shown up to approximately 75 per cent of the vehicle length.

Table 5.7. Parameters of the (laminar) flow past the HALIS configuration under reentry conditions at 72.0 km altitude [11].

M_∞	v_∞ $[m/s]$	T_∞ $[K]$	ρ_∞ $[kg/m^3]$	ω_{N_2}	ω_{O_2}	L_{ref} $[m]$	α [°]
24	7,027.54	212.65	$5.5 \cdot 10^{-5}$	0.738	0.262	32.77	40.0

Presented in Fig. 5.12 are computed data, and flight-measurement data from ordinary tiles, and from gauges with a catalytic coating. In general we see the drop of the wall temperature with increasing x as predicted quali-

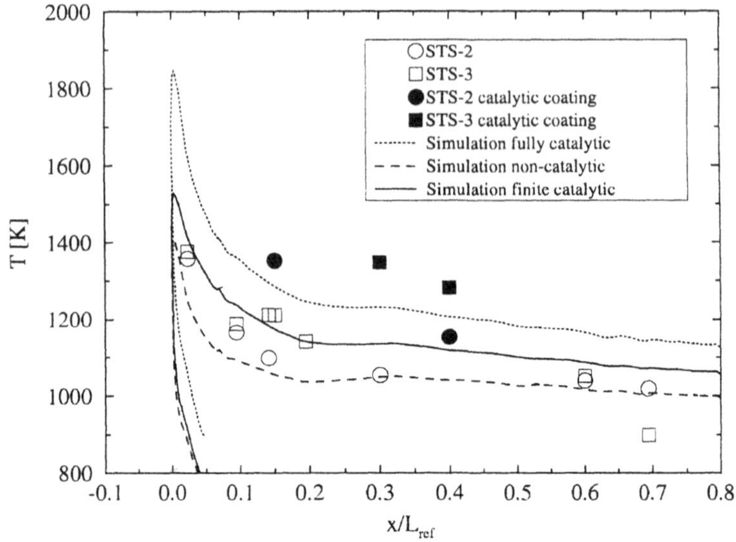

Fig. 5.12. Distributions of the radiation-adiabatic temperature ($T \equiv T_{ra}$) in the lower symmetry line of the HALIS configuration ($\epsilon = 0.85$) for different surface-catalytic recombination assumptions. Comparison of in-flight measurement data with CEVCATS data [11].

tatively in Sub-Section 3.2.1. The temperature difference between the fully catalytic and the non-catalytic case is strongest in the vehicle nose region with approximately 450.0 K. The results for the finite catalytic case lie between the results of the two limiting cases. The measured data are initially close to the computed finite catalytic data, and then to the non-catalytic data. This in contrast to the data shown for the heat flux in Fig. 5.11. The data measured on the catalytic coatings partly lie above the computed fully catalytic data.

Finally results from [17] are presented. The objective of that study was to determine the influence of the assumptions "fully catalytic" and "finite catalytic" wall on the wall temperature along the windward side of the X-38 with deflected body flap. The computations were made with the Navier-Stokes code URANUS with an axisymmetric representation of the windward symmetry-line contour. In Table 5.8 the flow parameters are given.

Fig. 5.13 shows that the computation with finite catalytic wall results in smaller temperatures than that with fully catalytic wall. The differences are about - 200.0 K in the nose region and at parts of the body, but small and reverse at the flap. The atomic nitrogen recombination coefficient γ_N is large at the nose and again at the flap, indicating strongly catalytic behaviour regarding N. The atomic oxygen recombination coefficient γ_O is moderate at

Table 5.8. Parameters of the (laminar) flow past the X-38 at 60.0 *km* altitude, and surface material (δ: downward deflection angle of flap) [17].

M_∞	v_∞ $[m/s]$	$Re_{\infty,L}$	L $[m]$	α [°]	δ [°]	ε	Nose cone	Body	Body flap
20	6,085.5	$6.2 \cdot 10^5$	9.14	40.0	20.0	0.87	SiC	SiO_2	SiC

the body and very large at the flap. There the finite-rate temperature $T_{w,fr}$ is even larger (about 50.0 K) than the fully-catalytic temperature $T_{w,fc}$. This is due to the very strong catalytic behaviour of the surface with regard to O and the resulting transport of atomic species in the boundary layer towards the wall. For the TPS design these data are the uncertainties range, which must be covered by appropriate design margins.

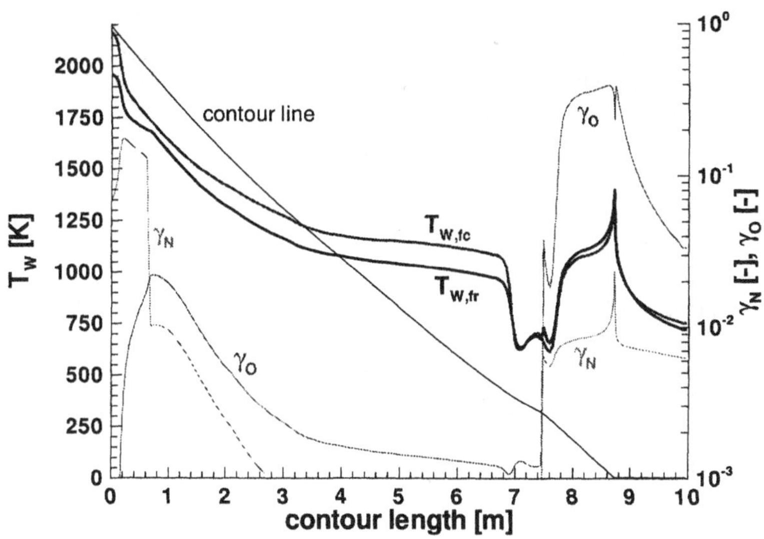

Fig. 5.13. Wall temperature ($T_w \equiv T_{ra}$) along the windward symmetry line of X-38 [17]. $T_{w,fc}$: wall temperature with fully catalytic wall (design assumption), $T_{w,fr}$: wall temperature with finite catalytic wall. Note the body contour in the figure. The hinge line lies at approximately 7.45 m.

5.7 A Few Remarks on Simulation Issues

Thermo-chemical freezing phenomena in the nozzles of hot hypersonic ground-simulation facilities seem not to be a major problem for the simulation of blunt-body (RV-type configuration) flows [21], [22]. The pressure is not much affected, but the bow shock stand-off distance may be wrong, which can be a problem in view of interaction phenomena on downstream configuration elements, Sub-Section 6.4.1. Also the determination of the test section Mach number can be problematic, Sub-Section 5.5.2. The possible large difference between the test section Mach number and the flight Mach number should be no problem, if the former is large enough that Mach-number independence is ensured, Section 10.3. However, the Space Shuttle experience regarding the pitching-moment anomaly, [23], must be taken very seriously, Section 10.3.

Another problem are thermal surface effects. For RV-type flight vehicles they concern at least catalytic surface recombination. In ground-simulation facilities we have in general cold model surfaces. In reality the surface is hot, $T_w \lessapprox 2{,}000.0\ K$, and thermal loads determined with a cold model will have deficiencies.

The situation is different for CAV-type flight vehicles. Thermo-chemical equilibrium or non-equilibrium will occur in the nose region, but downstream of it, where the body slope is small, freezing may set in. At a forebody with pre-compression and at the inlet ramps then again the situation will change. It is not clear what effect thermo-chemical freezing phenomena in the facility nozzle flow will have in this case. In [24] it was found that a species separation due to pressure-gradient induced mass diffusion can happen when the frozen free-stream flow passes the body (flat plate) induced oblique shock wave. Whether this can be of importance is not known.

Not much more is known about the influence of frozen nozzle flow on thermal surface effects, in this case regarding predominantly viscous flow phenomena. Boundary-layer instability and laminar-turbulent transition can be affected, Section 8.1.5, but for CAV-type vehicles the cold model surfaces will lead in any case to the major adverse effects.

Regarding computational simulation it depends on the flight speed, the altitude and the flight-vehicle type, and on the critical phenomenon/phenomena, whether an equilibrium or a non-equilibrium high-temperature real-gas model must be employed. This concerns not only the pressure field and near-wall/wall viscous and thermo-chemical phenomena, but via the shock stand-off distance also strong interaction effects, Section 9.2.2. Non-equilibrium thermo-chemical, but also radiation phenomena, however, must be regarded in any case at the very high speeds of, for instance, AOTV-type vehicles, see, e. g., [25].

5.8 Computation Models

For the computation of equilibrium or non-equilibrium flows different air models are available. If we can neglect ionization (Section 2.1), we work with the five species (N_2, O_2, N, O, NO) model with 17 reactions, and taking into account ionization with an eleven species model with 33 reactions [6]. In this section we summarize references to literature on detailed computation models for the thermo-chemical and transport properties of air in the temperature and density/pressure range of interest for both thermo-chemical equilibrium and non-equilibrium flow.

Equilibrium flow The thermo-chemical and transport properties of air can be determined with the models presented in Chapters 4 and 5. Characteristic data for thermo-chemical properties can be found e. g. in [6], and for transport properties in [26] to [29]. Regarding the general state of the art, including modeling problems, we refer to the review paper [8].

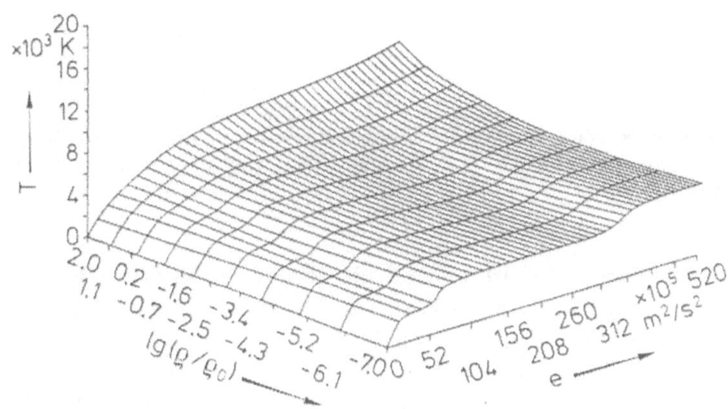

Fig. 5.14. State surface of the temperature T of equilibrium air as function of density ρ and internal energy e [33], database from [30] to [32] ($\rho_0 = 1.0\ kg/m^3$).

It is recommended to work in computation methods with state surfaces of thermodynamic and transport properties, because their use is computationally more efficient than that of basic formulations. As was mentioned in Sub-Section 4.2.5 such state surfaces are available in literature, e. g. [30] (only up to 6,000.0 K), [31], [32].

In [33] vectorized approximations of such state surfaces are given. We show as example in Fig. 5.14 the temperature T as function of the two variables density ρ (here in the form $lg(\rho/\rho_0)$), and internal energy e, and in Fig. 5.15 the ratio of specific heats γ as function of $lg(\rho/\rho_0)$ and $lg(p\rho_0/\rho p_0)$. Note that van der Waals effects, Section 5.1, are not included.

Examples of state surfaces of viscosity and thermal conductivity are given in Sub-Section 4.2.5.

Non-equilibrium flow Flows with thermo-chemical non-equilibrium must be modeled with help of the basic formulations. Characteristic data again can be found, e. g., in [6] and [10]. We refer also to the review paper [8] regarding the general state of the art, including modeling problems.

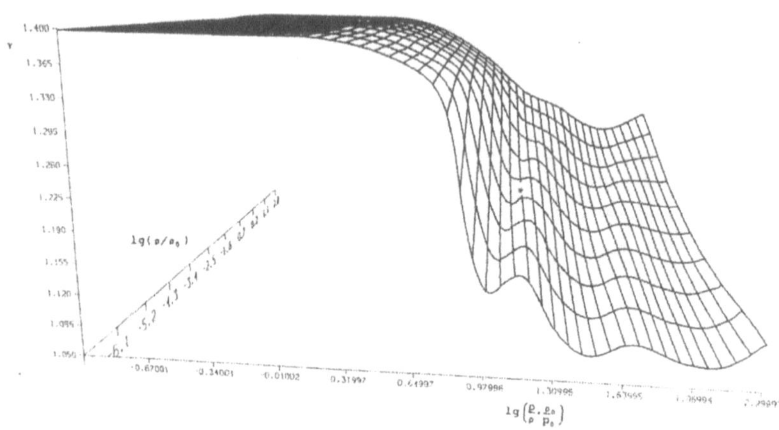

Fig. 5.15. State surface of the ratio of specific heats γ of equilibrium air as function of ρ and pressure/density p/ρ [33], database from [30] to [32] ($\rho_0 = 1.292\ kg/m^3$, $p_0 = 1.0133{\cdot}10^5\ Pa$).

Transport properties are treated like in equilibrium flow by taking into account the momentarily present thermo-chemical state of the gas with appropriate formulations, Sub-Section 4.2.5.

Special regard must be given to the boundary conditions of the species continuity equations, if finite catalytic surfaces are to be modeled. In [16] we find the formulations for the five and the eleven species gas model, and also general slip-flow boundary conditions.

5.9 Problems

Problem 5.1 A gas has the ratio of specific heats $\gamma = 1$. Show that this is equivalent to $f = \infty$.

Problem 5.2 The reservoir enthalpy of a ground-simulation facility with air as test gas is $h_t = 20.0\ MJ/kg$. What is a) the maximum possible speed at the exit of the nozzle, b) the speed, if the static temperature at the nozzle exit is $T_{exit} = 1{,}000.0\ K$ (assume a Lighthill gas at the exit), c) the speed, if in addition 20 per cent of the reservoir enthalpy is frozen?

Solution: $v =$ a) 6,324.5 m/s, b) 6,233.1 m/s, c) 5,554.5 m/s.

Problem 5.3 The stagnation temperature is the temperature at the stagnation point of a blunt body. For the inviscid flow of a perfect gas, it is the total temperature T_t. A RHPM-CAV-flyer flies with $v_\infty = 2.0\ km/s$ at 30.0 km altitude. What is a) T_t? What is T_t, if we mimic high-temperature real-gas effects with b) $\gamma_{eff} = 1.2$, and c) with $\gamma_{eff} = 1$?

Solution: $T_t =$ a) 2,217.5 K, b) 1,222.0 K, c) 226.51 K. Should or could it be 0.0 K in case c)? Discuss.

Problem 5.4 How does the uppermost curve of q in Fig. 5.11 scale with the x-dependence indicated in eq. (3.27)? Measure q at $x = 1.0\ m$, and compare with the measured value at $x = 6.0\ m$.

Solution: $\Delta q = +\ 0.0001\ MW/m^2$.

References

1. W. G. VINCENTI, C. H. KRUGER. "Introduction to Physical Gas Dynamics". John Wiley, New York and London/Sydney, 1965. Reprint edition, Krieger Publishing Comp., Melbourne, Fl., 1975
2. J. HILSENRATH, C. W. BECKETT, S. BENEDICT, L. FANO, J. HOGE, F. MASI, L. NUTTALL, S. TOULOUKIAN, W. WOOLLEY. "Tables of Thermodynamics and Transport Properties of Air, Argon, Carbon Dioxide, Carbon Monoxide, Hydrogen, Nitrogen, Oxygen and Steam". Pergamon Press, 1960.
3. E. H. HIRSCHEL. "Beitrag zur Beschreibung der Realgaseinflüsse bei gasdynamischen Problemen". DLR Mitt. 66-20, 1966.
4. G. SIMEONIDES. "Hypersonic Shock Wave Boundary Layer Interactions over Compression Corners". Doctoral Thesis, University of Bristol, U.K., 1992.
5. M. J. LIGHTHILL. "Dynamics of a Dissociating Gas. Part I: Equilibrium Flow". Journal of Fluid Mechanics, Vol. 2, 1957, pp. 1 - 32.
6. C. PARK. "Nonequilibrium Hypersonic Flow". John Wiley & Sons, New York, 1990.
7. G. DAMKÖHLER. "Einflüsse der Strömung, Diffusion und des Wärmeübergangs auf die Leistung von Reaktionsöfen". Zeitschrift für Elektrochemie, Vol. 42, No. 12, 1936, pp. 846 - 862.
8. G. S. R. SARMA. "Physico-Chemical Modelling in Hypersonic Flow Simulation". Progress in Aerospace Sciences, Vol. 36, No. 3-4, 2000, pp. 281 - 349.
9. S. SÉROR, E. SCHALL, M.-C. DRUGUET, D. E. ZEITOUN. "An Extension of CVCD model to Zeldovich Exchange Reactions for Hypersonic Non-Equilibrium Air Flows". Shock Waves, Vol. 8, 1998, pp. 285 - 298.
10. S. KANNE, H.-H. FRÜHAUF, E. W. MESSERSCHMID. "Thermochemical Relaxation Through Collisions and Radiation". AIAA Journal of Thermophysics and Heat Transfer, Vol. 14, No. 4, 2000, pp. 464 - 470.
11. S. BRÜCK. "Ein Beitrag zur Beschreibung der Wechselwirkung von Stössen in reaktiven Hyperschallströmungen (Contribution to the Description of the Interaction of Shocks in Reacting Hypersonic Flows)". Doctoral Thesis, Universität Stuttgart, Germany, 1998. Also DLR-FB 98-06, 1998.
12. K. HANNEMANN, V. HANNEMANN, S. BRÜCK, R. RADESPIEL, G. S. R. SARMA. "Computational Modelling for High Enthalpy Flows". SAMS, Vol. 34, No. 2, 1999, pp. 253 - 277.
13. K. HANNEMANN. "Computation of the Flow in Hypersonic Wind Tunnel Nozzles". *W. Kordulla, S. Brück (eds.), DLR Contribution to the Fourth European High Velocity Data Base Workshop.* DLR-FB 97-34, 1997.

14. L. M. G. WALPOT, G. SIMEONIDES, J. MUYLAERT, P. G. BAKKER. "High Enthalpy Nozzle Flow Insensitivity Study and Effects on Heat Transfer". Journal of Shock Waves, Vol. 16, 1996, pp. 197 - 204.
15. A. H. BOUDREAU. "Characterization of Hypersonic Wind-Tunnel Flow Fields for Improved Data Accuracy". AGARD CP 429, 1987, pp. 28-1 to 28-9.
16. C. D. SCOTT. "Wall Catalytic Recombination and Boundary Conditions in Nonequilibrium Hypersonic Flows - with Applications". *J. J. Bertin, J. Periaux, J. Ballmann (eds.), Advances in Hypersonics, Vol. 2, Modeling Hypersonic Flows.* Birkhäuser, Boston, 1992, pp. 176 - 250.
17. M. FERTIG, H.-H. FRÜHAUF. "Reliable Prediction of Aerothermal Loads at TPS-Surfaces of Reusable Space Vehicles". *Proc. 12th European Aerospace Conference, Paris, November 29 to Dezember 1, 1999.* 1999.
18. F. BERGEMANN. "Gaskinetische Simulation von kontinuumsnahen Hyperschallströmungen unter Berücksichtigung von Wandkatalye (Gas-Kinetic Simulation of Continuum-Near Hypersonic Flows with Regard to Surface Catalycity)". DLR-FB 94-30, 1994.
19. D. A. STEWARTT, J. V. RAKICH, M. J. LANFRANCO. "Catalytic Surface on Space Shuttle Thermal Protection System during Earth Entry of Flights STS-2 through STS-5". *J. P. Arrington, J. J. Jones (eds.), Shuttle Performance: Lessons Learned.* NASA-Report CP 2283, 1983, pp. 827 - 845.
20. K. J. WEILMUENSTER, P. A. GNOFFO, F. A. GREENE. "Navier-Stokes Simulations of the Shuttle Orbiter Aerodynamic Characteristics with Emphasis on Pitch Trim and Body Flap". AIAA-Paper 93-2814, 1993.
21. H. G. HORNUNG. "Non-Equilibrium Dissociating Nitrogen Flow over Spheres and Circular Cylinders". J. Fluid Mechanics, Vol. 53, Part 1, 1972, pp. 149 - 176.
22. M. N. MACROSSAN. "Hypervelocity Flow of Dissociating Nitrogen Downstream of a Blunt Nose". J. Fluid Mechanics, Vol. 217, 1990, pp. 167 - 202.
23. B. F. GRIFFITH, J. R. MAUS, J. T. BEST. "Explanation of the Hypersonic Longitudinal Stability Problem - Lessons Learned". *J. P. Arrington, J. J. Jones (eds.), Shuttle Performance: Lessons Learned.* NASA CP-2283, Part 1, 1983, pp. 347 - 380.
24. E. H. HIRSCHEL. "Hypersonic Flow of a Dissociated Gas over a Flat Plate". *L. G. Napolitano (ed.), Astronautical Research 1970.* North-Holland Publication Co., Amsterdam, 1971, pp. 158 - 171.
25. W. SCHNEIDER. "Effect of Radiation on Hypersonic Stagnation Flow at Low Density". Zeitschrift für Flugwissenschaften (ZFW), Vol. 18, No. 2/3, 1970, pp. 50 - 58.
26. R. B. BIRD, W. E. STEWART, E. N. LIGHTFOOT. "Transport Phenomena". John Wiley, New York and London/Sydney, 2nd edition, 2002 .
27. L. BIOLSI, D. BIOLSI. "Transport Properties for the Nitrogen System: N_2, N, N^+, and e". AIAA-Paper 83-1474, 1983.
28. L. BIOLSI. "Transport Properties for the Oxygen System: O_2, O, O^+, and e". AIAA-Paper 88-2657, 1988.

29. M. Fertig, A. Dohr, H.-H. Frühauf. "Transport Coefficients for High-Temperature Nonequilibrium Air Flows". AIAA Journal of Thermophysics and Heat Transfer, Vol. 15, No. 2, 2001, pp. 148 - 156.

30. S. Gordon, B. J. McBride. "Computer Progam for Calculation of Complex Chemical Equilibrium Compositions, Rocket Performance, Incident and Reflected Shocks, and Chapman-Jouguet Detonations". NASA SP-273, Interim Revision, 1976.

31. S. Srinivasan, J. C. Tannehill, K. J. Weilmuenster. "Simplified Curve Fits for the Thermodynamic Properties of Equilibrium Air". ISU-ERI-Ames-86401, Iowa State University, 1986.

32. S. Srinivasan, J. C. Tannehill, K. J. Weilmuenster. "Simplified Curve Fits for the Transport Properties of Equilibrium Air". ISU-ERI-Ames-88405, Iowa State University, 1987.

33. Ch. Mundt, R. Keraus, J. Fischer. "New, Accurate, Vectorized Approximations of State Surfaces for the Thermodynamic and Transport Properties of Equilibrium Air". Zeitschrift für Flugwissenschaften und Weltraumforschung (ZFW), Vol. 15, No. 3, 1991, pp. 179 - 184.

6 Inviscid Aerothermodynamic Phenomena

In Sub-Section 4.3.1 we have seen that if the Reynolds number characterizing a flow field is large enough, we can separate the flow field into inviscid and viscous portions. From Fig. 2.3, Section 2.1, we gather that the unit Reynolds numbers in the flight domain of interest are indeed sufficiently large. This does not mean that aerodynamic properties of hypersonic vehicles can be prescribed fully by means of inviscid theory. This is at best possible for the longitudinal movement of re-entry vehicles.

At transonic, supersonic and hypersonic flight we observe the important phenomenon of shock waves. The shock wave is basically a viscous phenomenon, but in general can be understood as a flow discontinuity embedded in an inviscid flow field. In this chapter we look at shock waves as compressibility phenomena occurring in the flow fields past hypersonic flight vehicles. We treat their basic properties, and also the properties of the isentropic Prandtl-Meyer expansion. Of importance in hypersonic flight-vehicle design is the stand-off distance of the bow-shock surface at blunt vehicle noses. We investigate this phenomenon as well as the effects of entropy-layer swallowing by the vehicle boundary layers.

Also of interest for the development of boundary layers is the change of the unit Reynolds number across shock waves. We will see that an increase of the unit Reynolds number will only occur, if the shock wave is sufficiently oblique. Then a boundary layer behind a shock wave close to the body surface will be thinned. This subsequently changes the thermal state of the surface, and rises the radiation-adiabatic temperature of a radiation-cooled surface.

Basics of Newtonian flow are considered then. Newtonian flow is an interesting limiting case, and the related computation method, with appropriate corrections, is an effective and cheap tool to estimate (inviscid) surface pressure and velocity fields. Related to Newtonian flow is the shadow effect, which can appear in hypersonic flow past flight vehicles, especially if they fly at large angle of attack. It is characterized by a concentration of the aerodynamic forces with increasing Mach number on the windward side of the vehicle, the "pressure side" of classical aerodynamics. The leeward side, the "suction side", is in the "shadow" of the body, and loses its role as a force-generating surface. We don't discuss this phenomenon, but refer to [1], where several examples are given, also regarding control-surface efficiency. In the context of

Newtonian flow finally an important principle for design and ground-facility simulation is treated, the Oswatitsch Mach-number independence principle.

6.1 Hypersonic Flight Vehicles and Shock Waves

Shock waves can occur if the speed in a flow field is larger than the speed of sound. Across shock waves the flow speed is drastically reduced, and density, pressure and temperature rise strongly. The phenomenon is connected to the finite propagation speed of pressure disturbances, the speed of sound. At sea level this is $a_\infty \approx 300.0\ m/s$ or $\approx 1{,}200.0\ km/h$. If the vehicle is flying slower, that is with subsonic speed $M_\infty < 1$, the pressure disturbances travel ahead of the vehicle: the air then gives gradually way to the approaching vehicle. If the vehicle flies faster than the speed of sound, with supersonic $M_\infty > 1$ or hypersonic $M_\infty \gg 1$, speed, the air gives way only very shortly ahead of the vehicle, and almost instantaneously, through a shock wave. In this case we speak about the (detached) bow shock, which envelopes the body at a certain distance with a generally convex shape. This bow shock moves in steady flight with the speed of the flight vehicle, i. e., it is fixed to the vehicle-reference frame, and (for steady flight) it is a steady phenomenon.

However, shock waves can also be embedded in the vehicle's flow field, if locally supersonic flow is present. Embedded shocks may occur already in the transonic flight regime, where the free-stream speed is still subsonic, but super-critical, and in the supersonic and hypersonic flight regimes. If the shock wave lies orthogonal to the local speed direction (normal shock), we have subsonic speed behind it, if it is sufficiently oblique, the speed behind it is supersonic. We treat these phenomena in Sub-Sections 6.3.1 and 6.3.2, but refer for a deeper treatment to the literature, for instance [2], [3], [4], [5].

Depending on the flight-vehicle configuration a flow field pattern with either subsonic or supersonic, or mixed, speed is resulting behind the bow shock, Fig. 6.1. At a blunt-nosed body, Fig. 6.1 a), we always have a detached bow shock with a small subsonic ($M < 1$) pocket behind it. The flow is then expanding to supersonic speed again. The bow shock is dissipating far downstream of the vehicle, while its inclination against the free-stream direction approaches asymptotically the free-stream Mach angle μ_∞:

$$sin\, \mu_\infty = \frac{1}{M_\infty}. \tag{6.1}$$

At a sharp-nosed cone[1], if the opening angle is small enough (sub-critical opening angle), the bow shock is attached, Fig. 6.1 b), the speed behind it is subsonic/supersonic or supersonic, Sub-Section 6.3.2. At a sharp-nosed cone of finite length with sufficiently large opening angle (super-critical opening

[1] At hypersonic speeds sharp-nosed cones would suffer untenable large thermal loads, so that cases b) and c) are hypothetical at such speeds.

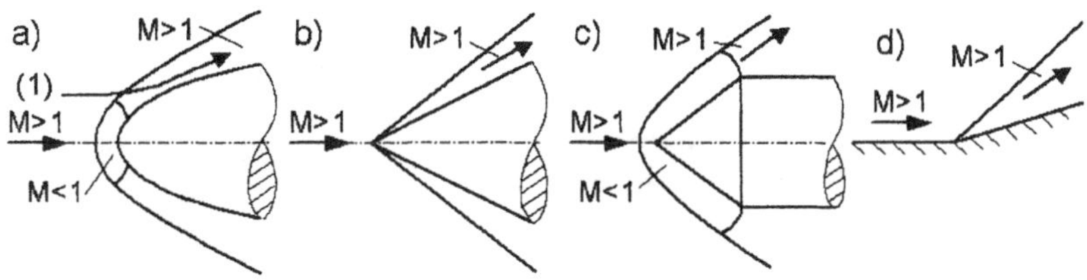

Fig. 6.1. Schematics of bow shocks and of an embedded shock wave: traces of shock surfaces in the symmetry plane of **a**) a blunt body, **b**) a sharp-nosed cone with sub-critical opening angle, **c**) a sharp-nosed cone with super-critical opening angle, and **d**) an embedded shock at a ramp with sub-critical ramp angle (two-dimensional or axisymmetric flow).

angle), Fig. 6.1 c), the bow shock is detached, the speed behind it is initially subsonic, and then, in the figure at the following cylinder, supersonic again. Lastly, at the ramp we have an embedded shock wave. If the ramp angle is sub-critical, the shock will be "attached", with supersonic speed behind the oblique shock, Fig. 6.1 d). If the ramp angle is sufficiently large (super-critical), the shock will be a "detached" normal shock, with a subsonic pocket behind it. Then the shock will bend around, like shown for the cone in Fig. 6.1 c).

The shape of the bow shock and the subsonic pocket are in principle different for RV-type flight vehicles and (airbreathing) CAV-type vehicles, Fig. 6.2. The slender CAV-type vehicle flies at small angle of attack, with a small subsonic pocket ahead of the small-bluntness nose, Fig. 6.2 a). The RV-type vehicle with its anyway blunt shape flies during re-entry at large angle of attack, and has a large subsonic pocket over the windward side[2], Fig. 6.2 b).

Why are bow-shock shapes of interest? They are of interest because the entropy rises across them, which is equivalent to a loss of total pressure. The consequence is the wave drag, which is a another form of aerodynamic drag besides induced drag, skin-friction drag, and viscosity-induced pressure drag (form drag). The entropy rise is largest where the shock is normal to the free-stream flow (normal shock), and becomes smaller with decreasing inclination of the shock against the free-stream (oblique shock). In this respect we note two fundamentally different cases:

---

[2] This is important also with regard to the proper choice of reference data in similarity parameters, Section 4.4. Over the CAV-type flight vehicle the boundary-layer edge Mach and unit Reynolds numbers for instance have approximately the order of magnitude of the free-stream values, Section 1.2, whereas on the RV-type vehicle especially the edge Mach number M_e is much lower: at an angle of attack of $\alpha \approx 40°$ at flight Mach numbers $25 \gtrapprox M_\infty \gtrapprox 10$ we have a large portion of subsonic flow, and then at most $M_e \approx 3$.

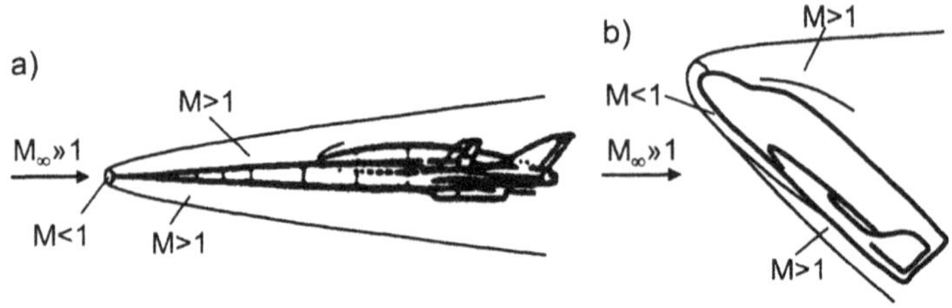

Fig. 6.2. Schematic of bow shocks and subsonic pockets: traces of shock surfaces in the symmetry plane of **a**) a CAV-type flight vehicle (typically small angle of attack), and **b**) of a RV-type vehicle (large angle of attack), see also Section 1.2.

1. For a symmetric body at zero angle of attack, see e. g. Fig. 6.1 a), the streamline on the axis carries the largest entropy rise $\triangle s$, because the bow shock is orthogonal to it.
2. For an asymmetric body at angle of attack, Fig. 6.3, the streamline through the locally normal bow-shock surface, at P_0, carries the largest entropy rise, $\triangle s_0 = \triangle s_{max}$. The stagnation-point streamline, however, which penetrates the bow-shock surface at P_1, which lies at a certain distance from P_0, carries a lower entropy rise, $\triangle s_1 < \triangle s_{max}$. Hence in this case the stagnation pressure is larger than in the symmetric case, because the total pressure loss is smaller. The boundary-layer edge flow, on the other hand, is not characterized, as in the symmetric case, by the maximum entropy state behind the bow-shock surface.

In general it can be stated: the larger the portion of the bow-shock surface with large inclination against the free stream, the larger is at a given free-stream Mach number the wave drag.

We understand now why a RV-type flight vehicle, which actually flies a braking mission, is blunt and flies at a large angle of attack. In addition, as we have seen in Section 3.2, large bluntness leads to thick boundary layers, and hence to large radiation-cooling efficiency. The blunt vehicle shape thus serves both large drag and low radiation-adiabatic surface temperatures, and hence thermal loads, which passive thermal protection systems can cope with, Fig. 3.2.

The situation is different with (airbreathing) CAV-type vehicles. Such vehicles must have a low total drag, therefore they are slender, have a small nose bluntness, and fly at small angles of attack. Here the small nose bluntness has an adverse effect regarding radiation cooling. Small nose radii lead to thin boundary layers, and hence to small radiation-cooling efficiency. In CAV-type vehicle design trade-offs are necessary to overcome the contradicting demands of small wave drag, and sufficient radiation-cooling efficiency.

Bow-shock shapes and shock locations in general are of interest also for another reason. They can lead to shock/boundary-layer, and shock/shock /bondary-layer interactions, which we will treat in Sub-Section 9.2. These

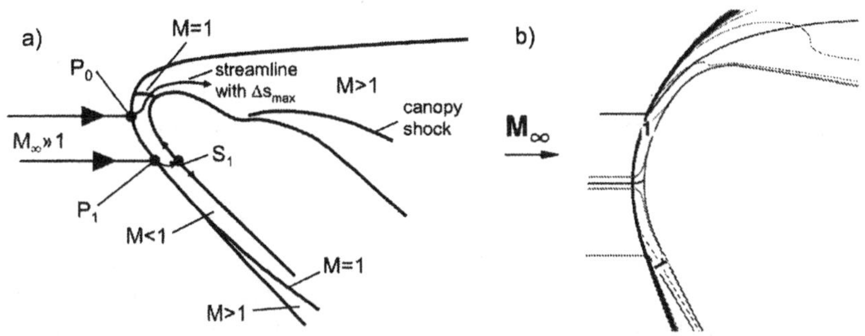

Fig. 6.3. Schematic of an asymmetric flow situation: **a**) traces of the bow shock surface and the streamlines in the symmetry plane of a RV-type flight vehicle at angle of attack: the streamline, which crosses the locally normal shock at P_0 carries the maximum entropy, the stagnation-point (S_1) streamline, which penetrates the bow-shock surface at P_1, carries a lower entropy (distance between P_0 and P_1 exaggerated), **b**) true situation computed for the X-38 (flow parameters see Table 3.2), [6]. Of the three streamlines in the middle of the graph, the lower goes through the normal portion of the bow shock, the one in the middle, very close to is, is the stagnation-point streamline, '1' denotes the upper and the lower trace of the sonic surface.

"strong" interactions can cause locally large mechanical and thermal loads with possible severe consequences for the airframe integrity. They go along also with local separation and unsteadiness phenomena. In Fig. 6.4 we show such a strong interaction situation, which can arise at both CAV-type and RV-type vehicle wings with a second delta part, which is needed to enhance the aspect ratio of the wing in order to assure proper flight characteristics at low-speed and high-angle of attack flight.

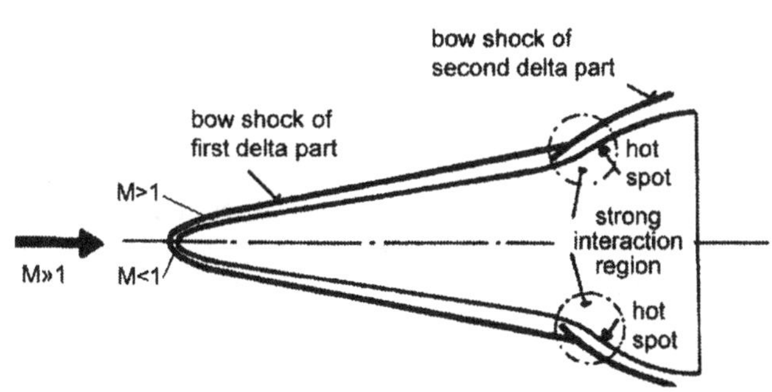

Fig. 6.4. Schematic of strong interaction of the vehicle bow shock with an embedded bow shock at the second delta part of the wing of a CAV-type flight vehicle (seen from above) [7].

Finally we look at the role, which shock waves play in hypersonic air-breathing propulsion and in aerothermodynamic airframe/propulsion integration of CAV-type flight vehicles, [8]. It is recalled that the flight-speed range of turbojet engines is up to $M_\infty = 3$ to 4, that of ramjet engines $3 \lessapprox M_\infty \lessapprox 6$, and of scramjet engines $6 \lessapprox M_\infty \lessapprox 12$.

Important in our context is the fact that the pre-compressor Mach number of turbojet engines, like the combustion-chamber Mach number of ramjet engines is $M = 0.4$ to 0.6, and the combustion-chamber Mach number of scramjet engines $M = 2$ to 3. These numbers show that the characteristic pre-engine Mach numbers are much lower than the actual flight Mach numbers.

The only way to attain these engine Mach numbers is to decelerate the air stream via shock waves [8]. Normal shock waves would give too large total-pressure losses, hence one or more oblique shock waves are employed. In Fig. 6.5 we show the schematic of a three-ramp inlet, which is typical for a ramjet engine.

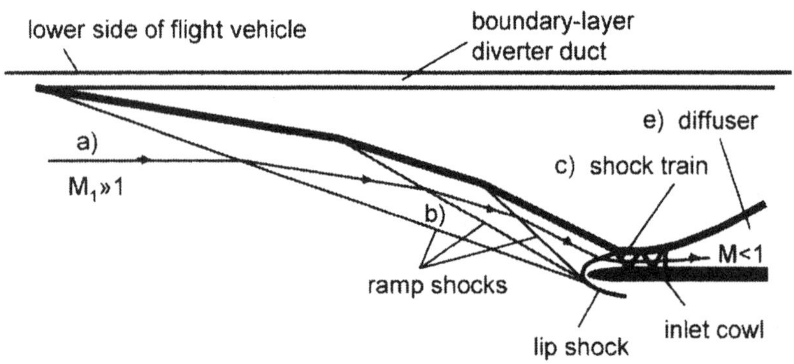

Fig. 6.5. Schematic of a three-ramp ramjet-engine inlet. The flow is from the left to the right. Not indicated is that at the lower side due to the forebody already a pre-compression takes place.

The inlet-onset flow a) will have been pre-compressed by the lower side of the forebody [7]. The forebody boundary layer must be diverted by the boundary-layer diverter, in order to avoid unwanted distortion of the inlet flow, and finally the flow entering the engine. In the outer compression regime b) the three oblique ramp shocks are centered on the lip of the inlet cowl.

The cowl lip, which is blunt in order to withstand mechanical and thermal loads, causes a bow shock, which interacts with the ramp shocks, Fig. 6.6. For practical reasons usually the exact shock-on-lip situation is avoided, because it may lead to adverse shock/shock/boundary-layer interaction effects.

In Fig. 6.5 the part of the inlet is indicated, where the internal compression by a train of oblique shocks, region c), finally leads to sub-sonic flow, which is

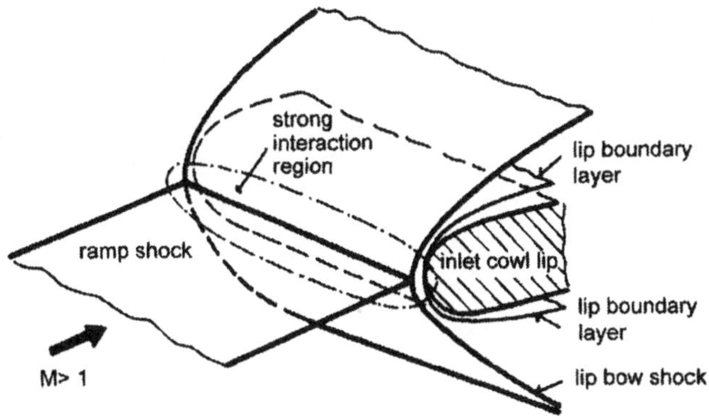

Fig. 6.6. Schematic of the shock-on-cowl-lip situation (only one ramp shock shown) [7].

further decelerated in the diffuser d) to the above mentioned small pre-engine Mach number of the flow entering the combustion chamber.

A fully oblique shock train is shown in Fig. 6.7 for a two-strut scramjet geometry [9] at a low (off-design) flight Mach number. In both the outer and in inner flow path we see shock reflections, shock intersections, Sub-Section 6.3.2, expansion waves, Sub-Section 6.5, and at the sharp strut-trailing edges slip surfaces.

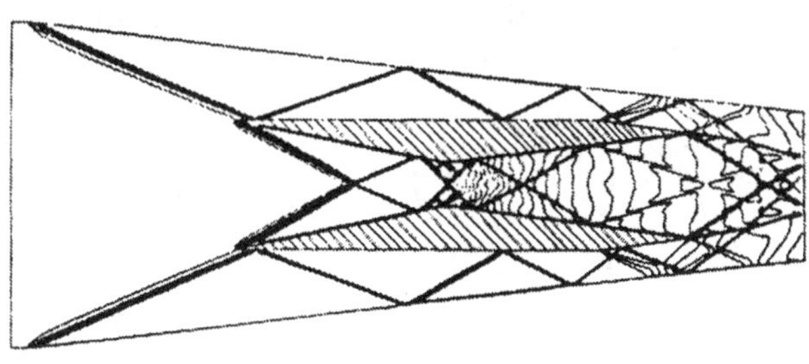

Fig. 6.7. Iso-Mach lines (and shock waves) predicted with an Euler code for a two-strut scramjet geometry at $M_\infty = 3$ [9].

6.2 One-Dimensional Shock-Free Flow

For this and the following sections we assume steady, inviscid, iso-energetic, one-dimensional[3] flow of a perfect gas. Used is the nomenclature (x, u) given

[3] This is to be distinguished from stream-tube flow, where in a tube with varying cross-section uniform distributions of the flow properties are assumed, see e. g. [3].

in Fig. 4.1, Section 4.1. We treat in this section continuous shock-free flow as a prerequisite, and then in the Sub-Sections 6.3.1 and 6.3.2 normal and oblique shock waves.

The governing equations in Section 4.3 for mass, momentum and energy transport, eqs. (4.83), (4.27), (4.63) reduce for one-dimensional flow to (we omit dx):

$$d(\rho u) = 0, \tag{6.2}$$

$$\rho u\, du = -dp, \tag{6.3}$$

$$\rho u\, dh = u\, dp. \tag{6.4}$$

The energy equation in differential form, eq. (6.4), can be rewritten by introducing the perfect-gas law, eq. (5.1), and relations (5.10):

$$\frac{\gamma}{\gamma - 1} d\left(\frac{p}{\rho}\right) + u\, du = 0. \tag{6.5}$$

This relation, combined with eq. (6.3), yields the pressure-density relation for isentropic flow:

$$\frac{p}{\rho^\gamma} = constant = \frac{p_t}{\rho_t^\gamma}. \tag{6.6}$$

The subscript 't' indicates the reservoir or 'total' condition. p_t is the total pressure. Eq. (6.6) can be generalized to:

$$\frac{p}{p_t} = \left(\frac{\rho}{\rho_t}\right)^\gamma = \left(\frac{T}{T_t}\right)^{\frac{\gamma}{\gamma-1}} = \left(\frac{a}{a_t}\right)^{\frac{2\gamma}{\gamma-1}}. \tag{6.7}$$

Here a is the speed of sound, the speed at which disturbances propagate through the fluid. Sound waves have so small amplitudes, that they can be considered as isentropic [3]: s = constant. Hence for perfect gas:

$$a^2 = \left(\frac{\partial p}{\partial \rho}\right)_s = \left(\frac{\partial(const.\, \rho^\gamma)}{\partial \rho}\right)_s = \gamma R T. \tag{6.8}$$

a_t is the speed of sound at the total temperature T_t:

$$a_t = \sqrt{\gamma R T_t}. \tag{6.9}$$

The entropy of a thermally perfect gas is defined by:

$$ds = c_v \frac{dp}{p} - c_p \frac{d\rho}{\rho}. \tag{6.10}$$

For a perfect gas we find:

$$s - s_{ref} = c_v \ln \frac{p}{p_{ref}} - c_p \ln \frac{\rho}{\rho_{ref}}. \tag{6.11}$$

In adiabatic flow the second law of thermodynamics demands:

$$s - s_{ref} \geq 0. \tag{6.12}$$

Isentropic flow processes are defined by:

$$s - s_{ref} \equiv 0 : \; s = constant. \tag{6.13}$$

The integrated energy equation eq. (6.5) in the familiar form for perfect gas reads, with c_p the specific heat at constant pressure[4]:

$$c_p T + \frac{1}{2} u^2 = c_p T_t. \tag{6.14}$$

From this relation we see, that with a given total temperature T_t by expansion only a maximum speed V_m can be reached. This is given when the "static" temperature T reaches zero:

$$V_m = \sqrt{2 c_p T_t}. \tag{6.15}$$

The maximum speed can be expressed as function of a_t and γ:

$$V_m = \sqrt{\frac{2}{\gamma - 1}}\, a_t. \tag{6.16}$$

Locally the "critical" or "sonic" condition is reached, when the speed u is equal to the speed of sound a:

$$u_* = a_*. \tag{6.17}$$

We can write eq. (6.14) in the form

$$\frac{1}{\gamma - 1} a^2 + \frac{1}{2} u^2 = c_p T_t, \tag{6.18}$$

and relate the critical speed of sound a_* to the total temperature T_t:

$$a_* = \sqrt{2 \frac{\gamma - 1}{\gamma + 1} c_p T_t}. \tag{6.19}$$

The relation for the "critical" Mach number M_* then is:

$$M_*^2 = \left(\frac{u}{a_*} \right)^2 = \frac{\frac{\gamma+1}{2} M^2}{1 + \frac{\gamma-1}{2} M^2}. \tag{6.20}$$

[4] The general formulation in terms of the enthalpy h is $h + \frac{1}{2}u^2 = h_t$, where h_t is the total enthalpy of the flow, see also Section 3.1.

Eq. (6.14) combined with eq. (6.7) gives the equation of Bernoulli for compressible isentropic flow of a perfect gas:

$$\frac{\gamma}{\gamma-1}\left(\frac{p_t}{\rho_t}\right)\left(\frac{p}{p_t}\right)^{\frac{\gamma-1}{\gamma}}+\frac{1}{2}u^2=\frac{\gamma}{\gamma-1}\left(\frac{p_t}{\rho_t}\right). \tag{6.21}$$

We remember Bernoulli's equation for incompressible flow. It can be found by integrating eq. (6.3) with the assumption of constant density ρ:

$$\frac{1}{2}\rho u^2+p=p_t. \tag{6.22}$$

The first term on the left-hand side is the dynamic pressure q (p is the "static" pressure):

$$q=\frac{1}{2}\rho u^2. \tag{6.23}$$

The concept of dynamic pressure is used also for compressible flow. There, however, it is no more simply the difference of dynamic and static pressure. Eq. (6.23) can be re-written for perfect gas as, for instance:

$$q=\frac{1}{2}\rho u^2=\frac{1}{2}\gamma p M^2. \tag{6.24}$$

Also in aerothermodynamics it is used to normalize pressure as well as aerodynamic forces and moments.

For perfect gas we can relate temperature, density and pressure to their total values, and to the flow Mach number:

$$T_t=T\left(1+\frac{\gamma-1}{2}M^2\right), \tag{6.25}$$

$$\rho_t=\rho\left(1+\frac{\gamma-1}{2}M^2\right)^{\frac{1}{\gamma-1}}, \tag{6.26}$$

$$p_t=p\left(1+\frac{\gamma-1}{2}M^2\right)^{\frac{\gamma}{\gamma-1}}. \tag{6.27}$$

Similarly we can relate temperature, density, pressure and the speed u to their free-stream properties '∞':

$$\frac{T}{T_\infty}=1-\frac{\gamma-1}{2}M_\infty^2\left[\left(\frac{u}{u_\infty}\right)^2-1\right], \tag{6.28}$$

$$\frac{\rho}{\rho_\infty}=\left\{1-\frac{\gamma-1}{2}M_\infty^2\left[\left(\frac{u}{u_\infty}\right)^2-1\right]\right\}^{\frac{1}{\gamma-1}}, \tag{6.29}$$

$$\frac{p}{p_\infty}=\left\{1-\frac{\gamma-1}{2}M_\infty^2\left[\left(\frac{u}{u_\infty}\right)^2-1\right]\right\}^{\frac{\gamma}{\gamma-1}}, \tag{6.30}$$

with $M_\infty = u_\infty / a_\infty$.

The pressure coefficient c_p is:

$$c_p = \frac{p - p_\infty}{q_\infty}, \tag{6.31}$$

where q_∞ is the dynamic pressure of the free-stream. For perfect gas the pressure coefficient reads with eq. (6.24):

$$c_p = \frac{2}{\gamma M_\infty^2} \left(\frac{p}{p_\infty} - 1 \right). \tag{6.32}$$

The pressure coefficient can be expressed in terms of the ratio local speed u to free-stream value u_∞, and the free-stream Mach number M_∞:

$$c_p = \frac{2}{\gamma M_\infty^2} \left\{ \left[1 + \frac{\gamma - 1}{2} M_\infty^2 [1 - \left(\frac{u}{u_\infty} \right)^2] \right]^{\frac{\gamma}{\gamma - 1}} - 1 \right\}, \tag{6.33}$$

as well as in terms of the local Mach number M and its free-stream value:

$$c_p = \frac{2}{\gamma M_\infty^2} \left[\left(\frac{1 + \frac{\gamma-1}{2} M_\infty^2}{1 + \frac{\gamma-1}{2} M^2} \right)^{\frac{\gamma}{\gamma - 1}} - 1 \right]. \tag{6.34}$$

With the help of eq. (6.27) c_p can be related to the total pressure p_t:

$$c_p = \frac{2}{\gamma M_\infty^2} \left[\frac{p}{p_t} \left(1 + \frac{\gamma - 1}{2} M_\infty^2 \right)^{\frac{\gamma}{\gamma - 1}} - 1 \right]. \tag{6.35}$$

In subsonic compressible flow we get for the stagnation point (isentropic compression) with $p = p_t$, $u = (M =)\ 0$:

$$c_p = \frac{2}{\gamma M_\infty^2} \left[\left(1 + \frac{\gamma - 1}{2} M_\infty^2 \right)^{\frac{\gamma}{\gamma - 1}} - 1 \right]. \tag{6.36}$$

In the case of supersonic flow of course the total-pressure loss across the shock must be taken into account (see next section).

For incompressible flow we find the pressure coefficient with the help of eq. (6.22) and constant p_t:

$$c_p = 1 - \frac{u^2}{u_\infty^2}, \tag{6.37}$$

and note finally that at a stagnation point c_p for compressible flow is always larger than that for incompressible flow with $c_p = 1$.

6.3 Shock Waves

In the following two sections we study normal and oblique shock waves and discuss the Rankine-Hugoniot conditions, which connect the flow properties upstream and downstream of them. We don't give tables and charts for the determination of the flow properties, and refer the reader instead to [10], [2], [3], [4].

6.3.1 Normal Shock Waves

Consider the shock surface in Fig. 6.1 a). On the axis, the symmetric shock surface lies orthogonal to the supersonic free stream. Its thickness is very small compared to its two principle radii. We approximate the shock surface locally by a plane surface. If the body lies very far downstream, we can consider the flow behind the normal portion of the shock wave as parallel and uniform, like the flow ahead of the shock wave.

With these assumptions we have defined the phenomenological model of the "normal shock wave", that is a plane shock surface, which lies orthogonal to a parallel and uniform supersonic stream. Behind it the flow is again parallel and uniform, but subsonic.

We study now the normal shock wave, like any aerodynamic object, in an object-fixed frame[6], Fig.6.8 (see also Fig. 5.4). We know that the shock wave is a viscous layer of small thickness with strong molecular transport across it. In continuum flow it usually can be considered approximately as a discontinuity in the flow field. The flow parameters, which are constant ahead of the shock surface, change across it instantly, Fig. 5.4 b), and are constant again downstream of it. How can we relate them to each other across the discontinuity?

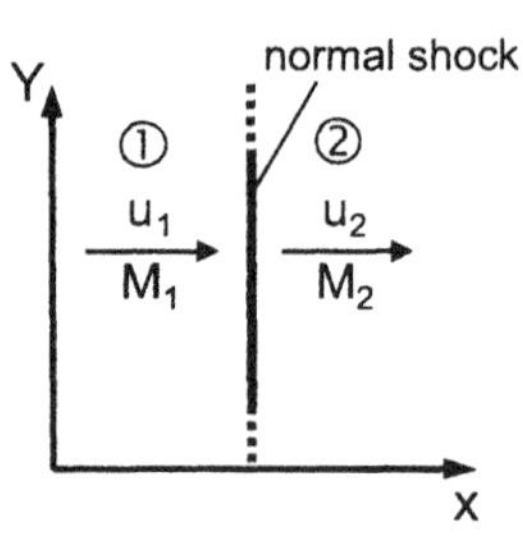

Fig. 6.8. Schematic of a normal shock wave and notation.

Obviously there are entities, which are constant across the discontinuity, as is shown in the following. These are the mass flux, the momentum flux and

[6] We use in this chapter for the convenience of the reader the notation from [10]. There a broad collection of important and useful relations and data concerning perfect-gas compressible air flow is given.

the total enthalpy flux. We find the describing equations by reduction of the conservative formulations of the governing equations, eq. (4.83), eq. (4.33), and eq. (4.64) to one-dimensional inviscid flow (we omit dx):

$$d(\rho u) = 0, \tag{6.38}$$

$$d(\rho u^2 + p) = 0, \tag{6.39}$$

$$d(\rho u h) - u dp = 0. \tag{6.40}$$

Eq. (6.40) can be combined with eqs. (6.3) and (6.38) to yield a form, which can be integrated immediately:

$$d[\rho u(h + \frac{1}{2}u^2)] = 0. \tag{6.41}$$

The integration then yields the constant entities postulated above:

$$\rho_1 u_1 = \rho_2 u_2, \tag{6.42}$$

$$\rho_1 u_1^2 + p_1 = \rho_2 u_2^2 + p_2, \tag{6.43}$$

$$\rho_1 u_1 \, (h_1 + \frac{1}{2}u_1^2) = \rho_2 u_2 \, (h_2 + \frac{1}{2}u_2^2). \tag{6.44}$$

The relation for the total enthalpy flux, eq. (6.44), reduces with eq. (6.42) to:

$$h_1 + \frac{1}{2}u_1^2 = h_2 + \frac{1}{2}u_2^2. \tag{6.45}$$

This relation says, that the total enthalpy h_t is constant across shock waves:

$$h_{t_1} = h_{t_2} = h_t. \tag{6.46}$$

This important result is in contrast to that for the total pressure. We will see below, that p_t is not constant across shock waves.

Relation (6.46) indicates also, that the maximum speed V_m is constant across the shock wave. For a perfect gas the total temperature T_t, the total speed of sound a_t, and the critical speed of sound a_* are constant across the shock wave, too.

For the following we assume perfect gas flow. We combine eqs. (6.42) and (6.45) and find:

$$\frac{p_1}{\rho_1 u_1} + u_1 = \frac{p_2}{\rho_2 u_2} + u_2. \tag{6.47}$$

After introducing the critical speed of sound[6] a_*, we arrive at Prandtl's relation:

$$u_1 u_2 = a_*^2 = \frac{p_2 - p_1}{\rho_2 - \rho_1} = \frac{\gamma - 1}{\gamma + 1} V_m^2, \tag{6.48}$$

and at the Rankine-Hugoniot relations:

$$\frac{p_2}{p_1} = \frac{(\gamma + 1)\rho_2 - (\gamma - 1)\rho_1}{(\gamma + 1)\rho_1 - (\gamma - 1)\rho_2}, \tag{6.49}$$

$$\frac{p_2 - p_1}{\rho_2 - \rho_1} = \gamma \frac{p_2 + p_1}{\rho_2 + \rho_1}. \tag{6.50}$$

Eq. (6.48) can be written in terms of a Mach number related to the critical speed of sound $M_* = u/a_*$ as:

$$M_{*_1} M_{*_2} = 1. \tag{6.51}$$

The Mach number M_* is related to the Mach number M by eq. (6.20). Both have similar properties, which are shown in Table 6.1.

Table 6.1. Relation of M_* and M.

M	0	$\sqrt{0.707}$	1	$\sqrt{2}$	∞
M_*	0	$\sqrt{\frac{\gamma+1}{\gamma+3}}$	1	$\sqrt{\frac{\gamma+1}{\gamma}}$	$\sqrt{\frac{\gamma+1}{\gamma-1}}$

Prandtl's relation in either form implies the two solutions, that the flow speed ahead of the normal shock is either supersonic or subsonic, and behind the normal shock accordingly subsonic or supersonic, i. e., that we have either a compression shock or an expansion shock.

We developed the phenomenological model "normal shock wave" starting from a supersonic upstream situation. In our mathematical model we did not make an assumption about the upstream Mach number. The question is now, whether both or only one of the two solutions is viable.

Before we decide this, we give some relations for the change of the flow parameters across the normal shock wave in terms of the upstream Mach number M_1, and the ratio of specific heats γ, see, e. g., [10]:

$$\frac{\rho_2}{\rho_1} = \frac{u_1}{u_2} = \frac{u_1^2}{a_*^2} = \frac{a_*^2}{u_2^2} = \frac{(\gamma + 1)M_1^2}{(\gamma - 1)M_1^2 + 2}, \tag{6.52}$$

---

[6] The location of the critical speed of sound is within the viscous shock layer, Fig. 5.4 a). If we consider the shock wave as a discontinuity, a_* is a fictitious entity, which is hidden in the discontinuity.

$$\frac{T_2}{T_1} = \frac{a_2^2}{a_1^2} = \frac{[2\gamma M_1^2 - (\gamma - 1)][(\gamma - 1)M_1^2 + 2]}{(\gamma + 1)^2 M_1^2}, \tag{6.53}$$

$$\frac{p_2}{p_1} = \frac{2\gamma M_1^2 - (\gamma - 1)}{\gamma + 1}, \tag{6.54}$$

$$M_2^2 = \frac{(\gamma - 1)M_1^2 + 2}{2\gamma M_1^2 - (\gamma - 1)}, \tag{6.55}$$

$$\frac{p_2}{p_{t_1}} = \frac{2\gamma M_1^2 - (\gamma - 1)}{\gamma + 1} \left[\frac{2}{(\gamma - 1)M_1^2 + 2}\right]^{\frac{\gamma}{\gamma-1}}, \tag{6.56}$$

$$\frac{p_2}{p_{t_2}} = \left[\frac{4\gamma M_1^2 - 2(\gamma - 1)}{(\gamma + 1)^2 M_1^2}\right]^{\frac{\gamma}{\gamma-1}}, \tag{6.57}$$

$$\frac{p_{t_2}}{p_{t_1}} = \frac{\rho_{t_2}}{\rho_{t_1}} = e^{-(s_2 - s_1)/R} = \left[\frac{(\gamma + 1)M_1^2}{(\gamma - 1)M_1^2 + 2}\right]^{\frac{\gamma}{\gamma-1}} \left[\frac{\gamma + 1}{2\gamma M_1^2 - (\gamma - 1)}\right]^{\frac{1}{\gamma-1}}. \tag{6.58}$$

The entropy increase can be expressed with eqs. (6.54) and (6.52) in terms of eq. (6.11):

$$\frac{s_2 - s_1}{c_v} = \ln\left[\frac{2\gamma M_1^2 - (\gamma - 1)}{\gamma + 1}\right] - \gamma \ln\left[\frac{(\gamma + 1)M_1^2}{(\gamma - 1)M_1^2 + 2}\right]. \tag{6.59}$$

The Rayleigh-Pitot formula reads [10]:

$$\frac{p_{t_2}}{p_1} = \left[\frac{(\gamma + 1)M_1^2}{2}\right]^{\frac{\gamma}{\gamma-1}} \left[\frac{\gamma + 1}{2\gamma M_1^2 - (\gamma - 1)}\right]^{\frac{1}{\gamma-1}}. \tag{6.60}$$

For practical purposes we note some pressure coefficients. Of course ahead of a shock wave in the free-stream we have:

$$c_{p_1} = \frac{p_1 - p_1}{q_1} = 0. \tag{6.61}$$

Behind the normal shock wave we find with eqs. (6.24) and (6.54):

$$c_{p_2} = \frac{p_2 - p_1}{q_1} = \frac{4(M_1^2 - 1)}{(\gamma + 1)M_1^2}. \tag{6.62}$$

Finally, assuming isentropic compression behind the normal shock towards the stagnation point (the situation, e. g., on the symmetry line between bow-shock and blunt-body surface in Fig. 6.1 a)), we find, with eqs. (6.24), (6.57) and (6.54), for the pressure coefficient at the stagnation point:

$$c_{p_{t_2}} = \frac{p_{t_2} - p_1}{q_1} = \frac{2}{\gamma M_1^2} \left\{ \left[\frac{(\gamma+1)^2 M_1^2}{4\gamma M_1^2 - 2(\gamma-1)} \right]^{\frac{\gamma}{\gamma-1}} \left[\frac{2\gamma M_1^2 - (\gamma-1)}{\gamma+1} \right] - 1 \right\}. \tag{6.63}$$

We make a plausibility check by putting $M_1 = 1$ into these relations. We find for the ratios of density, velocities, temperature, speed of sound, pressure, and total pressure, as well as for the Mach number M_2 the values "1", as was to be expected. The pressure quotient c_{p2} is zero, whereas the ratios of pressure to total pressure remain, also as to be expected, smaller than one. The pressure coefficient at the stagnation point, $c_{p_{t_2}}$, is the same as that for shock-free flow.

Now we come back to the question, which solution, that with supersonic or that with subsonic flow upstream of the shock location, is viable. To decide this, we consider the change of entropy across the shock wave. We note from eq. (6.58) that a total pressure decrease is equivalent to an entropy rise.

We just plug in numbers into eq. (6.58), choosing $\gamma = 1.4$. The results are given in Table 6.2. We find that only for Mach numbers $M_1 > 1$ the total pressure drops across the shock, i. e., only for supersonic flow ahead of the normal shock we observe a total pressure decrease, i. e. an entropy rise. Hence the above relations for the change across the normal shock wave are only valid, if the upstream flow is supersonic.

Expansion shocks are not viable, because they would, Table 6.2, lead to a rise of total pressure, i. e. an entropy reduction, which is ruled out by the second law of thermodynamics. Of course, we should have known this from the beginning, because we introduced the shock surface as a thin viscous layer.

Table 6.2. Change of total pressure $p_{t_2}/p_{t_1} = e^{-(s_2-s_1)/R}$ across a normal shock for selected upstream Mach numbers M_1 and $\gamma = 1.4$.

M_1	0	$\sqrt{\frac{1}{7}}$	0.5	0.75	1	2	5	10	∞
p_{t_2}/p_{t_1}	*undefined*	∞	2.256	1.037	1.0	0.721	0.062	0.003	0

The results given in Table 6.2 demand a closer inspection. We find from the second bracket of eq. (6.58) a singularity at

$$M_1^2 = \frac{\gamma - 1}{2\gamma}, \tag{6.64}$$

which with $\gamma = 1.4$ becomes $M_1^2 = 1/7$.

Below that value the total pressure ratio across the normal shock is undefined. From eq. (6.53) we see, that for $M_1 = \sqrt{(\gamma-1)/2\gamma}$ the temperature behind the shock wave becomes zero, i. e., that for this upstream Mach number the maximum velocity V_m, eq. (6.15), is reached.

The result thus is: if one assumes an expansion shock, the smallest possible upstream Mach number is that, for which behind the shock the largest possible speed is reached. Behind this result lies the fact that in our mathematical model the conservation of the total enthalpy is assured also for the physically not viable expansion shock.

Just plugging in numbers as we did above is somewhat unsatisfactory. The classical way to show that only the supersonic compression shock is viable, which however is restricted to Mach numbers not much different from $M_1 = 1$, is to make a series expansion, which yields:

$$\frac{p_{t_2}}{p_{t_1}} = e^{-(s_2-s_1)/R} = 1 - \frac{2\gamma}{3(\gamma+1)^2}(M_1^2-1)^3 + \text{terms of higher order.} \tag{6.65}$$

Because of the odd exponent of the bracket of the second term on the right-hand side of the equation we see that indeed a total-pressure decrease, and hence an entropy increase, across the normal shock happens only for $M_1 > 1$.

With this result we obtain from eqs. (6.52) to (6.55) that across the normal shock wave pressure p, density ρ, temperature T are increasing instantly, whereas speed u and Mach number M decrease instantly. The most important result is, that we always get subsonic flow behind the normal shock wave:

$$M_2 \mid_{normal\ shock} < 1. \tag{6.66}$$

For large upstream Mach numbers, i. e. $M_1 \to \infty$, we find some other interesting results[7]. They permit us, for instance, to judge in hypersonic flow the influence of a changing γ at large temperatures on the flow parameters downstream of a normal shock.

We obtain finite values for the ratios of velocities, and densities, as well as for the Mach number and the pressure coefficient behind the normal shock for $M_1 \to \infty$:

$$\frac{u_2}{u_1} \to \frac{\gamma-1}{\gamma+1}, \tag{6.67}$$

$$\frac{\rho_2}{\rho_1} \to \frac{\gamma+1}{\gamma-1}, \tag{6.68}$$

$$M_2^2 \to \frac{\gamma-1}{2\gamma}, \tag{6.69}$$

and for the pressure coefficients

$$c_{p_2} \to \frac{4}{\gamma+1}, \tag{6.70}$$

[7] Oswatitsch defines hypersonic flow past a body as flow with a free-stream Mach number $M_\infty \to \infty$. We come back to this in Section 6.8.

and

$$c_{p_{t_2}} \rightarrow \left[\frac{(\gamma+1)^2}{4\gamma}\right]^{\frac{\gamma}{\gamma-1}} \frac{4}{\gamma+1}. \tag{6.71}$$

Static temperature, speed of sound, and pressure ratios, however, for $M_1 \rightarrow \infty$ tend to infinite values:

$$\frac{T_2}{T_1} = \frac{a_2^2}{a_1^2} \rightarrow \infty, \tag{6.72}$$

$$\frac{p_2}{p_1} \rightarrow \infty, \tag{6.73}$$

whereas the for the total-pressure ratios, the entropy increase, and for the Rayleigh-Pitot formula the following results are obtained:

$$\frac{p_2}{p_{t_1}} \rightarrow 0, \tag{6.74}$$

$$\frac{p_2}{p_{t_2}} \rightarrow \left[\frac{4\gamma}{(\gamma+1)^2}\right]^{\frac{\gamma}{\gamma-1}}, \tag{6.75}$$

$$\frac{p_{t_2}}{p_{t_1}} = \frac{\rho_{t_2}}{\rho_{t_1}} = e^{-(s_2-s_1)/R} \rightarrow 0, \tag{6.76}$$

$$\frac{s_2 - s_1}{c_v} \rightarrow \infty, \tag{6.77}$$

$$\frac{p_{t_2}}{p_1} \rightarrow \infty. \tag{6.78}$$

6.3.2 Oblique Shock Waves

Consider streamline (1) in Fig. 6.1 a). It penetrates the bow-shock surface at a location, where the latter is inclined by approximately 45° against the free-stream direction. The streamline is deflected slightly behind the shock towards the shock surface.

We employ the same reasoning as for the normal shock wave, and define the phenomenological model "oblique shock wave", which is a plane shock surface, which lies inclined (shock angle θ) to a parallel and uniform supersonic stream. It is induced for example by a ramp with ramp angle δ, Fig. 6.9. The flow is inviscid, and the ramp surface is a streamline. Behind the oblique shock wave the flow is again parallel and uniform[8], but now it is deflected slightly towards the shock surface, Fig. 6.9.

For a shock wave, like for any wave phenomenon, changes of flow parameters happen only in the direction normal to the shock wave. This means,

[8] This holds for plane flow. At a cone, Fig. 6.12 b), the situation is different.

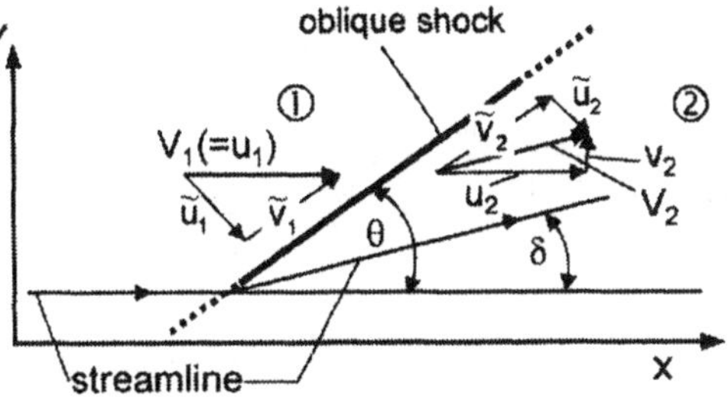

Fig. 6.9. Schematic of an oblique shock wave and notation.

that the governing entity at any point of a shock surface is the normal Mach number M_{1_N}, i. e., the Mach number of the flow velocity component normal to the shock surface $\tilde{u}_1$ at that point, Fig. 6.9:

$$M_{1_N} = \frac{\tilde{u}_1}{a_1} = \frac{V_1\, sin\theta}{a_1} = M_1 sin\theta. \tag{6.79}$$

The shock relations (again for perfect gas) are in principle the same as those for the normal shock, only that now $\tilde{u}_1$ replaces u_1. Prandtl's relation now reads:

$$\tilde{u}_1\tilde{u}_2 = a_*^2 - \frac{\gamma-1}{\gamma+1}\tilde{v}^2, \tag{6.80}$$

where $\tilde{v}$ is the component of V_1 tangential to the oblique shock wave, Fig. 6.9, which does not change across it[9]:

$$\tilde{v} = \tilde{v}_1 = \tilde{v}_2 = V_1 cos\theta. \tag{6.81}$$

The Rankine-Hugoniot relations are identical to those for the normal shock wave.

We give now, as we did for the normal shock, some relations for the change of the flow parameters across the oblique shock, Fig. 6.9, in terms of the upstream Mach number M_1, the shock-inclination angle θ, and the ratio of specific heats γ, again see, e. g., [10]:

$$\frac{\rho_2}{\rho_1} = \frac{\tilde{u}_1}{\tilde{u}_2} = \frac{(\gamma+1)M_1^2\sin^2\theta}{(\gamma-1)M_1^2\sin^2\theta+2}, \tag{6.82}$$

$$\frac{T_2}{T_1} = \frac{a_2^2}{a_1^2} = \frac{[2\gamma M_1^2\sin^2\theta-(\gamma-1)][(\gamma-1)M_1^2\sin^2\theta+2]}{(\gamma+1)^2M_1^2\sin^2\theta}, \tag{6.83}$$

[9] At a general three-dimensional shock surface, locally the angle between the shock-surface normal and the free-stream direction is 90°- θ. We then define two tangential velocity components at the shock surface.

$$\frac{p_2}{p_1} = \frac{2\gamma M_1^2 \sin^2\theta - (\gamma - 1)}{\gamma + 1}, \tag{6.84}$$

$$M_2^2 = \frac{(\gamma+1)^2 M_1^4 \sin^2\theta - 4(M_1^2 \sin^2\theta - 1)(\gamma M_1^2 \sin^2\theta + 1)}{[2\gamma M_1^2 \sin^2\theta - (\gamma - 1)][(\gamma - 1)M_1^2 \sin^2\theta + 2]}, \tag{6.85}$$

$$\frac{\tilde{u}_2}{V_1} = \frac{(\gamma - 1)M_1^2 \sin^2\theta + 2}{(\gamma + 1)M_1^2 \sin^2\theta} \sin\theta, \tag{6.86}$$

$$\frac{\tilde{v}_2}{V_1} = \frac{\tilde{v}_1}{V_1} = \cos\theta, \tag{6.87}$$

$$\frac{u_2}{V_1} = 1 - \frac{2(M_1^2 \sin^2\theta - 1)}{(\gamma + 1)M_1^2}, \tag{6.88}$$

$$\frac{v_2}{V_1} = \frac{2(M_1^2 \sin^2\theta - 1)}{(\gamma + 1)M_1^2} \cot\theta, \tag{6.89}$$

$$\frac{V_2^2}{V_1^2} = 1 - 4\frac{(M_1^2 \sin^2\theta - 1)(\gamma M_1^2 \sin^2\theta + 1)}{(\gamma + 1)^2 M_1^4 \sin^2\theta}, \tag{6.90}$$

$$\frac{p_2}{p_{t_1}} = \frac{2\gamma M_1^2 \sin^2\theta - (\gamma - 1)}{\gamma + 1} \left[\frac{2}{(\gamma - 1)M_1^2 + 2}\right]^{\frac{\gamma}{\gamma-1}}, \tag{6.91}$$

$$\frac{p_2}{p_{t_2}} = \left\{\frac{[4\gamma M_1^2 \sin^2\theta - 2(\gamma - 1)][(\gamma - 1)M_1^2 \sin^2\theta + 2]}{(\gamma + 1)^2 M_1^2 \sin^2\theta[(\gamma - 1)M_1^2 + 2]}\right\}^{\frac{\gamma}{\gamma-1}}, \tag{6.92}$$

$$\begin{aligned} \frac{p_{t_2}}{p_{t_1}} &= \frac{\rho_{t_2}}{\rho_{t_1}} = e^{-(s_2 - s_1)/R} = \\ &= \left[\frac{(\gamma + 1)M_1^2 \sin^2\theta}{(\gamma - 1)M_1^2 \sin^2\theta + 2}\right]^{\frac{\gamma}{\gamma-1}} \left[\frac{\gamma + 1}{2\gamma M_1^2 \sin^2\theta - (\gamma - 1)}\right]^{\frac{1}{\gamma-1}}, \end{aligned} \tag{6.93}$$

The entropy rise is found in analogy to that of the normal shock surface:

$$\frac{s_2 - s_1}{c_v} = \ln\left[\frac{2\gamma M_1^2 \sin^2\theta - (\gamma - 1)}{\gamma + 1}\right] - \gamma \ln\left[\frac{(\gamma + 1)M_1^2 \sin^2\theta}{(\gamma - 1)M_1^2 \sin^2\theta + 2}\right]. \tag{6.94}$$

The Rayleigh-Pitot formula for the oblique shock reads:

$$\begin{aligned} \frac{p_{t_2}}{p_1} &= \left\{\frac{(\gamma + 1)M_1^2 \sin^2\theta[(\gamma - 1)M_1^2 + 2]}{2[(\gamma - 1)M_1^2 \sin^2\theta + 2]}\right\}^{\frac{\gamma}{\gamma-1}} \cdot \\ &\cdot \left[\frac{(\gamma + 1)}{2\gamma M_1^2 \sin^2\theta - (\gamma - 1)}\right]^{\frac{1}{\gamma-1}}. \end{aligned} \tag{6.95}$$

Concerning the pressure coefficients we note first c_{p_2}:

$$c_{p_2} = \frac{4(M_1^2 \sin^2 \theta - 1)}{(\gamma + 1) M_1^2}. \tag{6.96}$$

We remember then the asymmetric stagnation-point situation shown in Fig. 6.3. The stagnation-point streamline, which crosses the bow-shock surface at P_1 obviously leads to a larger p_{t_2} than the normal-shock value. We find now with eqs. (6.24), (6.92) and (6.84) the pressure coefficient at the stagnation point:

$$c_{p_{t_2}} = \frac{2}{\gamma M_1^2} \left\{ \left[\frac{(\gamma+1)^2 M_1^2 \sin^2\theta [(\gamma-1) M_1^2 + 2]}{[4\gamma M_1^2 \sin^2\theta - 2(\gamma - 1)][(\gamma-1) M_1^2 \sin^2\theta + 2]} \right]^{\frac{\gamma}{\gamma-1}} \cdot \left[\frac{2\gamma M_1^2 \sin^2\theta - (\gamma - 1)}{\gamma + 1} \right] - 1 \right\}. \tag{6.97}$$

We can at once take over qualitatively the results for the normal shock wave, when we deal with an oblique compression shock. This means that also across the oblique shock pressure p, density ρ, temperature T are increasing instantly, whereas speed V decreases instantly. The important difference is, that we get subsonic flow behind the shock too, but only for the flow component normal to the shock surface:

$$M_{2_N} \,|_{oblique\ shock} < 1. \tag{6.98}$$

Then the question arises, whether the *resultant* flow behind the oblique shock is subsonic or supersonic, i. e., whether the Mach number $M_2 = V_2/a_2$ < 1 or > 1. In fact, as we have indicated in Figures 6.1 to 6.4, behind a detached shock surface at a blunt or a sharp nosed body we find a subsonic pocket, which can be rather large, Fig. 6.2 b). This implies, that in the vicinity of the point, where the shock surface is just a normal shock, the flow behind the oblique shock surface is subsonic, i. e., $M_2 < 1$, while farther away it becomes supersonic.

With the help of eq. (6.85) we determine, at what angle θ this happens. We set $M_2^2 = 1$ and obtain the critical angle θ_* in terms of the upstream Mach number M_1, and the ratio of specific heats γ:

$$\sin\theta_* = \pm \frac{1}{4\gamma M_1^2} \cdot \left[(\gamma+1) M_1^2 + (\gamma - 3) + \sqrt{(\gamma+1)[(\gamma+1) M_1^4 + 2(\gamma-3) M_1^2 + (\gamma + 9)]} \right]. \tag{6.99}$$

For $-\theta_* \leqq \theta \leqq +\theta_*$ we have subsonic, and above these values supersonic flow behind the oblique shock surface. Of course we cannot determine on this basis the shape of the sonic surface, i. e., the shape of the surface on which between a detached shock and the body surface $M = 1$.

For large upstream Mach numbers, i. e. $M_1 \to \infty$, we find the limiting values of θ_*:

$$\sin \theta_* |_{M_1 \to \infty} = \pm \sqrt{\frac{\gamma + 1}{2\gamma}}. \tag{6.100}$$

For a perfect gas with $\gamma = 1.4$ they are $\theta_*|_{M_1 \to \infty} = \pm\, 67.79°$. In a large Mach number range θ_* varies only weakly, Fig. 6.10. For $\gamma = 1.4$ the value of θ_* drops from 90° at $M_1 = 1$ to a minimum of approximately 61° at $M_1 = 2$, and rises then monotonously to the limiting value $\theta_*|_{M_1 \to \infty} = 67.79°$.

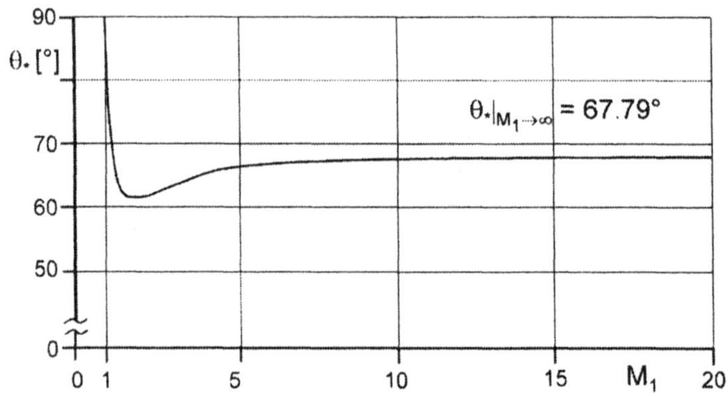

Fig. 6.10. Variation of the sonic shock-wave angle $+\theta_*$ with M_1.

Because, for a given Mach number M_∞, $90° \geqq \theta_* \geqq \mu_\infty$, Fig. 6.17, the sonic line can meet the bow shock, especially at low Mach numbers, far away from the body surface, Fig. 6.11.

Now to the flow parameters at large upstream Mach numbers, i. e. $M_1 \to \infty$. Generally we obtain the same limiting values across the oblique shock for the ratios of densities, temperatures et cetera in the same way as those across the normal shock. We note here only the following exceptions:

$$M_2^2 \to \frac{(\gamma+1)^2 - 4\gamma \sin^2 \theta}{2\gamma(\gamma - 1)\sin^2 \theta}, \tag{6.101}$$

$$\frac{p_2}{p_{t_2}} \to \left[\frac{4\gamma \sin^2 \theta}{(\gamma+1)^2}\right]^{\frac{\gamma}{\gamma-1}}, \tag{6.102}$$

and finally:

$$c_{p_2} \to \frac{4 \sin^2 \theta}{\gamma + 1}, \tag{6.103}$$

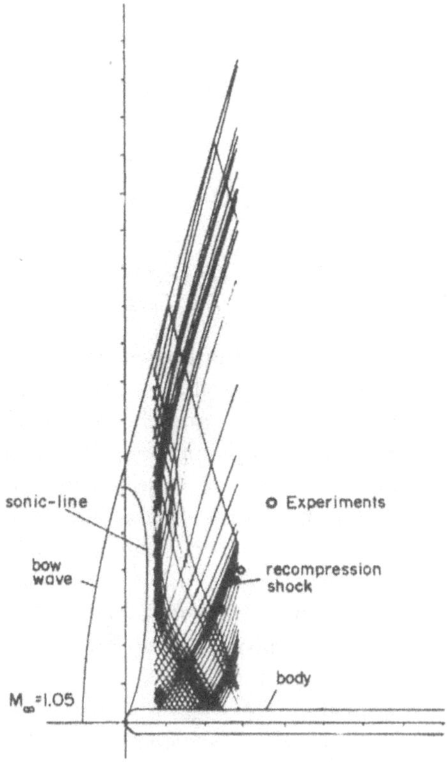

Fig. 6.11. Computed hemisphere-cylinder flow. $M_\infty = 1.05$, $\alpha = 0°$, inviscid flow, Euler solution, [11], with shock fitting.

$$c_{p_{t_2}} \to \left[\frac{(\gamma+1)^2}{4\gamma \sin^2\theta}\right]^{\frac{\gamma}{\gamma-1}} \frac{4\sin^2\theta}{\gamma+1}. \tag{6.104}$$

Comparing eq. (6.104) with eq. (6.71) we see that the stagnation point pressure in the limit $M_1 \to \infty$ for the asymmetric situation of Fig. 6.3 indeed is larger than the normal shock value. This holds for all finite Mach numbers $M_1 > 1$, too.

The relations for the velocity components at $M_1 \to \infty$ behind the oblique shock are:

$$\frac{\tilde{u_2}}{\tilde{u_1}} \to \frac{\gamma-1}{\gamma+1}, \tag{6.105}$$

$$\frac{\tilde{u_2}}{V_1} \to \frac{\gamma-1}{\gamma+1}\sin\theta, \tag{6.106}$$

$$\frac{\tilde{v}_2}{V_1} \to \cos\theta, \tag{6.107}$$

$$\frac{u_2}{V_1} \to 1 - \frac{2\sin^2\theta}{\gamma + 1}, \tag{6.108}$$

$$\frac{v_2}{V_1} \to \frac{2\sin^2\theta}{\gamma + 1}\cot\theta, \tag{6.109}$$

$$\frac{V_2^2}{V_1^2} \to 1 - \frac{4\gamma\sin^2\theta}{(\gamma + 1)^2}. \tag{6.110}$$

We note finally that the shape of the bow-shock surface, either detached or attached, depends for a perfect-gas flow only on the free-stream Mach number M_1, the ratio of specific heats γ, and the effective body shape with its embedded shock wave and expansion phenomena[10]. All shock relations given above for normal and oblique shocks are valid always locally on bow shock and embedded shock surfaces. If real-gas effects, especially high-temperature effects, are present, they influence the bow-shock shape, Sub-Section 6.4.1, but also the shock layer, Section 5.4.

Explicit relations, which connect the shock properties to the body shape, are available only for the two-dimensional inviscid ramp flow. These are the oblique-shock relations discussed above.

In Fig. 6.9 we have indicated the angle δ, by which the streamlines are deflected behind the oblique shock. If we select such a streamline, we obtain the two-dimensional ramp flow, Fig. 6.1 d). We find for it the relation between shock-wave angle θ and ramp angle δ, i. e. the deflection angle, in terms of the upstream Mach number M_1, and the ratio of specific heats γ:

$$\tan\delta = \frac{v_2}{u_2} = \frac{2\cot\theta(M_1^2\sin^2\theta - 1)}{2 + M_1^2(\gamma + 1 - 2\sin^2\theta)}. \tag{6.111}$$

This equation can be rearranged to yield:

$$M_1^2\sin^2\theta - 1 = \frac{\gamma + 1}{2}M_1^2\frac{\sin\delta\sin\theta}{\cos(\delta - \theta)}. \tag{6.112}$$

For small ramp angles δ it becomes:

$$M_1^2\sin^2\theta - 1 \approx \left(\frac{\gamma + 1}{2}M_1^2\tan\theta\right)\delta, \tag{6.113}$$

and we obtain for θ at large Mach numbers $M_1 \to \infty$ (θ then is small anyway):

$$\theta \to \frac{\gamma + 1}{2}\delta. \tag{6.114}$$

[10] The effective body shape is the shape, which the flow "sees", i. e., the configuration of the flight vehicle at angle of attack, yaw, et cetera.

If the ramp angle δ is reduced to zero, we find from eq. (6.112) the Mach angle μ, which we had already noted in Section 6.1:

$$\theta|_{\delta\to 0} = \mu = sin^{-1}(\frac{1}{M}). \tag{6.115}$$

In Figs. 6.1 b) and c) we have noted that at a given free-stream Mach number a critical cone angle exists, at which the bow-shock surface becomes detached. Such an angle, δ_{max}, exists also for ramp flow. The (largest possible) shock-wave angle related to it is:

$$\sin\theta_{*\delta_{max}} = \pm\frac{1}{4\gamma M_1^2}\cdot \cdot\left[(\gamma+1)M_1^2 - 4 + \sqrt{(\gamma+1)[(\gamma+1)M_1^4 + 8(\gamma-1)M_1^2 + 16]}\right]. \tag{6.116}$$

Cone flow, of course, is different from ramp flow, because it has, in a sense, three-dimensional aspects. For a given upstream Mach number M_1 and equal values of ramp angle δ and (circular) cone half-angle σ we find a larger shock-wave angle θ for the ramp flow, Fig. 6.12.

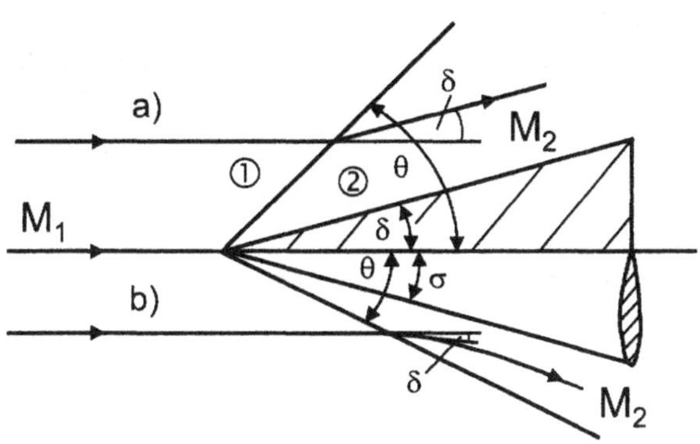

Fig. 6.12. Schematic of the flow past **a**) a two-dimensional ramp, and **b**) a circular cone.

The flow behind the attached oblique ramp shock is uniform and parallel to the ramp surface. The flow behind the attached shock surface of the cone at zero angle of attack of the cone is also circular, see, e. g., [3], [10]. It is first, i. e. directly behind the shock surface, deflected according to eq. (6.111). The streamlines then approach asymptotically the direction of the cone surface, i. e., the semi-vertex angle σ. All flow quantities are constant on each concentric conical surface, which lies between the shock surface and the body surface. The Mach number on the body surface, and hence also away from it, can be subsonic or supersonic, depending on the pairing M_∞ and σ. Graphs of the flow parameters for circular cone flow can be found in [10].

In closing this sub-section we have a look at shock intersections and reflections. Such phenomena can occur, see Chapter 9, in viscous flows past ramps, at the interaction of embedded shocks and the bow shock but also in flows in propulsion inlets et cetera, Section 6.1.

Intersections of shock waves belonging to opposite families[11] in uniform and parallel flow are symmetric, if the incoming shocks are symmetric, i. e. of equal strength, Fig. 6.13 a), and asymmetric, if not, Fig. 6.13 b). In the symmetric case the streamlines are first deflected towards the shock waves. Behind the intersected shocks they are again parallel and uniform. Indeed, this is the governing condition for the shock intersection.

This also holds for the asymmetric case. However, because the involved shocks are of different strength, we get behind the intersection two parallel and uniform streams with different total-pressure loss. Since the static pressures are the same, these two streams have different speeds. The result is, in terms of inviscid flow, a slip surface, which in reality is a vortex sheet, Fig. 6.13b). This vortex sheet can become unstable via the Kelvin-Helmholtz mechanism [12], [13].

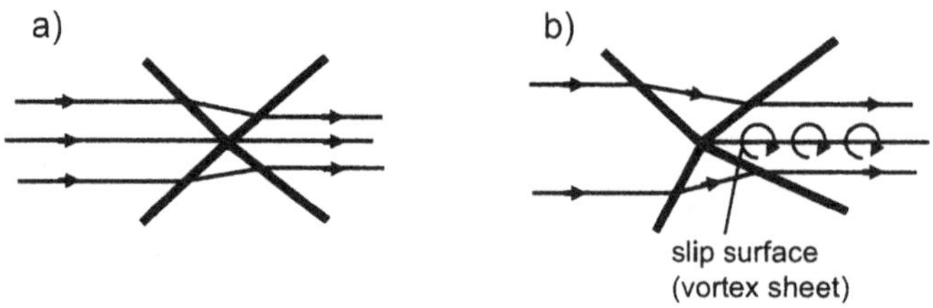

Fig. 6.13. Schematic of intersections of shocks belonging to opposite families : **a**) symmetric intersection, **b**) asymmetric intersection.

Shock reflection at a wall can be interpreted, in inviscid flow, as, for instance, one half of a symmetric shock intersection (the upper half of Fig. 6.13 a)). In viscous flow the interaction with the boundary layer must be taken into account. If the effective shock strength is small, a thickening of the boundary layer will happen, otherwise boundary-layer separation will occur, Section 9.1.

We note finally that intersecting shocks of the same family form in a triple point T a single stronger shock [3]. In T also a slip line and an expansion fan originate, see Fig. 9.6 c) in Sub-Section 9.2.1.

[11] Shock waves belong to the same family, if their shock angles θ have the same sign. This sign is the same as that of the angles (Mach angles μ) of the respective Mach lines (characteristics) of the flow in which they are imbedded [3], [2]. The shock waves belong to opposite families, if their shock angles have opposite signs.

6.3.3 Treatment of Shock Waves in Computational Methods

Shock waves have a definite structure with a thickness of only a few mean free paths. This would mean that for their computation the Boltzmann equation must be employed [14]. In general the Boltzmann equation is valid in all four flow regimes, Section 2.3, i. e. from continuum to free molecular flow. In the free-molecular flow regime it is used as the so called collisionless Boltzmann equation. However, due to the very large computational effort to solve it (recent developments in solution algorithms and computer power have improved the situation), discrete numerical continuum methods for the solution of the Euler and the Navier-Stokes equations and their derivatives will be employed in aerothermodynamics whenever possible. For computational cases, where a gap exists between the applicability range of the latter, and that of the Boltzmann equation, the bridging-function concept can be used [15].

The question is now how to solve the computational problem of shock waves in the continuum regime. In [16] it is shown that the structure of a normal shock wave can be computed with a continuum method, i. e. the Navier-Stokes equations, for pre-shock Mach numbers $M_1 \lessapprox 2$. This is done by comparing solutions of the Boltzmann equation, which employ the Bhatnagar-Gross-Krook model, with solutions of the Navier-Stokes equations.

Since it is the Mach number M_N normal to the front of the shock wave, which matters, also the structures of oblique shock waves can be treated with the Navier-Stokes equations, if $M_N \lessapprox 2$. We show this in Fig. 6.14 with the computation case of a reflecting shock wave at a surface without boundary layer [17]. It was set up such that the normal Mach number of the incoming shock wave, $M_N = M_1 \sin\beta = 1.5$, was equal to that of a case treated in [16]. In Fig. 6.15 the solution of the Boltzmann equation (curve 1) [16] is compared at some station ahead of the reflection location, with solutions of a space-marching, shock-structure resolving scheme of a derivative of the Navier-Stokes equations [17].

For the coarse discretization of the latter solution with about ten nodes in the shock wave in stream-wise direction (curve 2) the comparison is not satisfactory. The fine discretization with about twenty nodes (curve 3) yields a good agreement on the high-pressure side (behind the shock wave). On the low-pressure side (ahead of the shock wave) the agreement is only fair. This could be due to the fact, that in [16] the Prandtl number $Pr = 1$ was used, and in [17] the actual value.

This brings us back to the original question of how to treat shock waves in flow fields past large flight vehicles, which globally are fully in the continuum regime. Obviously a discretization with twenty nodes across the shock structure is beyond the performance range of computers for a long time to come. Even if this would be possible, the problem remains, that shock-wave structures can be described by means of the Navier-Stokes equations only if the relevant shock-wave Mach number is small enough.

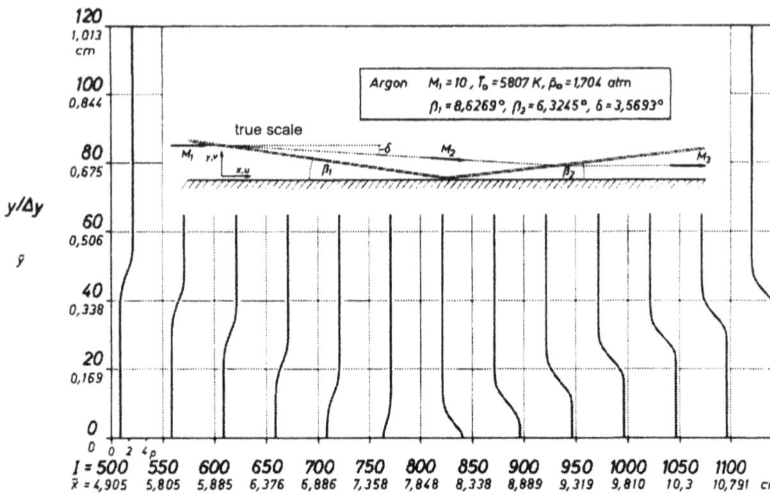

Fig. 6.14. Pressure profiles in the shock-wave reflection zone at a surface without boundary layer, inset: reflection process in true scale [17].

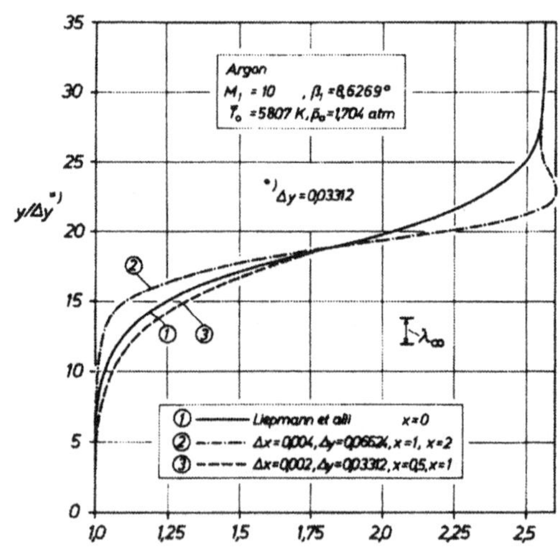

Fig. 6.15. Comparison of local pressure profiles upstream of the shock-wave reflection location shown in Fig. 6.14. On the left side is the region ahead of the oblique shock wave, on the right side the region behind it [17].

In Euler and Navier-Stokes solutions shock waves are therefore treated as discontinuity surfaces in the frame of the "weak-solution concept", see, e. g., [18]. This "shock-capturing" approach demands to employ the governing equations in conservative formulation, Chapter 12. The "shock-fitting" approach, see, e. g., [19], which is applicable more or less only to describe the bow-shock surface of a vehicle, is now seldom used.

The shock-capturing approach is well proven. In the slip-flow regime, however, where over a blunt body a "thick" shock wave can even merge with the also "thick" boundary layer (merged-layer regime [20]) the shock-capturing approach is somewhat questionable, also in view of the fact that the shock wave has a "secondary" structure due to thermochemical relaxation effects.

6.4 Blunt-Body Flow

6.4.1 Bow-Shock Stand-Off Distance at a Blunt Body

We have noted in the preceding sub-section that the shape of the bow-shock surface is governed by the free-stream properties and the effective body shape. These also govern the stand-off distance Δ_0 of the bow-shock surface from the body surface. Especially the shape of the body nose has a strong influence on Δ_0, Fig. 6.16 [3], where $\delta/d \equiv \Delta_0/2R_b$. The smallest stand-off distance is observed for the sphere, the largest for the two-dimensional cases. In general Δ_0 is smaller at axisymmetric bodies, because there the stagnation-point flow is relieved three-dimensionally, in contrast to two-dimensional flow. With increasing free-stream Mach number all curves approach asymptotically limiting values of the bow-shock stand-off distance Δ_0.

For the bow-shock stand-off distance at spheres and circular cylinders, approximate analytical relations are available for sufficiently large Mach numbers, say, $M_\infty \gtrapprox 5$, see, e. g., [21]. We consider first a sphere with radius R_b, Fig. 6.17. It is assumed that the bow-shock surface in the vicinity of the flow axis lies close and parallel to the body surface, and thus is a sphere with the radius R_s. With the assumption of constant density in the thin layer between body surface and bow-shock surface it is found for the shock stand-off distance Δ_0 at a sphere with radius R_b [21]:

$$\Delta_0 = R_s - R_b = \frac{\varepsilon\, R_s}{1 + \sqrt{\frac{8\varepsilon}{3}}}, \tag{6.117}$$

with ε being the inverse of the density rise across the normal shock portion, Fig. 6.17:

$$\varepsilon = \frac{\rho_\infty}{\rho_s} = \frac{\rho_1}{\rho_2} = \frac{(\gamma - 1)M_1^2 + 2}{(\gamma + 1)M_1^2}. \tag{6.118}$$

Eq. (6.117) can be expanded in a power series to yield:

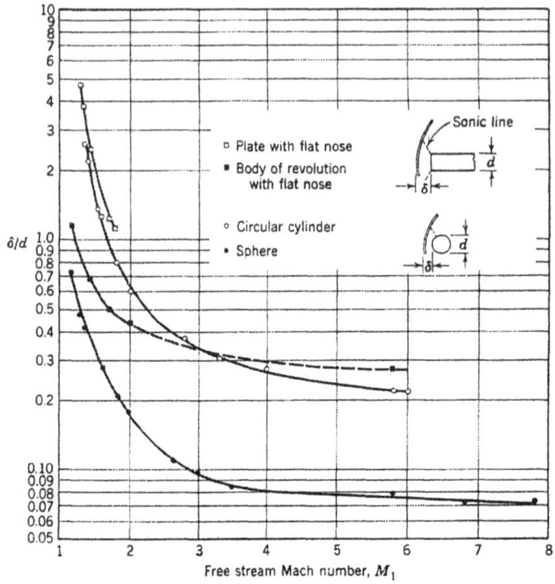

Fig. 6.16. Bow-shock stand-off distances δ/d at two-dimensional (plate, circular cylinder) and axisymmetric (flat-nosed body of revolution, sphere ($\delta/d \equiv \Delta_0/2R_b$) bodies as function of the free-stream Mach number M_1. Experimental data, cold hypersonic wind tunnels, air, $\gamma = 1.4$, [3].

$$\Delta_0 = \varepsilon \, R_s \left[1 - \sqrt{\frac{8\varepsilon}{3}} + \frac{8}{3}\varepsilon + O(\varepsilon^{3/2}) \right] . \tag{6.119}$$

A similar relation for the circular cylinder reads [21]:

$$\Delta_0 = R_s - R_b = \frac{1}{2}\varepsilon \, R_s \left[ln\frac{4}{3\varepsilon} + \frac{\varepsilon}{2}(ln\frac{4}{3\varepsilon} - 1) + O(\varepsilon^3) \right] . \tag{6.120}$$

For sufficiently large free-stream Mach numbers ε can be approximated for both normal and oblique shock waves, eqs. (6.68) and (6.82), by the value for $M_1 \rightarrow \infty$:

$$\varepsilon|_{M_{1\rightarrow\infty}} = \frac{\gamma - 1}{\gamma + 1}, \tag{6.121}$$

where we must demand $\gamma > 1$. For our purposes this is a fair approximation for $M_1 \gtrapprox 6$ at least for $\gamma = 1.4$, as can be seen from Fig. 6.18, where the inverse of ε is shown as function of M_1, and for three values of γ.

In a flow with large total enthalpy the temperature behind the bow-shock surface in the vicinity of the stagnation point region will rise strongly. Accordingly the ratio of specific heats γ will decrease. If we take an effective

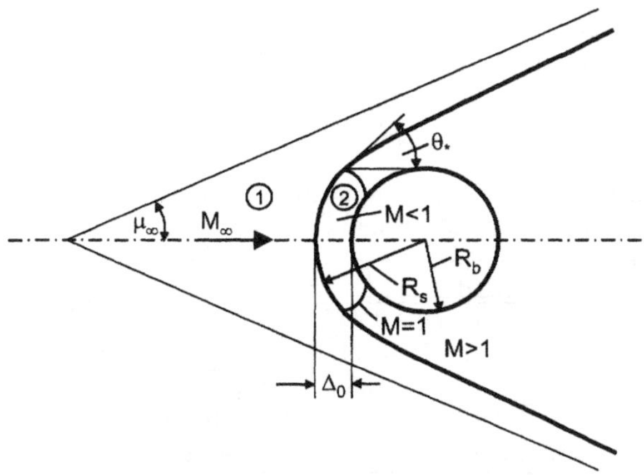

Fig. 6.17. Schematic and notation of bow-shock stand-off distance Δ_0 at a sphere of radius R_b.

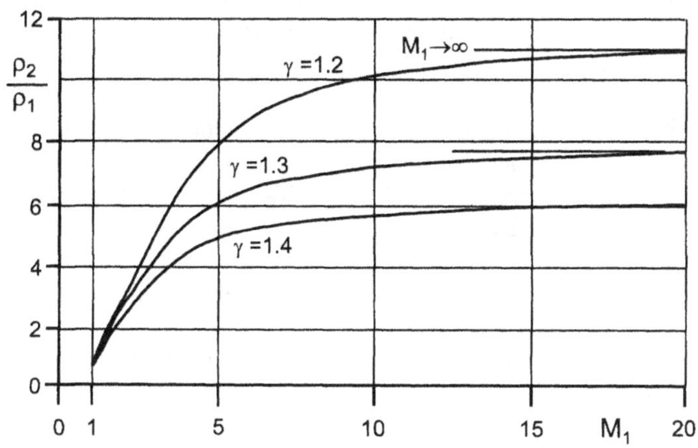

Fig. 6.18. ρ_2/ρ_1, the inverse of ε across a normal shock as function of M_1, and for three values of γ.

mean ratio of the specific heats γ as a measure for the bow shock stand-off distance at large free-stream Mach numbers, we obtain for instance for the sphere, eq. (6.119), to first order:

$$\frac{\Delta_0}{R_s} = \frac{\gamma - 1}{\gamma + 1}\Big|_{mean,shock\ layer}. \tag{6.122}$$

This result tells us that the bow-shock stand-off distance Δ_0 at a body in hypersonic flight is largest for perfect gas, and with increasing high-temperature real-gas effects will become smaller as the effective mean γ is reduced. A rigorous treatment of this phenomenon can be found in, e. g., [22], [23].

We illustrate this phenomenon with data computed by means of a coupled Euler/second-order boundary-layer method for a hyperbola of different dimensions at hypersonic Mach numbers, Table 6.3.

Table 6.3. Flow parameters of computation cases of a hyperbola with laminar boundary layer at hypersonic Mach numbers.

Case	L [m]	R_b [m]	α [°]	M_∞	$Re_{\infty R_b}$	T_∞ [K]	T_w	Gas	Reference
1	0.075	0.015488	0.0	10	$0.247 \cdot 10^4$	220.0	adiabatic	air	[24]
2	0.075	0.015488	0.0	10	$0.247 \cdot 10^4$	220.0	adiabatic	air	[24]
3	0.075	0.015488	0.0	25	$0.0356 \cdot 10^4$	198.4	adiabatic	N_2	[25]
4	0.75	0.15488	0.0	25	$0.356 \cdot 10^4$	198.4	adiabatic	N_2	[25]
5	75.0	15.488	0.0	25	$35.6 \cdot 10^4$	198.4	adiabatic	N_2	[25]

The bow-shock stand-off distance in case 1 (perfect gas computation), is more than two times larger than that in case 2 (equilibrium gas computation), Fig. 6.19 a), Table 6.4.

In reality, however, the situation can be more complex, depending on the size of the body, and hence on the first Damköhler number of the problem, Section 5.4. This is studied by means of true non-equilibrium computations for three different body sizes, cases 3 to 5 in Table 6.3. For the small body with $L = 0.075\ m$ (case 3) only a small degree of dissociation is found, Fig. 6.19 b). The reason is that due to the small absolute bow-shock stand-off distance and the large flow speed not enough time is available for dissociation to occur, $DAM1 \to 0$.

For the large body with $L = 75.0\ m$, case 5, the situation is different. The absolute bow-shock stand-off distance is large, and the degree of dissociation too, Fig. 6.19 b). At the stagnation point, $X1S = 0$, it compares well with the degree of dissociation $m^*_{N_2}$ ($\equiv \omega^*_{N_2}$) found behind an isolated normal shock wave with the same flow conditions, hence we have an equilibrium situation, $DAM1 \to \infty$.

Table 6.4. Thermodynamic models of computation cases for a hyperbola at hypersonic Mach numbers (see Table 6.3), and results. The ratio of boundary-layer to shock-layer thickness δ/Δ is that at half the body length ($X1S = 0.375$ in Fig. 6.19).

Case	Thermodyn. model	Surface	$DAM1$	Thermodyn. result	Δ_0/R_b	δ/Δ
1	perfect gas	-	-	-	≈0.59	≈0.14
2	equilibrium	-	-	-	≈0.25	≈0.16
3	non-equilibrium	non-catalytic	$\to 0$	frozen	≈0.33	≈0.31
4	non-equilibrium	non-catalytic	$O(1)$	non-equilibrium	≈0.26	≈0.33
5	non-equilibrium	non-catalytic	$\to \infty$	equilibrium	≈0.14	≈0.05

That m_{N_2} at the stagnation point is a little smaller than $m^*_{N_2}$, is due to the fact that between the bow-shock surface and the body surface a further compression happens, and hence a rise of the temperature from 5,600.0 K behind the (isolated) normal shock to approximately 5,900.0 K at the stagnation point, which increases the degree of dissociation.

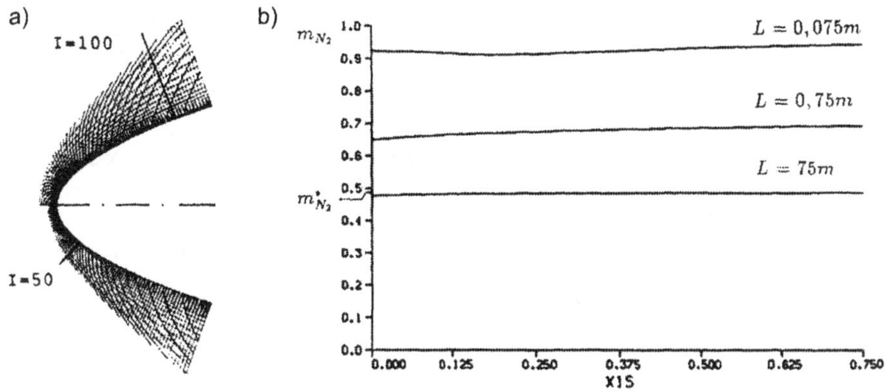

Fig. 6.19. Results of coupled Euler/second-order boundary-layer computations of hypersonic flow past hyperbolas. **a**) bow-shock shapes [24], upper part: perfect gas (case 1), lower part: equilibrium real gas (case 2). **b**) mass fraction m_{N_2} ($\equiv \omega_{N_2}$) at hyperbola surface for different body dimensions, cases 3 to 5 [25], $X1S$ is the respective location on the body axes.

Case 4, $L = 0.75\ m$, finally lies between the two other cases, and is a typical non-equilibrium case, $DAM1 = \text{O}(1)$, Table 6.4. Regarding the almost constant mass fractions for the cases 3 to 5 in Fig. 6.19 b), we note that an adiabatic wall was assumed throughout. Therefore the levels of surface temperature remain approximately the same along the hyperbola contours, and with them the respective mass fractions.

The results prove that with a coupled Euler/second-order boundary-layer computation formulated for non-equilibrium flow also the limiting cases of "frozen flow" and "equilibrium flow" can be handled. For pure inviscid flow computations of this kind, problems have been observed in the vicinity of the stagnation point of blunt bodies, where in any case the flow locally approaches equilibrium, which makes a special post-processing necessary, [26].

The differences in the bow-shock stand-off distances between case 1 and case 3, as well as between case 2 and case 5, are due to the fact that we have viscous-flow cases. The approximate ratio of boundary-layer to shock-layer thickness δ/Δ at 0.5 body length L is given for each case in Table 6.4. These ratios are approximately the same for cases 1 and 2, and also for cases 3 and 4, because the Reynolds numbers are the same, respectively comparable for each pair. Of course the surface temperature levels are much higher in cases

3 and 4, because of the much larger free-stream Mach number, and hence the boundary layers are thicker, Sub-Section 7.1. Case 5 finally, with a very large Reynolds number, has a very small boundary-layer thickness compared to the shock-layer thickness.

The results given in Table 6.4 point to the potential problem of a wrong determination of the shock stand-off distance compared to free flight. This can happen in a cold hypersonic tunnel, in a high-enthalpy facility, or with a computation method, because of the strong sensitivity of the shock stand-off distance on the high-temperature thermo-chemical behaviour of the flow. Such a wrong determination can lead, for instance, to an erroneous localization of intersections of the bow-shock surface with the aft part of the flight vehicle - we discuss examples in Sub-Section 9.2.1 - but also to errors in drag prediction et cetera, related to the interaction of the bow-shock surface with embedded shock waves and expansion phenomena[12].

The absolute value of a potential error in the shock stand-off distance, of course, depends on the effective radius of the stagnation-point area, e. g., the blunt shape of a re-entry vehicle at large angle of attack. If we take as example a body at high Mach number with an effective nose radius of $R_b = 1.5\ m$, we would get in flight, with an assumed mean $\gamma = 1.2$ in the shock layer, a shock stand-off distance of approximately $0.1\ m$. For perfect gas with $\gamma = 1.4$ this distance would be approximately $0.2\ m$, which is wrong by a factor of two.

On the basis of the above relations we obtain an overview of parameters in the shock layer in the nose region of a body, i. e. near the stagnation point, which are affected, if high-temperature real-gas effects are not properly taken into account there. The pressure coefficients c_{p_2} and $c_{p_{2_t}}$ in that region for large free-stream Mach numbers M_1 become independent of Mach number, eqs. (6.70) and (6.71), see also Section 6.8, and also do not vary strongly with γ. Hence we can assume that for given large Mach numbers the pressure also does not vary strongly, and that the mean density in the layer between the bow shock and the body surface in the nose region will be inverse to the mean temperature there:

$$\rho|_{mean_{shock\ layer}} \approx \frac{1}{T|_{mean_{shock\ layer}}}. \qquad (6.123)$$

With that result we get qualitatively the differences of flow parameters in the shock layer at the nose-region between the flight situation and the perfect-gas situation, i. e. with an inadequate simulation of high-temperature real-gas effects, Table 6.5. The actual differences, of course, depend on the nose shape, and speed and altitude of flight compared to the parameters and the flow situation in ground-simulation. Whether they are of relevance in vehicle design, depends on the respective design margins, Section 10.1.

[12] Actually the body contour *and* these phenomena together with the free-stream properties determine the shape of the bow-shock surface.

However, our goal is to have uncertainties in ground-simulation data small in order to keep design margins small, [27].

Table 6.5. Flow parameters in the nose-region shock layer of a blunt body at large free-stream Mach number M_1. Qualitative changes near the stagnation point from the flight situation with high-temperature real-gas effects (baseline) to the perfect-gas situation.

Parameter	Flight	Perfect gas	Remarks
Pressure	baseline	$\approx$ equal	Oswatitsch independ. principle
Speed	baseline	$\approx$ equal	$u \sim \Delta p$
Temperature	baseline	larger	gas absorbs less heat
Density	baseline	smaller	$\rho \sim p/T$
Viscosity	baseline	larger	$\mu \sim T^{\omega_\mu}$
Thermal conductivity	baseline	larger	$k \sim T^{\omega_k}$
Shock stand-off distance	baseline	larger	$\Delta_0/R \sim \rho_1/\rho_2$
Unit Reynolds number	baseline	smaller	$\rho_1 u_1 \approx \rho_2 u_2$: $Re_2^* \sim 1/T_2^{\omega_\mu}$
Boundary-layer thickness	baseline	larger	$\delta \sim 1/\sqrt{Re}$

6.4.2 The Entropy Layer at a Blunt Body

We have seen in Sub-Chapters 6.3.1 and 6.3.2 that in inviscid iso-energetic flow, i. e. adiabatic flow without friction and heat conduction, the entropy increases across normal and oblique shock waves. Behind these shock waves the entropy is constant again along the streamlines. If the entropy is constant also from streamline to streamline, we call this flow a homentropic flow. This situation is found in the uniform free stream ahead of a shock wave, which we have assumed for our investigations, and behind it, if the shock wave is not curved[13]. If we have flow, where the entropy is constant only along streamlines, we call this isentropic flow. This is the situation behind a curved shock wave.

We have seen also that for a given free-stream Mach number the normal shock wave leads to the largest entropy increase, eq. (6.58). Across an oblique shock the entropy increase is smaller, eq. (6.93), and actually becomes smaller and smaller with decreasing shock angle θ, until the shock angle approaches the Mach angle $\theta \rightarrow \mu$.

The total pressure decreases if the entropy increases:

$$\frac{p_{t_2}}{p_{t_1}} = e^{-(s_2 - s_1)/R}, \tag{6.124}$$

[13] We have introduced both the normal and the oblique shock as phenomenological models. They are not curved, and hence we find constant entropy behind them from streamline to streamline.

and hence behind a curved shock surface we get a total pressure, which changes from streamline to streamline. The lowest total pressure is found on the streamline, which has passed the locally normal shock surface. If the total pressure changes from streamline to streamline, we have, with a given pressure field, a change of the flow velocity from streamline to streamline, see Bernoulli's equation for compressible flow, eq. (6.21): the flow behind a curved shock surface is rotational.

The law which relates the vorticity $\underline{\omega}$ of this rotational inviscid flow to the entropy gradient across the streamlines is Crocco's theorem [3]:

$$\underline{V} \times rot\, \underline{V} = -\, T\, grad\, s, \tag{6.125}$$

which reads in terms of the vorticity $\underline{\omega} = rot\, \underline{V}$

$$\underline{V} \times \underline{\omega} = -\, T\, grad\, s. \tag{6.126}$$

We illustrate the velocity gradient in the inviscid flow behind a curved bow-shock surface in Fig. 6.20 a) for the symmetric case, and in Fig. 6.20 b) for the asymmetric case, see also Section 6.1. The latter case we do not consider further, because it is much more complicated than the symmetric case, which completely permits to study the phenomenon of entropy-layer swallowing.

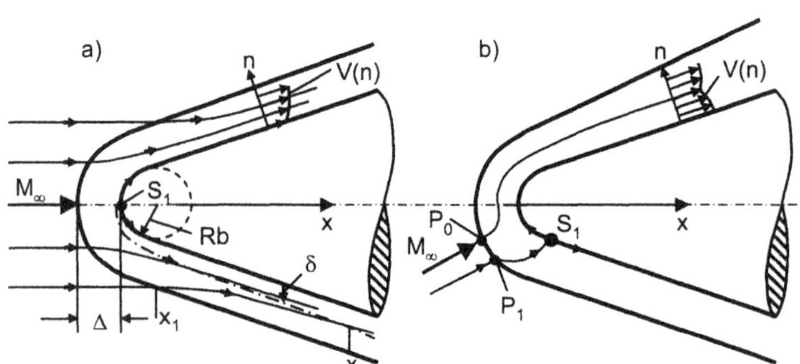

Fig. 6.20. Schematic of velocity gradients across streamlines due to the entropy layer in the inviscid flow field behind a curved bow-shock surface in **a**) the symmetric, and **b**) the asymmetric situation.

Why is the entropy layer of interest? We consider the development of the boundary layer at the blunt body in the symmetrical case. We assume that the characteristic Reynolds number is so large, that we can clearly distinguish between the inviscid flow field and the boundary layer, Sub-Section 4.3.1. We assume further laminar flow throughout, and neglect the Mangler effect, [28], as well as the influence of the surface temperature, which with radiation cooling decreases very fast from its maximum value in the stagnation-point area[14].

The boundary layer has a thickness δ, which grows with increasing running length along the body surface in the downstream direction[15]. The edge of the boundary layer is not coincident with a streamline. On the contrary, it is crossed by the streamlines of the inviscid flow, i. e., it entrains, or swallows, the inviscid flow[16].

We have indicated at the lower side of Fig. 6.20 a) the point x_1 on the bow-shock surface. It is the point downstream of which the bow-shock surface can be considered as straight. This means that all streamlines crossing the shock downstream of x_1 have the same constant entropy behind the shock surface. The streamline, which crosses the shock surface at x_1, enters the boundary layer at x_2. Thus the boundary layer swallows the entropy-layer of the blunt body between the stagnation point S_1 and the point x_2. The length on the body surface between the stagnation point and x_2 is called the entropy-layer swallowing distance, see, e. g. [29], in which the body bluntness, i. e., the entropy-layer swallowing, can have an effect on the boundary-layer parameters.

Entropy-layer swallowing is an interesting and, for hypersonic vehicle design, potentially important phenomenon. It leads to a decrease of the boundary-layer thicknesses[17], and subsequently to an increase of the heat flux in the gas at the wall q_{gw}, as well of the wall shear stress. The changes in the characteristic boundary-layer thicknesses, i. e., of the form of the boundary-layer velocity profile, also influence the laminar-turbulent transition behaviour of the boundary layer, see Section 8.1.5.

The effects of entropy layer swallowing have found attention quite early, see, e. g., [30], [31]. In the following we consider entropy-layer swallowing from a purely qualitative and illustrative point of view.

We assume a blunt body in a given hypersonic free stream with Mach number M_∞, and unit Reynolds number Re^u_∞. In Fig. 6.20 a) we have indicated the stand-off distance Δ of the bow-shock surface. It is approximately a function of the normal-shock density ratio ε, and the nose diameter R_s, eq. (6.119). We consider two limiting cases, assuming a comparable boundary-layer development.

1. The nose radius R_s is very large, and hence the boundary-layer thickness is very small compared to the shock stand-off distance: $\delta \ll \Delta$. The

[14] For the influence of the Mangler effect and the surface temperature on the boundary-layer thickness see Sub-Section 7.2.1.

[15] For the problem of defining this thickness see also Sub-Section 7.2.1.

[16] Indeed, it is only in first-order boundary-layer theory, where the boundary layer is assumed to be infinitely thin, that the inviscid streamline at the body surface is considered as the edge streamline of the boundary layer.

[17] We will see below, that the boundary layer first of all must be thick enough that the effect can happen.

boundary layer development is governed by the inviscid flow, which went through the normal-shock portion of the bow-shock surface. The entropy layer is not swallowed or only to a very small degree.

2. The nose radius R_s is small, and hence the boundary-layer thickness is not small compared to the shock stand-off distance: $\delta \approx O(\Delta)$. This situation is shown in the lower part of Fig. 6.20 a). The entropy layer is swallowed, and this means that increasingly inviscid flow with higher speed and lower temperature[18] is entrained into the boundary layer. This increases the effective local unit Reynolds number. As will be shown in Sub-Section 6.6, the unit Reynolds number directly behind the bow shock is smallest for the normal-shock part, and increases with decreasing shock angle θ. This effect is directly significant for the boundary layer only away from the stagnation-point region, where the inviscid flow is turned weakly. The increased effective unit Reynolds number then increases the heat flux in the gas at the wall and the wall shear stress, as was mentioned above.

It is clear that we can make a similar consideration assuming constant M_∞ and constant nose radius R_s, and a changing unit Reynolds number Re^u_∞. Then we observe entropy-layer swallowing with decreasing Re^u_∞, because this increases the boundary-layer thickness.

We illustrate the results found qualitatively above with results of flow-field computations made with a first-order and a second-order coupled Euler/boundary-layer method for the second case, instead with solutions for two different nose radii.

A first-order boundary-layer scheme takes into account only the properties of the stagnation-point streamline, and hence does not capture entropy-layer swallowing, which in reality happens. This is done only if we employ either a one-layer computation scheme, like a viscous shock-layer or a Navier-Stokes method, or a suitable two-layer computation scheme, i. e. a second-order boundary-layer method, which is coupled to an Euler method. We show results in Fig. 6.21 and Fig. 6.22 from the application of a second-order boundary-layer method with perturbation coupling, see Sub-Section 7.2.1, to the flow past a hyperbola at zero angle of attack [32], see also the symmetric case in Fig. 6.20 a).

In Fig. 6.21 a) we find for a given location the rotational inviscid flow profile at the wall (see also Fig. 6.20 a), upper part), and the boundary-layer velocity profiles. The boundary-layer profile found with the first-order solution, i. e., without entropy-layer swallowing, is governed by the inviscid flow at the body surface, and hence by the stagnation-point entropy. The profile found with the second-order solution blends into the inviscid flow

[18] We assume isoenergetic inviscid flow, where the total enthalpy is constant in the whole flow field. Hence for a given free-stream Mach number the (static) temperature behind the shock wave is the smaller the smaller the shock angle θ is, eq. (6.83).

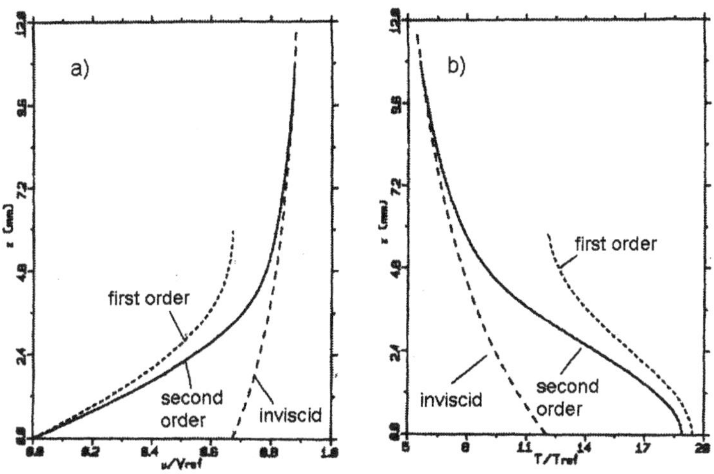

Fig. 6.21. Illustration of the effect of entropy-layer swallowing at a hyperbola, case 1 of Table 6.3: **a**) typical velocity profiles of the inviscid flow and the boundary layer without (first-order solution), and with (second-order solution) entropy-layer swallowing, **b**) typical temperature profiles of the inviscid flow and the boundary layer without (first-order solution), and with (second-order solution) entropy-layer swallowing [32].

profile at a finite distance from the surface and hence swallows the entropy-layer as the boundary layer develops from the stagnation-point region in downstream direction.

In the second-order case, i. e., the case with entropy-layer swallowing, the boundary-layer edge velocity u_e/V_{ref} is larger than in the first-order case, i. e., without entropy-layer swallowing. Consequently we have with entropy-layer swallowing a larger velocity gradient $\partial(u/V_{ref})/\partial z|_w$ and because of approximately the same wall temperatures, Fig. 6.21 b), and hence the same viscosity at the wall, a larger wall shear stress τ_w.

Fig. 6.21 b) shows the temperature profiles for the adiabatic wall situation. In this case the adiabatic temperature T_r is a little smaller with entropy-layer swallowing than without. This is in contrast to the usually observed behaviour of the wall-heat flux q_{gw}, see also Fig. 6.22, where entropy-layer swallowing increases the heat flux at the wall. The reason for this probably is the overall higher temperature level in the case without entropy-layer swallowing, because of the already much higher boundary-layer edge temperature T_e/T_{ref}.

A typical cold-wall case is sketched in Fig. 6.22. There is indicated the much larger temperature gradient near the body surface, where dissipation work is larger in the case with entropy-layer swallowing due to the larger velocity gradient, Fig. 6.21 a). Consequently the temperature gradient at the wall is larger than in the case without entropy-layer swallowing, and with the

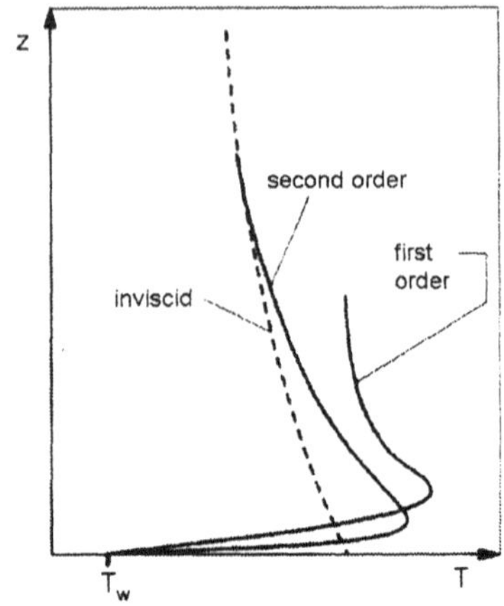

Fig. 6.22. Illustration of the effect of entropy-layer swallowing at a hyperbola. Cold wall with fixed temperature T_w: typical temperature profiles of the inviscid flow and the boundary layer without (first-order solution), and with entropy-layer swallowing (second-order solution) [32].

same wall temperature T_w in both cases, we find a larger q_{gw} with entropy-layer swallowing than without.

6.5 Supersonic Turning: Prandtl-Meyer Expansion and Isentropic Compression

The basics of fluid mechanics tell us, that supersonic flow will be accelerated (expanded), if the characteristic cross-section of the flow-domain is enlarged, and decelerated (compressed), if it is reduced. We call this expansion or compression by supersonic flow turning, see, e. g., [3].

Supersonic flow turning implies that small turning rates are involved, with basically isentropic changes of the flow. The Prandtl-Meyer expansion is an isentropic flow phenomenon in which the flow speed is increased. In a sense it can be considered as the counterpart of an oblique shock wave, where the flow speed is reduced. However, across the shock wave we have a discontinuity of speed, whereas with the Prandtl-Meyer expansion a gradual increase of the flow speed occurs.

The Prandtl-Meyer expansion is a "centered" expansion phenomenon, Fig. 6.23. Downstream of the expansion fan the flow again is parallel to the body surface. If a streamline (a) in the expansion fan is selected as a body contour, we have above it a "simple" expansion. However, in such a case the hidden Prandtl-Meyer corner can be used to describe the flow. The Laval-nozzle expansion between nozzle throat and contour inflection point, Fig. 5.7, can be considered as an example for such a simple expansion.

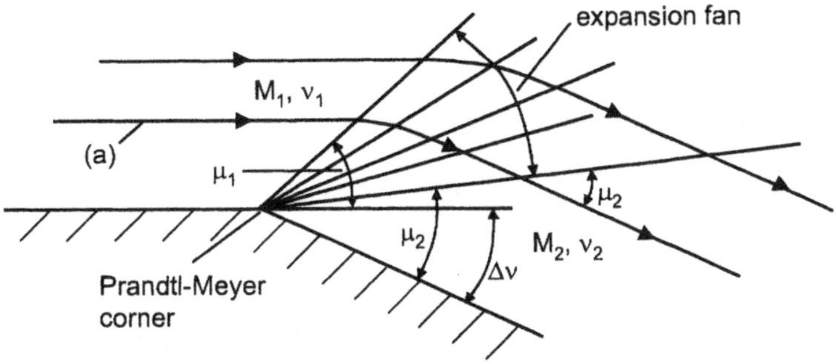

Fig. 6.23. Schematic of Prandtl-Meyer expansion with $\Delta\nu$ being the Prandtl-Meyer angle. If streamline (a) would be the body contour, the flow above it would be a "simple" expansion flow.

The real counterpart of the Prandtl-Meyer expansion is the isentropic compression of supersonic flow, which can be envisaged, for example, as the inverse flow in a Laval nozzle. Investigations have been made whether isentropic compression can be used in supersonic engine flow inlets [8]. Such a flow can be imagined as a compression by an infinite number of very weak shocks. In practice, however, such flow would be very sensitive to contour disturbances - the boundary layer already is such a disturbance - so that isentropic compression is considered today as not feasible.

Supersonic flow turning of a perfect gas is described by an expression obtained from eq. (6.113), which was derived for small ramp angles [3]:

$$-d\delta = \sqrt{M^2 - 1}\frac{dV}{V}. \tag{6.127}$$

Here δ is the deflection angle and V the resultant flow velocity. Integrating this expression gives the Prandtl-Meyer angle $\nu(M)$:

$$-\delta + c = \int \sqrt{M^2 - 1}\frac{dV}{V} = \nu(M). \tag{6.128}$$

After further manipulation we get the following relation for $\nu(M)$:

$$\nu(M) = \int \frac{\sqrt{M^2 - 1}}{1 + \frac{\gamma-1}{2}M^2}\frac{dM}{M}. \tag{6.129}$$

This relation integrated yields the Prandtl-Meyer function, [3], [10], [5]:

$$\Delta\nu = \nu - \nu_1 = \left[\sqrt{\frac{\gamma+1}{\gamma-1}}\tan^{-1}\sqrt{\frac{\gamma-1}{\gamma+1}(M^2-1)} - \tan^{-1}\sqrt{M^2-1}\right]_{M_1}^{M}. \tag{6.130}$$

It describes supersonic isentropic flow turning, Fig. 6.24, which is either expansion turning:

$$\nu = \nu_1 + |\delta - \delta_1| \tag{6.131}$$

or compression turning:

$$\nu = \nu_1 - |\delta - \delta_1|. \tag{6.132}$$

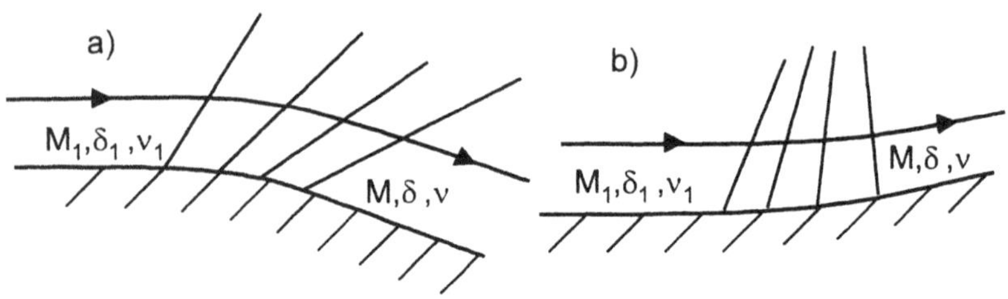

Fig. 6.24. Supersonic isentropic turning: **a)** expansion turning $\nu = \nu_1 + |\delta - \delta_1|$, **b)** compression turning $\nu = \nu_1 - |\delta - \delta_1|$.

$\Delta\nu$ can be computed as a function of M_1 and γ directly from eq. (6.130). No explicit relation is available to find the resulting Mach number for a given Prandtl-Meyer angle. An iterative solution procedure can be programmed. Tabulated data for M and ν for $\gamma = 1.4$ can be found in , e. g., [10], [3], [4]. Since both expansion and compression turning is isentropic, the relations given in Section 6.2 can be used to determine pressure, temperature and the like.

We consider now supersonic isentropic flow turning at very large Mach numbers. In such cases $\sqrt{M^2-1} \approx M$, and eq. (6.130) reduces to

$$\Delta\nu = \nu - \nu_1 = \left[\sqrt{\frac{\gamma+1}{\gamma-1}} \tan^{-1} \sqrt{\frac{\gamma-1}{\gamma+1}} M - \tan^{-1} \sqrt{M}\right]_{M_1}^{M}. \tag{6.133}$$

A Taylor expansion yields:

$$\Delta\nu = \left\{\sqrt{\frac{\gamma+1}{\gamma-1}} \left[\frac{\pi}{2} - \sqrt{\frac{\gamma+1}{\gamma-1}} \frac{1}{M} + O(M^{-3})\right] - \left[\frac{\pi}{2} - \frac{1}{M} + O(M^{-3})\right]\right\}_{M_1}^{M}. \tag{6.134}$$

Omitting the higher-order terms we obtain then:

$$\Delta\nu = \nu - \nu_1 = \frac{2}{\gamma-1}\left(\frac{1}{M_1} - \frac{1}{M}\right). \tag{6.135}$$

The Prandtl-Meyer angle itself for $M_1 \to \infty$ is:

$$\nu \rightarrow \frac{\pi}{2}\left(\sqrt{\frac{\gamma+1}{\gamma-1}}-1\right). \tag{6.136}$$

The pressure coefficient for large Mach numbers is found with the help of eq. (6.27) for large Mach numbers:

$$\frac{p}{p_1} = \left(\frac{M_1}{M}\right)^{\frac{2\gamma}{\gamma-1}}, \tag{6.137}$$

and finally with eq. (6.32)

$$c_p = \frac{p-p_1}{q_1} = \frac{2}{\gamma M_1^2}\left[\left(1-\frac{\gamma-1}{2}\Delta\nu M_1\right)^{\frac{2\gamma}{\gamma-1}} - 1\right]. \tag{6.138}$$

The largest possible expansion turning angle is that, for which $V_2 = V_m$. One has to keep in mind, that V_m is finite, eq. (6.15). If it is reached, $T_2 = 0.0$ K, and hence $M_2 \rightarrow \infty$. Because the Prandtl-Meyer relation is formulated in terms of the Mach number, we cannot determine from it the maximum expansion turning angle. The largest possible compression turning angle is that which leads to $M_2 = 1$.

With these results we have attained insight into the two turning phenomena, of which the Prandtl-Meyer expansion is the more important one. However, it does not find a direct application in computation schemes, except for the shock-expansion scheme [3], which is used in approximate computation methods for aerothermodynamic loads determination, e. g., [34] and also Chapter 11.

We give finally examples for expansion and compression turning in Table 6.6. It is interesting to observe, that for $M_1 = 20$ despite the large Mach number changes (M_2) with the given turning angles only small velocity (V_2) and temperature (T_2) changes occur. The static pressure p_2 however reacts strongly.

Table 6.6. Supersonic isentropic turning: examples of expansion and compression turnings.

M_1	ν_1	Turning	$\Delta\nu$	ν_2	M_2	V_2/V_1	p_2/p_1	T_2/T_1
6	84.955°	expansion	+10.0°	94.955°	7.84	1.026	0.184	0.617
6	84,955°	compression	-10.0°	74.955°	4.80	0.967	3.779	1.461
20	116.20°	expansion	+10.0°	126.20°	67.0	1.006	$2.2 \cdot 10^{-4}$	0.090
20	116.20°	compression	-10.0°	106.20°	11.66	0.988	40.2	2.872

6.6 Change of Unit Reynolds Number Across Shock Waves

We have seen that the shock angle θ of a bow-shock surface decreases from 90° at its normal shock portion to asymptotically the Mach angle μ_∞, which corresponds to the free-stream Mach number M_∞. At a two-dimensional ramp, which represents either a hypersonic inlet ramp, or an approximation of a control surface, we have, if we neglect viscous effects, shock angles, which for perfect gas are functions of the free-stream Mach number M_∞, the ratio of specific heats γ, and the ramp angle δ only. Eq. (6.111) connects these entities.

We consider now a plane oblique shock surface at shock angles $90° \geq \theta \geq \mu_\infty$ regardless of the geometry that is inducing it, Fig. 6.25.

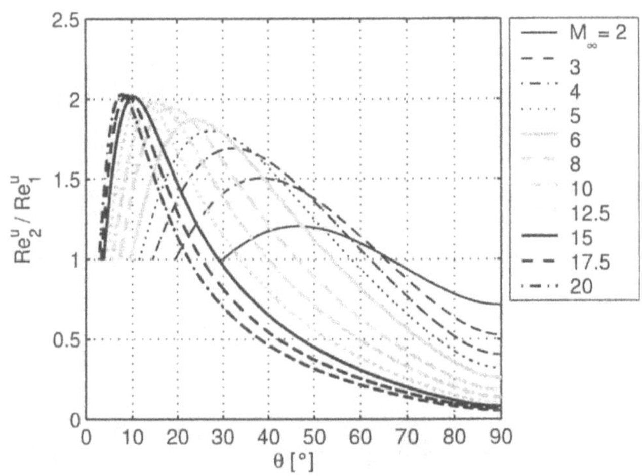

Fig. 6.25. Ratio of unit Reynolds numbers ($Re^u = \rho u/\mu$) across shock surfaces as function of the shock angle θ for different free-stream Mach numbers M_∞ ($\equiv M_1$) and the ratio of the specific heats $\gamma = 1.4$ [35]. The viscosity μ in all cases was computed with the power-law relation eq. (4.15).

We observe first, that for $\theta = 90°$ at all Mach numbers $M_1 > 1$ the unit Reynolds number Re^u behind the shock is smaller than ahead of it:

$$\theta = 90° : \; Re^u_2/Re^u_1 < 1. \tag{6.139}$$

This is obvious, because $\rho_2 u_2 = \rho_1 u_1$, and $T_2 > T_1$, and hence the viscosity $\mu_2 > \mu_1$.

If we now reduce with a given Mach number M_1 the shock angle θ, we find that the ratio Re^u_2/Re^u_1 increases, and that a shock angle exists, below which $Re^u_2/Re^u_1 > 1$. This holds for all Mach numbers and ratios of specific heat [35].

This behaviour is due to the fact that u_2/u_1 increases with decreasing shock angle θ, Fig. 6.26. The density ratio ρ_2/ρ_1, however, decreases first weakly and then strongly with decreasing θ, Fig. 6.27, while the temperature ratio T/T decreases first weakly and for small θ weakly again, Fig. 6.28.

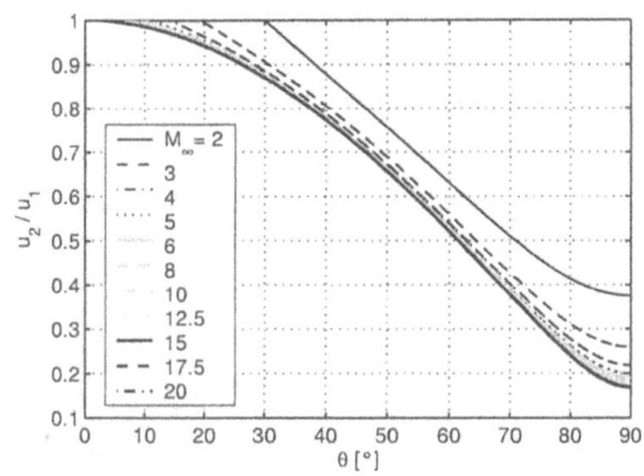

Fig. 6.26. Behaviour of u_2/u_1 as function of the shock angle θ and the Mach number M_∞ ($\equiv M_1$), $\gamma = 1.4$ [35].

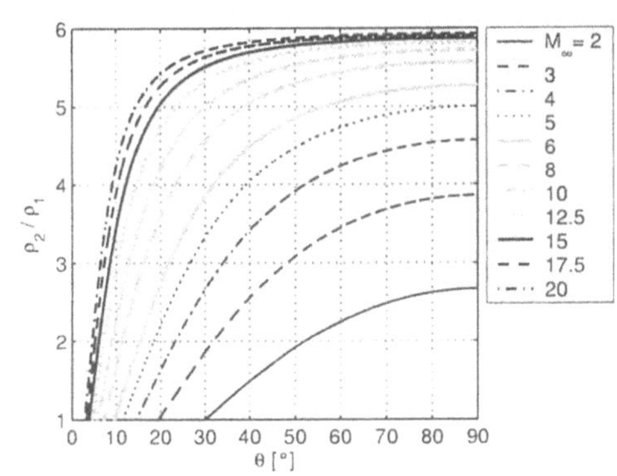

Fig. 6.27. Behaviour of ρ_2/ρ_1 as function of the shock angle θ and the Mach number M_∞ ($\equiv M_1$), $\gamma = 1.4$ [35].

The behaviour of the ratio of the unit Reynolds numbers across shock surfaces is of interest insofar as the development of the boundary layer behind an oblique shock wave, induced for instance by an inlet ramp or a control surface, to a large degree is governed by it, provided that the flow is sufficiently two-dimensional. This holds for attached boundary layers, but in

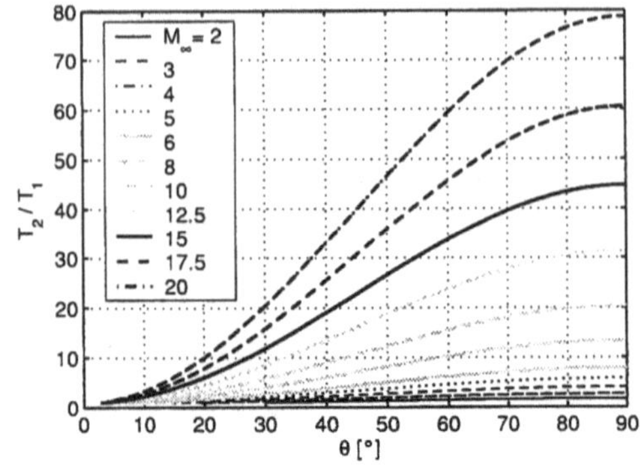

Fig. 6.28. Behaviour of T_2/T_1 as function of the shock angle θ and the Mach number M_∞ ($\equiv M_1$), $\gamma = 1.4$ [35].

principle also at the foot of the shock surface, where strong-interaction phenomena may occur. We understand now that an increase of, for instance, the radiation-adiabatic temperature of the body surface behind an oblique shock surface happens only, if the shock angle θ is below a certain value, which causes an increase of the unit Reynolds number. Otherwise a decrease will be observed.

We illustrate our findings in Figs. 6.29 and 6.30 with results from a study with an approximate method of the flow past a generic three-ramp inlet [36]. The free-stream Mach-number is $M_\infty = 7$, the altitude $H = 35.0\ km$, and the emissivity of the radiation-cooled ramp surfaces $\varepsilon = 0.85$. The flow is considered two-dimensional and fully turbulent.

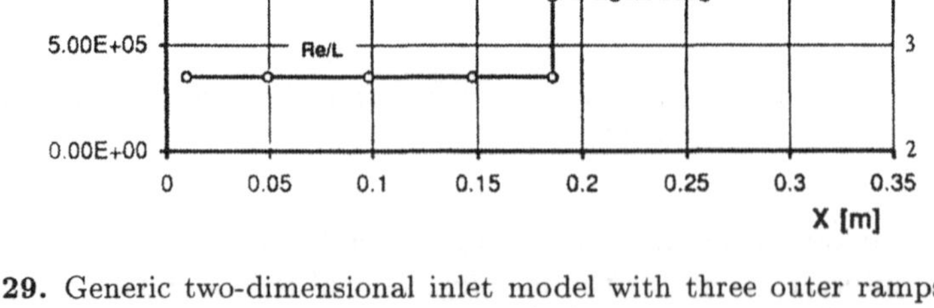

Fig. 6.29. Generic two-dimensional inlet model with three outer ramps (for the ramp geometry see Fig. 6.30): Mach number (M) and unit Reynolds number ($Re/L \equiv Re^u$) on the ramps [36].

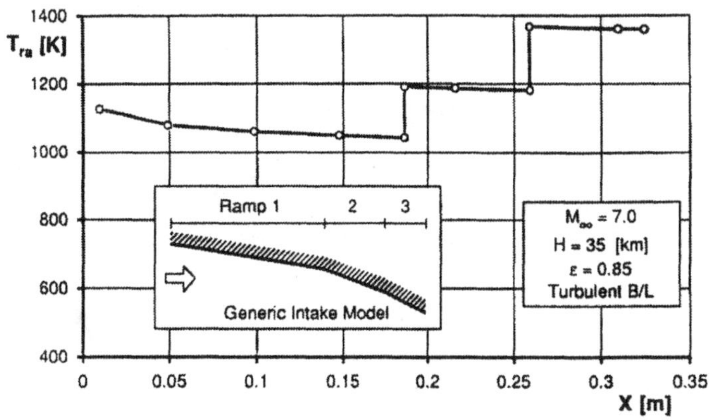

Fig. 6.30. Generic two-dimensional inlet model with three outer ramps: radiation-adiabatic temperature T_{ra} on the ramps [36].

We find in Fig. 6.29 a stepwise[19] reduction of the Mach number, and a stepwise increase of the unit Reynolds number. With the increase of the unit Reynolds number on each ramp the thickness of the boundary layer decreases. Therefore the radiation-adiabatic temperature rises on each ramp stepwise by almost 200.0 K to a higher level, Fig. 6.30, and then always decreases slightly with the boundary-layer growth, see Sub-Section 3.2.1, on each ramp.

6.7 Newton Flow

6.7.1 Basics of Newton Flow

At flight with very large speed, respective Mach number, the enthalpy h_∞ of the undisturbed atmosphere is small compared to the kinetic energy of the flow relative to the moving flight vehicle. The relation for the total enthalpy, eq. (3.2), can be reduced to:

$$M_\infty \to \infty : \; h_t = \frac{1}{2} v_\infty^2. \tag{6.140}$$

If we neglect the flow enthalpy also in the flow field at the body surface, the forces exerted on the body would be due only to the kinetic energy of the fluid particles. A corresponding mathematical flow model was proposed by Newton [21]. His model assumes that the fluid particles, coming as a parallel stream, hit the body surface, exert a force, and then move away parallel to the body surface, Fig. 6.31 a). With the notation given there we obtain for

[19] This accentuated step behaviour is due to the properties of the employed approximate method.

the force $R^\star$ acting on an infinitely thin flat plate with the reference surface A:

$$R^\star = \rho v^2 \sin^2 \alpha A. \tag{6.141}$$

This is in contrast to specular reflection, Fig. 6.31 b) (free-molecular flow), which would lead to a force twice as large, however, it models surprisingly well the actual forces in the continuum flow regime.

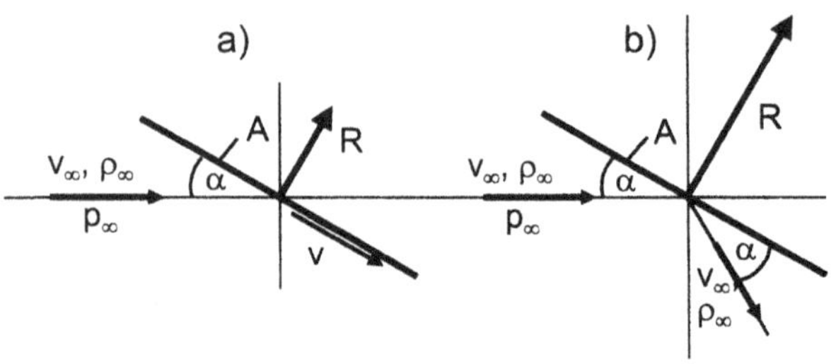

Fig. 6.31. Schematic of Newton flow at an infinitely thin flat plate with reference surface A: **a)** Newton´s model, **b)** specular reflection model.

If we also take into account the ambient pressure p_∞, we get the aerodynamic force acting on the windward side of the flat plate:

$$R = (\rho_\infty v_\infty^2 \sin^2 \alpha + p_\infty) A. \tag{6.142}$$

The surface-pressure coefficient on the windward side, $c_{p_{ws}}$, then is:

$$c_{p_{ws}} = \frac{p - p_\infty}{q_\infty} = \frac{1}{q_\infty}(\frac{R}{A} - p_\infty) = 2 \sin^2 \alpha. \tag{6.143}$$

On the lee-side of the flat plate only the ambient pressure p_∞ acts, because this side is not hit by the incoming free-stream[20]. The surface-pressure coefficient on this side, $c_{p_{ls}}$, hence is

$$c_{p_{ls}} = 0. \tag{6.144}$$

With the usual definitions we find the lift and the drag coefficient for the infinitely thin flat plate:

$$C_L = (c_{p_{ws}} - c_{p_{ls}}) \cos \alpha = 2 \sin^2 \alpha \cos \alpha, \tag{6.145}$$

and

$$C_D = (c_{p_{ws}} - c_{p_{ls}}) \sin \alpha = 2 \sin^3 \alpha, \tag{6.146}$$

[20] This side in Newton flow is the "shadow" side of the flat plate.

and also the resulting force coefficient, which has the same value as the pressure coefficient on the windward side

$$C_R = (c_{p_{w_s}} - c_{p_{l_s}}) = 2\sin^2\alpha. \tag{6.147}$$

These relations obviously are valid only for large Mach numbers. Take for instance an airfoil in subsonic flow. Its lift at angles of attack up to approximately maximum lift is directly proportional to the angle of attack α, and not to $\cos\alpha\sin^2\alpha$.

Newton's model is the constitutive element of "local surface inclination" or "impact" methods for the determination of aerodynamic forces and even flow fields at large Mach numbers. In design work, especially with incremental approaches, it can be used down to flight Mach numbers as low as $M_\infty \approx 3$ to 4, [37].

If we compare pressure coefficients on the windward side of the RHPM-flyer, Chapter 11, we find at all angles of attack α, that the Newton value is below the exact value, Fig. 6.32, [1], where results for two angles of attack are given. The difference increases with increasing α, and decreases with increasing M_∞.

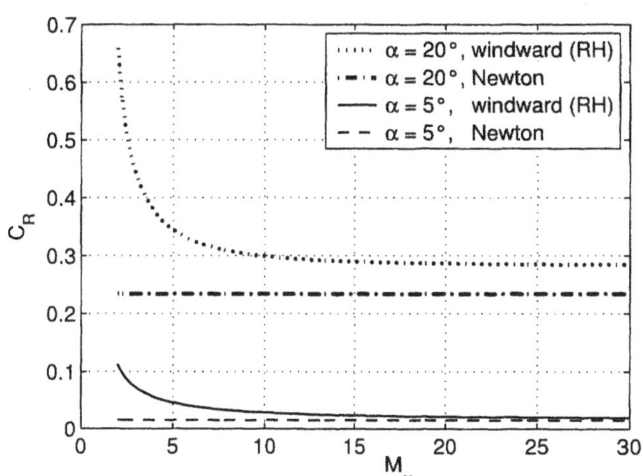

Fig. 6.32. Comparison of total force coefficients C_R at the windward side of a flat plate (RHPM-flyer) found with eq. (6.147) (Newton), and eq. (6.96) (RH, exact), two different angles of attack α, $\gamma = 1.4$ [1].

The exact (RH) value obviously becomes "Mach-number independent" (Section 6.8) for $M_\infty \gtrsim 8$. For a sharp-nosed cone Mach-number independence is found already at $M_\infty \gtrsim 4$, and the Newton result lies closer to, but still below, the exact result.

Certainly there is a physical meaning behind Newton's model. It is Mach-number independent per se, and the difference to the exact value becomes smaller with decreasing ratio of specific heats γ. We find an explanation, when we reconsider the pressure coefficient behind the oblique shock for very large Mach numbers, eq. (6.103):

$$M_\infty \to \infty : \ c_p \to \frac{4\sin^2\theta}{\gamma+1}, \tag{6.148}$$

and also the relation between the shock angle θ and the ramp angle δ for very large Mach numbers, eq. (6.114):

$$M_\infty \to \infty : \ \theta \to \frac{\gamma+1}{2}\delta. \tag{6.149}$$

For $\gamma \to 1$ we obtain that the shock surface lies on the ramp surface (infinitely thin shock layer):

$$\theta \to \delta, \tag{6.150}$$

and hence

$$c_p \to 2\sin^2\delta. \tag{6.151}$$

Since the ramp angle δ is the angle of attack α of the flat plate, we have found the result, see eq. (6.143), that in the limit $M_\infty \to \infty$, and $\gamma \to 1$, Newton's theory is exact. This is illustrated in Fig. 6.33, where with increasing Mach number and decreasing ratio of specific heats finally the exact solution meets Newton's result.

6.7.2 Modification Schemes, Application Aspects

We have noted above, that already at free-stream Mach numbers above $M_\infty \approx 3$ to 4 useful results can be achieved with Newton's theory. However, for both the flat plate and the sharp-nosed cone results of Newton's theory lie somewhat below exact results, as long as we have finite Mach numbers and realistic ratios of specific heats. What about the situation then at generally shaped, and especially blunt-nosed bodies?

Two modification schemes to the original formulation have been proposed. The first is Busemann's centrifugal correction [38], [21], which, however is not very effective.

The second scheme is important especially for blunt-nosed bodies. In [39] it is observed from experimental data, that for such bodies a pressure coefficient, corrected with the stagnation-point value $c_{p_{max}}$, gives satisfactory results for $M_\infty \gtrapprox 2$.

With $c_{p_{max}} = c_{p_{t_2}}$ for the stagnation point, eq. (6.63), writing p_e for the pressure on the body surface, respectively the boundary-layer edge, and using

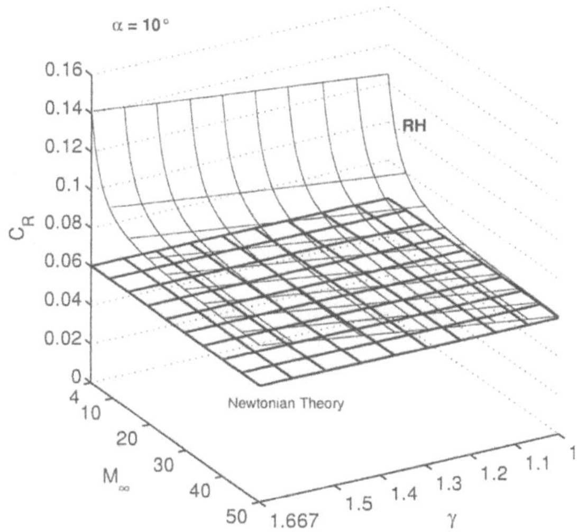

Fig. 6.33. The total force coefficient C_R at the windward side of a flat plate (RHPM-flyer) at $\alpha = 10.0°$ found with eq. (6.147) (Newtonian Theory, constant value, lower surface), and eq. (6.96) (exact (RH), upper surface), for $4 \leqq M_\infty \leqq 50$, and $1 \leqq \gamma \leqq 1.66\bar{6}$ [1].

θ_b, Fig. 6.34, for the local body slope, instead of α, we find the "modified" Newton relation:

$$c_{p_e} = \frac{p_e - p_\infty}{q_\infty} = c_{p_{max}} \sin^2 \theta_b =$$
$$= \frac{2}{\gamma M_\infty^2} \left\{ \left[\frac{(\gamma+1)^2 M_\infty^2}{4\gamma M_\infty^2 - 2(\gamma-1)} \right]^{\frac{\gamma}{\gamma-1}} \left[\frac{2\gamma M_\infty^2 - (\gamma-1)}{\gamma+1} \right] - 1 \right\} \sin^2 \theta_b. \tag{6.152}$$

This modified Newton relation, in contrast to the original one, depends on the free-stream Mach number M_∞.

For the investigation of boundary-layer properties, Section 7.2, at the stagnation point of blunt bodies or at the attachment line of blunt swept leading edges, et cetera, we need the gradient of the external inviscid velocity u_e there.

Assuming either a spherical or a cylindrical contour, Fig. 6.34 a), we find for both of them, see, e. g., [33], the Euler equation for the ψ direction:

$$\rho_e u_e \frac{1}{R} \frac{\partial u_e}{\partial \psi} = - \frac{1}{R} \frac{\partial p_e}{\partial \psi}. \tag{6.153}$$

At the stagnation point it holds:

$$\psi = 0: \; p_e = p_s = p_{max}, \; \frac{\partial p_s}{\partial \psi} = 0, \; u_e = 0, \; \frac{\partial u_e}{\partial \psi} > 0. \tag{6.154}$$

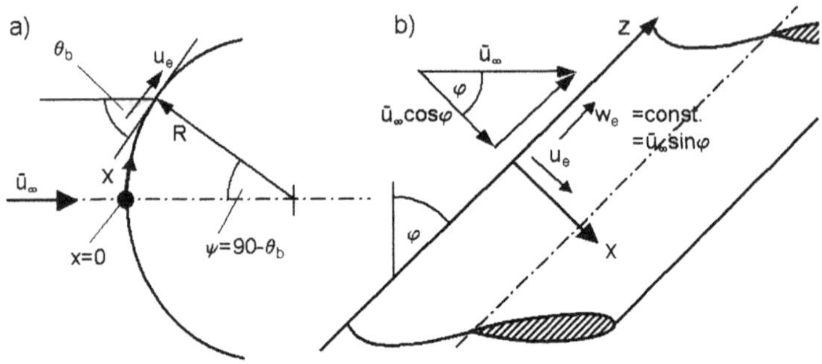

Fig. 6.34. Schematic and notation of flow past **a**) sphere and circular cylinder (2-D case), **b**) infinite swept circular cylinder with sweep angle φ.

For small angles ψ it can be written:

$$\frac{x}{R} = \sin\psi = \psi, \; \cos\psi = 1, \tag{6.155}$$

and

$$\frac{d\psi}{dx} = \frac{1}{R}. \tag{6.156}$$

To derive $du_e/d\psi$, we reformulate eq. (6.152) in terms of ψ, Fig. 6.34:

$$p_e = q_\infty c_{p_{max}} \cos^2\psi + p_\infty, \tag{6.157}$$

and find the gradient:

$$\frac{\partial p_e}{\partial \psi} = -q_\infty c_{p_{max}} 2 \sin\psi \cos\psi. \tag{6.158}$$

Combining this with eqs. (6.153) and (6.156), and writing it in the notation found usually in the literature, du_e/dx instead of $\partial u_e/\partial \psi$, we obtain:

$$\frac{du_e}{dx} = \frac{2 c_{p_{max}} q_\infty}{\rho_e u_e} \sin\psi \cos\psi \frac{1}{R}. \tag{6.159}$$

In the vicinity of the stagnation point with $x = 0$ it holds for $u_e(x)$:

$$u_e(x) = u_e|_{x=0} + \frac{du_e}{dx}|_{x=0} x + ..., \tag{6.160}$$

and hence:

$$\frac{u_e(x)}{x}|_{x>0} \approx \frac{du_e}{dx}|_{x=0}. \tag{6.161}$$

We substitute now u_e in eq. (6.159) with the help of eq. (6.161) and q_∞ with the help of eq. (6.157), and find with eqs. (6.155) and (6.156):

$$\left(\frac{du_e}{dx}\right)^2|_{x=0} = \left[\frac{2(p_s - p_\infty)}{\rho_s x}\frac{x}{R}\frac{1}{R}\right]_{x=0}, \tag{6.162}$$

where p_s and ρ_s are pressure and density in the stagnation point. Note that $c_{p_{max}}$ did cancel out.

The gradient of the inviscid velocity at the stagnation point in x - direction, Fig. (6.34), finally is:

$$\frac{du_e}{dx}|_{x=0} = \frac{k}{R}\sqrt{\frac{2(p_s - p_\infty)}{\rho_s}}. \tag{6.163}$$

This result holds, in analogy to potential theory, with $k = 1$ for the sphere, and $k = 1.33$ for the circular cylinder[21] (2-D case), and for both Newton and modified Newton flow.

At the infinite swept cylinder with sweep angle φ, it is the component of the free-stream vector normal to the cylinder axis, $u_\infty \cos\varphi$, Fig. 6.34 b), which matters. It can be written in terms of the unswept case and $\cos\varphi$:

$$\frac{du_e}{dx}|_{x=0,\varphi>0} = \cos\varphi \left[\frac{1.33}{R}\sqrt{\frac{2(p_s - p_\infty)}{\rho_s}}\right]_{\varphi=0}. \tag{6.164}$$

The important result in all cases is that the velocity gradient du_e/dx at the stagnation point or at the stagnation line, is $\sim 1/R$. The larger R, the smaller is the gradient. At the attachment line of the infinite swept cylinder it reduces also with increasing sweep angle, i. e. $du_e/dx|_{x=0,\varphi>0} \sim \cos\varphi/R$.

We note finally that recent experimental work has shown, that the above result has deficiencies, which are due to high-temperature real-gas effects [41]. With increasing density ratio across the shock, i. e. decreasing shock stand-off distance, Sub-Section 6.4.1, $du_e/dx|_{x=0}$ becomes larger. The increment grows nearly linearly up to 30 per cent for density ratios up to 12. If $du_e/dx|_{x=0}$ is needed with high accuracy, this effect must be taken into account.

We can determine now, besides the pressure at the stagnation point and on the body surface, also the inviscid velocity gradient at the stagnation point or across the attachment line of infinite swept circular cylinders. With the relations for isentropic expansion, starting from the stagnation point, it is possible to compute the distribution of the inviscid velocity, as well as the temperature and the density on the surface of the body. With these data, and with the help of boundary-layer relations, wall shear stress, heat flux in the gas at the wall et cetera can be obtained.

---

[21] For the circular cylinder $k = \sqrt{2}$ is given in [40] as more adequate value.

This is possible also for surfaces of three-dimensional bodies, with a stepwise marching downstream, depending on the surface paneling [34]. Of course, surface portions lying in the shadow of upstream portions, for example the upper side of a flight vehicle at angle of attack, cannot be treated. In the approximate method HOTSOSE [34] the modified Newton scheme is combined with the Prandtl-Meyer expansion and the oblique-shock relations[22] in order to overcome this shortcoming.

However, entropy-layer swallowing cannot be taken into account, because the bow-shock shape is not known. Hence the streamlines of the inviscid flow field between the bow-shock shape and the body surface and their entropy values cannot be determined.

This leads us to another limitation of the modified Newton method. It is exact only if the stagnation point is hit by the streamline which crossed the locally normal shock surface, Fig. 6.1 a). If we have the asymmetric situation shown in Fig. 6.3, we cannot determine the pressure coefficient in the stagnation point, because we do not know where P_1 lies on the bow-shock surface. Hence we cannot apply eq. (6.97). In such cases the modified Newton method, eq. (6.152), will give acceptable results only if P_1 can be assumed to lie in the vicinity of P_0.

We discuss finally a HOTSOSE result for the axisymmetric biconic reentry capsule shown in Fig. 6.35. The scheme was applied for free-stream Mach numbers $M_\infty = 4$ to 10 and a large range of angles of attack $0 \leqq \alpha \leqq 60°$. We show here only the results for $M_\infty = 4$, which are compared in Fig. 6.36 with Euler results.

The lift coefficient C_L found with HOTSOSE compares well with the Euler data up to $\alpha \approx 20°$. The same is true for the pitching-moment coefficient C_M. For larger angles of attack the stagnation point moves away from the spherical nose cap and the solution runs into the problem that P_1 becomes different from P_0, Fig. 6.3. In addition three-dimensional effects play a role, which are not captured properly by the scheme, if the surface discretisation remains fixed [34]. The drag coefficient C_D compares well for the whole range of angle of attack, with a little, inexplicable difference at small angles.

6.8 Mach-Number Independence Principle of Oswatitsch

We have found in Sub-Sections 6.3.1 and 6.3.2 that for $M_\infty \to \infty$ velocity components, density, Mach number and pressure coefficient behind normal and oblique shocks, and hence behind a bow-shock surface, become independent of M_∞.

We consider now the situation behind a bow-shock surface, for convenience only in two dimensions. We use the notation from Fig. 6.9 for the

[22] These are the elements of the shock-expansion method, which is employed for the RHPM-flyer.

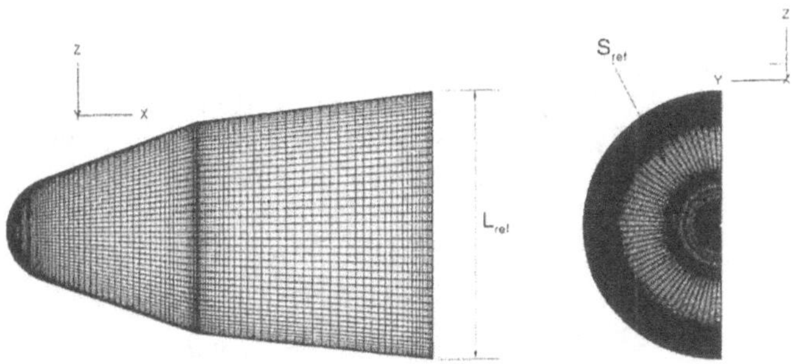

Fig. 6.35. Configuration and HOTSOSE mesh of the biconic re-entry capsule (BRC) [34].

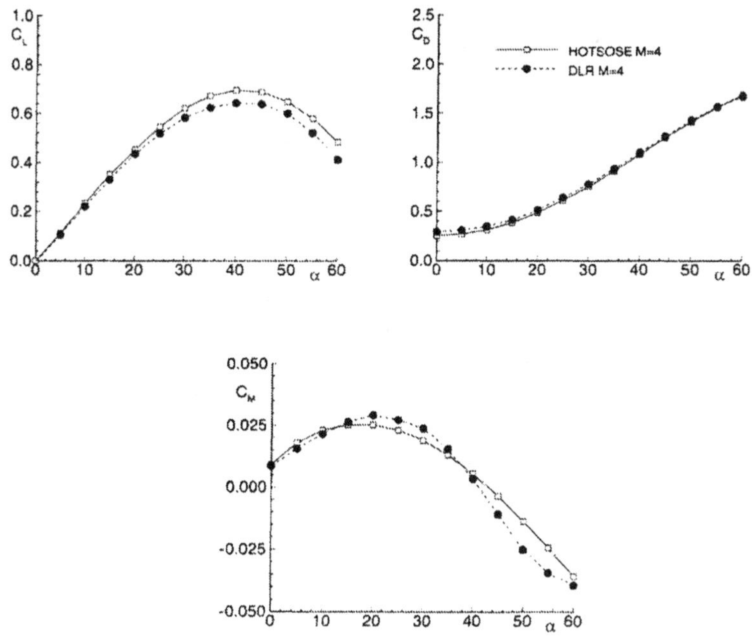

Fig. 6.36. Aerodynamic coefficients (inviscid flow, perfect gas) for the biconic re-entry capsule at $M_\infty = 4$ obtained with HOTSOSE (∘) and compared with Euler results (•) [34].

Cartesian velocity components behind a shock surface. We replace in eqs. (6.108) to (6.110), (6.68), (6.101), and (6.103) the subscript 1 with ∞, write u_∞ instead of V_1, omit the subscript 2, and introduce the superscript $'$ for all flow parameters non-dimensionalised with their free-stream parameters [23]:

$$\frac{u_2}{u_1} = \frac{u}{u_\infty} = u' = 1 - \frac{2\sin^2\theta}{\gamma+1}, \tag{6.165}$$

$$\frac{v_2}{u_1} = \frac{v}{u_\infty} = v' = \frac{2\sin^2\theta}{\gamma+1}\cot\theta, \tag{6.166}$$

$$\frac{V_2^2}{u_1^2} = \frac{V^2}{u_\infty^2} = V'^2 = u'^2 + v'^2 = 1 - \frac{4\gamma\sin^2\theta}{(\gamma+1)^2}, \tag{6.167}$$

$$\frac{\rho_2}{\rho_1} = \frac{\rho}{\rho_\infty} = \rho' = \frac{\gamma+1}{\gamma-1}, \tag{6.168}$$

$$M_2^2 = M^2 = \frac{(\gamma+1)^2 - 4\gamma\sin^2\theta}{2\gamma(\gamma-1)\sin^2\theta}, \tag{6.169}$$

$$c_{p_2} = c_p = \frac{4\sin^2\theta}{\gamma+1}. \tag{6.170}$$

Experimental observations, and these relations have led to what is now called Oswatitsch's independence principle [21]. In his original work [42] Oswatitsch defines flow at very large Mach numbers $M \gg 1$ as hypersonic flow. All flows at large Mach numbers have already hypersonic properties in the sense that certain force coefficients become independent of the Mach number. This happens, depending on the body form, already for free-stream Mach numbers as low as $M_\infty = 4$ to 5, see for instance Fig. 6.37.

We consider here especially the case of blunt bodies. This relates mainly to RV-type flight vehicles, but also to high flight Mach number CAV-type vehicles, which always have a nose with finite bluntness in order to withstand the thermal loads at large flight speeds.

The shape of the bow-shock surface is governed by the body shape and the free-stream parameters. The flow properties just behind the bow-shock surface are the upstream boundary conditions for the flow past the body. For $M_\infty \to \infty$ they are those given in eqs. (6.165) to (6.168). They are functions of the shape of the bow-shock surface, i. e. the shock angle θ, and the ratio of specific heats γ.

In the following we consider still only the two-dimensional case. This suffices completely to show the essence of the Mach-number independence principle. The derivation of the general three-dimensional case is straight forward.

---

[23] The normal-shock values are included with $\theta = 90°$.

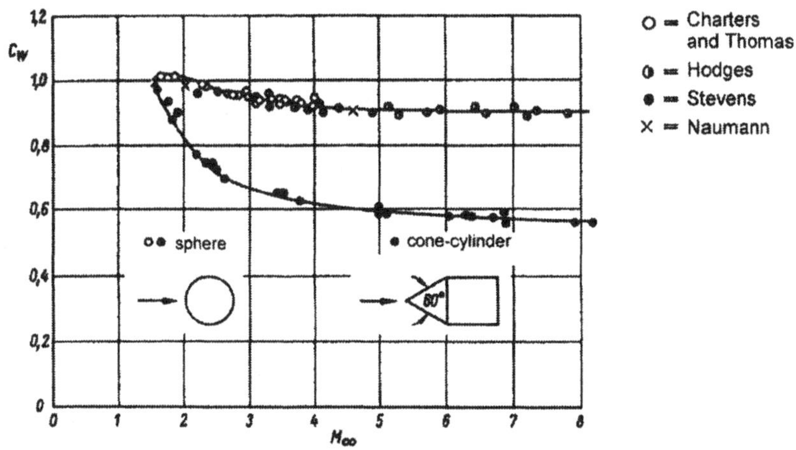

Fig. 6.37. Measured drag coefficients ($C_W \equiv C_D$) of sphere and cone-cylinder configuration as function of M_∞ from several sources [4].

We derive first the gas dynamic equation. We consider steady inviscid flow of a perfect gas. We re-formulate the pressure-gradient terms of the Euler equations, see in two dimensions eqs. (4.27) and (4.28) without the molecular and mass-diffusion transport terms, with the speed of sound, eq. (6.8):

$$\partial p = \left(\frac{\partial p}{\partial \rho}\right)_s \partial \rho = a^2 \partial \rho. \tag{6.171}$$

We combine now the Euler equations with the continuity equation eq. (4.83), neglect the time derivatives, and find the gas dynamic equation in two-dimensional Cartesian coordinates:

$$(u^2 - a^2)\frac{\partial u}{\partial x} + (v^2 - a^2)\frac{\partial v}{\partial y} + uv\left(\frac{\partial u}{\partial y} + \frac{\partial v}{\partial x}\right) = 0. \tag{6.172}$$

Together with Crocco´s theorem

$$\underline{V} \times rot\, \underline{V} = -\, T\, grad\, s, \tag{6.173}$$

we have three equations for the determination of u, v, and the entropy s.

To eliminate the speed of sound a and the temperature T we use eq. (6.14). We rewrite that equation and find with $V^2 = u^2 + v^2$ (see Fig. 6.9):

$$a^2 = \frac{(\gamma - 1)u_\infty^2}{2}\left(1 + \frac{2}{\gamma - 1}\frac{1}{M_\infty^2} - \frac{V^2}{u_\infty^2}\right), \tag{6.174}$$

and

$$T = \frac{u_\infty^2}{2c_p}\left(1 + \frac{2}{\gamma - 1}\frac{1}{M_\infty^2} - \frac{V^2}{u_\infty^2}\right). \tag{6.175}$$

For $M_\infty \to \infty$ these relations become

$$a^2 = \frac{(\gamma - 1)u_\infty^2}{2}\left(1 - \frac{V^2}{u_\infty^2}\right), \tag{6.176}$$

and

$$T = \frac{u_\infty^2}{2c_p}\left(1 - \frac{V^2}{u_\infty^2}\right). \tag{6.177}$$

The question is whether here the terms with $1/M_\infty^2$ can be omitted, because at slender bodies $V \approx u_\infty$, and hence $1 - (V^2/u_\infty^2)$ is small, too. It is argumented that V/u_∞ depends only on the shock angle of the bow shock, eq. (6.167), and not on M_∞ [42]. Therefore eqs. (6.177) and (6.176) are sufficiently exact, as long as M_∞ is sufficiently large.

It remains the problem to express in eq. (6.173) the entropy s for $M_\infty \to \infty$. From eq. (6.77) for the normal shock surface we gather that in this case $s \to \infty$, which also holds for an oblique shock surface. This problem is circumvented by introducing the difference $s' = s_{oblique\ shock} - s_{normal\ shock}$[24]:

$$\begin{aligned} s' = {} & \ln\left[\frac{2\gamma M_\infty^2 \sin^2\theta - (\gamma - 1)}{\gamma + 1}\right] - \gamma \ln\left[\frac{(\gamma + 1)M_\infty^2 \sin^2\theta}{(\gamma - 1)M_\infty^2 \sin^2\theta + 2}\right] - \\ & - \ln\left[\frac{2\gamma M_\infty^2 - (\gamma - 1)}{\gamma + 1}\right] + \gamma \ln\left[\frac{(\gamma + 1)M_\infty^2}{(\gamma - 1)M_\infty^2 + 2}\right]. \end{aligned} \tag{6.178}$$

Now we find for $M_\infty \to \infty$ a finite value of s', which depends also only on the shape of the bow-shock surface:

$$s' = 2\ln(\sin\theta). \tag{6.179}$$

We introduce eqs. (6.177), (6.176), and (6.179) into eqs. (6.172) and (6.173), non-dimensionalize and obtain:

$$\begin{aligned} & (\frac{2}{\gamma - 1}u'^2 - 1 + V'^2)\frac{\partial u'}{\partial x} + (\frac{2}{\gamma - 1}v'^2 - 1 + V'^2)\frac{\partial v'}{\partial y} + \\ & + \frac{2}{\gamma - 1}u'v'\left(\frac{\partial u'}{\partial y} + \frac{\partial v'}{\partial x}\right) = 0, \end{aligned} \tag{6.180}$$

and

[24] This is permitted, because a constant term is subtracted from the argument of *grad* s in eq. (6.173).

$$\underline{V}' \times rot\, \underline{V}' = -\frac{1}{2\gamma}(1 - V'^2)\; grad\, \frac{s'}{c_v}. \tag{6.181}$$

These equations describe the flow between the bow-shock and the body surface. They are supplemented by the boundary conditions behind the bow-shock surface, eqs. (6.165), (6.166) (6.179), and the boundary conditions at the body surface. The latter are kinematic conditions, which demand vanishing flow-velocity component normal to the body surface, i. e. the flow to be tangential to it.

The system of equations is independent of the free-stream Mach number M_∞. Assuming uniqueness, we deduce from it that u', v' and s' in the flow field between bow-shock surface and body surface do not depend on M_∞, but only on the body shape, and the ratio of specific heats γ. This holds also for the shape of the bow-shock surface, and the patterns of the streamlines, the sonic line, and the Mach lines in the supersonic part of the flow field. Density ρ', eq. (6.168) and pressure coefficient c_p, eq. (6.170), and with the latter the force and moment coefficients, are also independent of M_∞[25].

Oswatitsch called his result a similarity law. In [21] it is argued that to call it "independence principle" would be more apt, because it is a special type of similitude, being stronger than a general similitude. It is valid on the windward side of a body, i. e. in the portions of the flow past a body, where the bow-shock surface lies close to the body surface, see, e. g., Fig. 6.2 b). It is not valid on the lee-side of a body, where, due to the hypersonic shadow effect, no forces are exerted on the body surface. Nothing is stated about transition regimes between windward side and lee-side surface portions.

In [42] no solution of the system of equations (6.180) and (6.181) is given. In [21] several theories are discussed, some of which base directly on the findings of Oswatitsch. We have used results of two of these theories in Sub-Section 6.4.1 to discuss the influence of high-temperature real-gas effects on the bow-shock stand-off distance at a blunt body.

Oswatitsch's independence principle is an important principle for applied aerothermodynamics. If we obtain for a given body shape experimentally for instance the force coefficients at a Mach number M'_∞ large enough, they are valid then for all larger Mach numbers $M_\infty > M'_\infty$, Fig. 6.37, provided, however, that we have perfect-gas flow. In [21] it is argued that Oswatitsch's independence principle also holds for non-perfect gas flow. In any case we can introduce an effective ratio of specific heats γ_{eff}, and thus study the influence of high-temperature real-gas effects. In [21] it is claimed too, that the principle holds for boundary layers in hypersonic flows, as long as the external inviscid flow follows the independence principle. We will come back to the topic in Section 10.3.

For slender, sharp-nosed bodies at very large Mach numbers M_∞ the Mach angle μ_∞ may be of the same order of magnitude as the maximum deflection

[25] This result is analogous to that which we obtain for incompressible flow by means of the Laplace equation.

angle β, which the flow undergoes at the body surface. This class of high Mach number flow is characterized by the hypersonic similarity parameter, see, e. g., [2], [3]:

$$K = M_\infty \sin\beta \gg 1, \tag{6.182}$$

which was introduced by Tsien[26] [43].

In terms of the thickness ratio τ (body thickness/body length) of such bodies it reads:

$$K = M_\infty \tau \gg 1. \tag{6.183}$$

Not going into details of the theory we note the results [42], [3], which are important for aerodynamic shape definition:

- the surface pressure coefficient of a body with thickness ratio τ follows

$$c_p \sim \tau^2, \tag{6.184}$$

- and the wave drag coefficient

$$C_{D_w} \sim \tau^3. \tag{6.185}$$

It can be expected that these results give also the right increments in the case of slender, blunt-nosed configurations at small angle of attack, i. e. CAV-type flight vehicles. Of course at a blunt nose (the nose bluntness is be the major driver of the wave drag) the pressure coefficient is not covered by relation (6.184).

6.9 Problems

Problem 6.1 Consider expansion of air as perfect gas with a total temperature T_t = 1,500.0 K. How large is the maximum possible speed V_m?
Solution: V_m = 1,037.5 m/s.

Problem 6.2 What is the critical speed for the condition from Problem 6.1?
Solution: u_* = 599.0 m/s.

Problem 6.3 How large is the pressure coefficient c_p (perfect gas) in a stagnation point for subsonic flow with M_∞ = a) 0.01, b) 0.1, c) 0.5, d) 1?
Solution: c_p = a) ∞, b) 1.0025, c) 1.064, d) 1.276. Why do you find $c_p = \infty$ for $M_\infty = 0$? Remember Problem 4.5.

[26] Tsien neglects in this work the occurrence of shock waves and hence obtains results of restricted validity [42].

Problem 6.4 How large is the pressure coefficient c_p (perfect gas) in a stagnation point for supersonic flow (normal shock) with M_∞ = a) 1, b) 5, c) 10, d) ∞?

Solution: c_p = a) 1.276, b) 1.809, c) 1.832, d) 1.839.

Problem 6.5 Estimate the shock stand-off distance Δ_0 at a sphere with the radius $R_b = 1.0\ m$ for perfect-gas flow with $\gamma = 1.4$ for M_∞ = a) 5, b) 10, c) ∞. Use eq. (6.117), for convenience simplified to $\Delta_0 = \varepsilon\, R_s$.

Solution: c_p = a) 0.25 m, b) 0.21 m, c) 0.2 m. What do you observe for increasing Mach number?

Problem 6.6 Estimate like in Problem 6.5 the shock stand-off distance Δ_0 at a sphere with the radius $R_b = 1.0\ m$, but now for flow with $\gamma_{eff} = 1.2$ for M_∞ = a) 5, b) 10, c) ∞.

Solution: c_p = a) 0.12 m, b) 0.11 m, c) 0.1 m. What do you observe compared to Problem 6.5?

Problem 6.7 Estimate like in Problem 6.5 the shock stand-off distance Δ_0 at a sphere with the radius $R_b = 1.0\ m$, but now for flow with $\gamma_{eff} = 1$ for M_∞ = a) 5, b) 10, c) ∞.

Solution: c_p = a) 0.042 m, b) 0.01 m, c) 0.0 m. What do you observe compared to Problems 6.5 and 6.6?

Problem 6.8 Determine for a RHPM-CAV-flyer flying with $M_\infty = 6$ at 30.0 km altitude with the help of Newton's relations the aerodynamic (i. e. without g-reduction) lift L, drag D, lift to drag ratio L/D, and pitching moment M. The parameters are $\alpha = 6°$, $A_{ref} = 1860.0\ m^2$, $L_{ref} = 40.0\ m$, the reference point for the moment is the nose of the vehicle[27].

Solution: $L = 1.220{\cdot}10^6\ N$, $D = 0.128{\cdot}10^6\ N$, $L/D = 9.51$, $M = 48.801{\cdot}10^6\ Nm$.

Problem 6.9 Determine like in Problem 6.8 for the RHPM-CAV-flyer flying with $M_\infty = 6$, but now at 40.0 km altitude, lift L, drag D, lift to drag ratio L/D, and pitching moment M.

Solution: $L = 0.239{\cdot}10^6\ N$, $D = 0.031{\cdot}10^6\ N$, L/D =9.51, $M = 11.71{\cdot}10^6\ Nm$.

Problem 6.10 Compare the results from Problem 6.8 and 6.9. What is the main reason for the smaller aerodynamic forces at 40.0 km altitude? The aerodynamic lift there could be corrected by increasing the angle of attack α, or by enlarging the flight vehicle planform to get a larger A_{ref}. Do both and discuss that a larger α means also a larger drag (disregard the trimming issue), and a larger A_{ref} a larger structural weight.

---

27 Remember that a force coefficient C_F is defined by the actual force F divided by the free-stream dynamic pressure q_∞ times the reference area A_{ref}: $C_F = F/q_\infty A_{ref}$. The reference area is either the wing surface (planform surface), or for the flight vehicles considered here, the overall planform surface.

References

1. B. THORWALD. "Demonstration of Configurational Phenomena Exemplified by the RHPM-Hypersonic-Flyer". Diploma Thesis, University Stuttgart, Germany, 2003.
2. K. OSWATITSCH. "Gas Dynamics". Academic Press, New York, 1956.
3. H. W. LIEPMANN, A. ROSHKO. "Elements of Gasdynamics". John Wiley & Sons, New York/ London/ Sidney, 1966.
4. J. ZIEREP. "Theorie der schallnahen und der Hyperschallströmungen". Verlag G. Braun, Karlsruhe, 1966.
5. J. D. ANDERSON. "Modern Compressible Flow: with Historical Perspective". McGraw-Hill, New York, 1982.
6. W. ZEISS. Personal communication, München 2003.
7. E. H. HIRSCHEL. "Aerothermodynamic Phenomena and the Design of Atmospheric Hypersonic Airplanes". *J. J. Bertin, J. Periaux, J. Ballmann (eds.), Advances in Hypersonics, Vol. 1, Defining the Hypersonic Environment.* Birkhäuser, Boston, 1992, pp. 1 - 39.
8. W. H. HEISER, D. T. PRATT. "Hypersonic Airbreathing Propulsion". AIAA Education Series, Washington, 1994.
9. A. EBERLE, M. A. SCHMATZ, N. C. BISSINGER. "Generalized Flux Vectors for Hypersonic Shock Capturing". AIAA-Paper 90-0390, 1990.
10. AMES RESEARCH STAFF. "Equations, Tables, and Charts for Compressible Flow". NACA R-1135, 1953.
11. C. WEILAND. "A Contribution to the Computation of Transonic Supersonic Flows over Blunt Bodies". Computers and Fluids, Vol. 9, 1981, pp. 143 - 162.
12. H. J. LUGT. "Introduction to Vortex Theory". Vortex Flow Press, Potomac, Maryland, 1996.
13. P. J. SCHMID, D. S. HENNINGSON. "Stability and Transition in Shear Flows". Springer-Verlag, New York/Berlin/Heidelberg, 2001.
14. J. N. MOSS. "Computation of Flow Fields for Hypersonic Flight at High Altitudes". *J. J. Bertin, J. Periaux, J. Ballmann (eds.), Advances in Hypersonics, Vol. 3, Computing Hypersonic Flows.* Birkhäuser, Boston, 1992, pp. 371 - 427.
15. J. K. HARVEY. "Rarefied Gas Dynamics for Spacecraft". *J. J. Bertin, R. Glowinski, J. Periaux (eds.), Hypersonics, Vol. 1, Defining the Hypersonic Environment.* Birkhäuser, Boston, 1989, pp. 483 - 509.
16. H. W. LIEPMANN, R. NARASIMHA, M. T. CHAHINE. "Structure of a Plane Shock Layer". J. Physics of Fluids, Vol. 5, No. 11, 1962, pp. 1313 - 1324.
17. E. H. HIRSCHEL. "The Structure of a Reflecting Oblique Shock Wave". *H. Cabannes, R. Temann (eds.), Proc. of the Third International Conference on Numerical Methods in Fluid Mechanics, Vol. II, Problems of Fluid Mechanics.* Lecture Notes in Physics 19, Springer, Berlin/Heidelberg/New York, 1973, pp. 153 - 160.
18. A. EBERLE, A. RIZZI, E. H. HIRSCHEL. "Numerical Solutions of the Euler Equations for Steady Flow Problems". *Notes on Numerical Fluid Mechanics, NNFM 34.* Vieweg, Braunschweig/Wiesbaden, 1992.

19. C. WEILAND, M. PFITZNER, G. HARTMANN. "Euler Solvers for Hypersonic Aerothermodynamic Problems". Notes of Numerical Fluid Mechanics, NNFM 20, Vieweg, Braunschweig/Wiesbaden, 1988, pp. 426 - 433.

20. G. KOPPENWALLNER. "Rarefied Gas Dynamics". *J. J. Bertin, R. Glowinski, J. Periaux (eds.), Hypersonics, Vol. 1, Defining the Hypersonic Environment.* Birkhäuser, Boston, 1989, pp. 511 - 547.

21. W. D. HAYES, R. F. PROBSTEIN. "Hypersonic Flow Theory, Volume 1, Inviscid Flows". Academic Press, New York/London, 1966.

22. H. G. HORNUNG. "Non-Equilibrium Dissociating Nitrogen Flow Over Spheres and Circular Cylinders". J. Fluid Mechanics, Vol. 53, Part 1, 1972, pp. 149 - 176.

23. H. OLIVIER. "A Theoretical Model for the Shock Stand-Off Distance in Frozen and Equilibrium Flows". J. Fluid Mechanics, Vol. 413, 2000, pp. 345 - 353.

24. CH. MUNDT, M. PFITZNER, M. A. SCHMATZ. "Calculation of Viscous Hypersonic Flows Using a Coupled Euler/Second-Order Boundary-Layer Method". Notes of Numerical Fluid Mechanics, NNFM 29, Vieweg, Braunschweig/Wiesbaden, 1990, pp. 422 - 433.

25. CH. MUNDT. "Rechnerische Simulation reibungsbehafteter Strömungen im chemischen Nichtgleichgewicht (Computational Simulation of Viscous Flows in Chemical Non-Equilibrium)". Doctoral Thesis, Technische Universität München, 1992.

26. J.-A. DÉSIDÉRI, M.-V. SALVETTI. "Inviscid Non-Equilibrium Flow in the Vicinity of a Stagnation Point". *Mathematical problems in Mechanics (Mathematical Analysis).* C. R. Académie des Sciences, Paris, t. 316, Série I, 1993, pp.525 - 530.

27. E. H. HIRSCHEL. "Thermal Surface Effects in Aerothermodynamics". *Proc. Third European Symposium on Aerothermodynamics for Space Vehicles, Noordwijk, The Netherlands, November 24 - 26, 1998.* ESA SP-426, 1999, pp. 17 - 31.

28. H. SCHLICHTING. "Boundary Layer Theory". 7th edition, McGraw-Hill, New York, 1979.

29. V. ZAKKAY, E. KRAUSE. "Boundary Conditions at the Outer Edge of the Boundary Layer on Blunted Conical Bodies". AIAA Journal, Vol. 1, 1963, pp. 1671 - 1672.

30. A. FERRI. "Some Heat Transfer Problems in Hypersonic Flow". Aeronautics and Astronautics, Pergamon Press, 1960, pp. 344 - 377.

31. N. R. ROTTA, V. ZAKKAY. "Effects of Nose Bluntness on the Boundary Layer Characteristics of Conical Bodies at Hypersonic Speeds". Astronautica Acta, Vol. 13, 1968, pp. 507 - 516.

32. F. MONNOYER. "Hypersonic Boundary-Layer Flows". *H. Schmitt (ed.), Advances in Fluid Mechanics.* Computational Mechanics Publications, Southampton, 1997, pp. 365 - 406.

33. R. B. BIRD, W. E. STEWART, E. N. LIGHTFOOT. "Transport Phenomena". John Wiley, New York and London/Sydney, 2nd edition, 2002.

34. U. Reisch, Y. Anseaume. "Validation of the Approximate Calculation Procedure HOTSOSE foe Aerodynamic and Thermal Loads in Hypersonic Flow with Existing Experimental and Numerical Results". DLR-FB 98-23, 1998.

35. M. Mharchi. "Demonstration of Hypersonic Thermal Phenomena and Viscous Effects with the RHPM-Flyer". Diploma Thesis, University Stuttgart, Germany, 2003.

36. H.-U. Georg. Personal communication, München 1996.

37. J. J. Bertin. "State-of-the-Art Engineering Approaches to Flow Field Computations". *J. J. Bertin, R. Glowinski, J. Periaux (eds.), Hypersonics, Vol. 2, Computation and Measurement of Hypersonic Flows.* Birkhäuser, Boston, 1989, pp. 1 - 91.

38. A. Busemann. "Flüssigkeits- und Gasbewegung". Handwörterbuch der Naturwissenschaften, Vol. IV, 2nd Edition, G. Fischer, Jena, 1933, pp. 244 - 279.

39. L. Lees. "Hypersonic Flow". IAS-Reprint 554, 1955.

40. E. Reshotko, C. B. Cohen. "Heat Transfer at the Forward Stagnation Point". NACA TN-3513, 1955.

41. H. Olivier. "Influence of the Velocity Gradient on the Stagnation Point Heating in Hypersonic Flow". Shock Waves, Vol. 5, 1995, pp. 205 - 216.

42. K. Oswatitsch. "Ähnlichkeitsgesetze für Hyperschallströmung". ZAMP, Vol. II, 1951, pp. 249 - 264. Also: "Similarity Laws for Hypersonic Flow". Royal Institute of Technology, Stockholm, Sweden, KTH-AERO TN 16, 1950.

43. H. S. Tsien. "Similarity Laws of Hypersonic Flows". J. Math. Phys., Vol. 25, 1946, pp. 247 - 251.

7 Attached High-Speed Viscous Flow

At the begin of Chapter 6 we have noted that in the hypersonic flight domain, like in the lower speed regimes, flow fields can be separated into inviscid and viscous portions. At large altitudes, in general at the limit of the continuum regime, this separation becomes questionable. However, the central topic of this chapter is attached viscous flow, whose basic properties we describe with help of the phenomenological model "boundary layer".

We look first at the typical phenomena arising in viscous flows and at their consequences for aerothermodynamic vehicle design. Attached viscous flow is characterized, in general, by the molecular transport of momentum, energy and mass, Chapter 4, towards the vehicle surface, with wall-shear stress, the thermal state of the surface, thermo-chemical wall phenomena, et cetera, as consequences. We treat the boundary-layer equations, consider their limits in hypersonic flow, examine the implications of radiation cooling of vehicle surfaces, and define integral properties and surface parameters, including viscous thermal-surface effects. Finally we give simple relations for laminar and turbulent flow, with extensions to compressible flow by means of the reference temperature/enthalpy concept, for the estimation of boundary-layer thicknesses, wall-shear stress, and the thermal state of a surface (wall heat flux and wall temperature) for planar surfaces, spherical noses and cylindrical swept edges. A case study closes the chapter.

Laminar-turbulent transition and turbulence in attached viscous flow, which we have to cope with at altitudes below approximately 40.0 to 60.0 km, are considered in Chapter 8. We discuss there the basic phenomena, their dependence on flow and surface parameters, and on the thermal state of the surface. The state of the art regarding transition prediction and turbulence models is also reviewed.

Connected too to viscous flow are the flow-separation phenomenon and other strong-interaction phenomena. In Chapter 9 we treat separation and the typical supersonic and hypersonic strong interaction phenomena including hypersonic viscous interaction, and finally rarefaction effects, which are related to the latter phenomenon. All these phenomena are of large importance in aerothermodynamic design, because they may influence not only the aerodynamic performance, flyability and controllability of a flight vehicle, but also, and that very strongly, mechanical and thermal loads on the airframe and its components.

7.1 Attached Viscous Flow

7.1.1 Attached Viscous Flow as Flow Phenomenon

The flow past a body exhibits, beginning at the forward stagnation point, a thin layer close to the body surface, where viscous effects play a role. They are due to the fact that the fluid in the continuum regime sticks fully to the surface: no-slip boundary condition, eq. (4.47), or only partly in the slip-flow regime: slip-flow boundary condition, eq. (4.48). We speak about attached viscous flow. Away from this layer the flow field is inviscid, i. e. viscous effects can be neglected there. Of course the inviscid flow field behind the flight vehicle, and at large angle of attack also above it, contains vortex sheets and vortices, which are viscous phenomena, together with shock waves, see Section 6.1.

Attached viscous flow must always be seen in connection with the flow past the body as a whole. The body surface with either no-slip or slip boundary condition, together with the thermal and thermo-chemical boundary conditions, Section 4.3, is the causa prima for that flow. If the layer of attached viscous flow is sufficiently thin, the flow in it is nearly parallel to the body surface and the gradient of the pressure in it in direction normal to the surface vanishes. Attached viscous flow in this case is governed by the pressure field of the external inviscid flow field, and, of course, its other properties, and the surface boundary conditions. We call such an attached viscous flow sheet a "boundary layer"[1]. The boundary layer can be considered as the phenomenological model of attached viscous flow. In the following we use the terms "attached viscous flow" and "boundary layer" synonymously.

Of special interest in flight vehicle design are the flow boundary-layer thicknesses, the wall shear stress, and the thermal state of the surface, which encompasses both the wall temperature and the heat flux in the gas at the wall. They will be discussed in detail in the following sections.

For these discussions we assume in general two-dimensional boundary-layer flows. In reality the flows past hypersonic flight vehicles are three-dimensional, see, e. g., Figs. 7.7, 7.8, 9.4, and 9.5. However, large portions of the flow can be considered as quasi two-dimensional, and can, with due reservations, be treated with the help of two-dimensional boundary-layer models for instance for initial qualitative considerations, and for the approximate determination of boundary-layer parameters. This does not hold in regions with attachment and separation phenomena et cetera, which are present in flight with a large angle of attack, Figs. 9.4 and 9.5.

[1] We remember that in fact three kinds of boundary layers can be present simultaneously: flow, thermal, and mass-concentration boundary layer, Section 4.3.

7.1.2 Some Properties of Three-Dimensional Attached Viscous Flow

To apply a simplification or an approximation makes only sense, if the real situation is sufficiently known. Therefore we consider now some properties of attached viscous flow, i. e. three-dimensional boundary layers. No comprehensive description of three-dimensional attached viscous, and of separated and vortical flows is available. We refer the reader for details to, e. g., [1], [2], [3].

A three-dimensional boundary layer is governed by the two-dimensional external inviscid flow at or very close to the surface of the body. This external inviscid flow is two-dimensional in the sense that its streamlines are curved tangentially to the body surface. As a consequence of this curvature all boundary-layer streamlines are stronger curved than the external streamline [1]. The skin-friction line, finally, thus can have quite another direction than the external inviscid streamline. In practice this means, that, for instance, the surface oil-flow pattern on a wind-tunnel model is not necessarily representative for the pattern of the external inviscid flow field close to surface.

A stream surface, i. e. a boundary-layer profile, in a three-dimensional boundary layer defined in direction normal to the body surface, will become skewed in downstream direction due to the different curvatures of its streamlines. This is in contrast to a two-dimensional boundary layer, where the stream surface keeps its shape in the downstream direction. Hence the boundary layer profiles look different in the two cases, Fig. 7.1 a) and b).

If in a three-dimensional boundary layer on an arbitrary body surface the coordinate system locally is oriented at the external stream line, we call it a locally monoclinic orthogonal external-streamline oriented coordinate system [4][2]. In such a coordinate system the skewed boundary-layer profile can be decomposed into the main-flow profile and the cross-flow profile, Fig. 7.1 a). The main-flow profile resembles the two-dimensional profile, Fig. 7.1 b). The cross-flow profile by definition has zero velocity $v^{*2} = 0$ at both the surface, $x^3 = 0$, and the boundary-layer edge, $x^3 = \delta$.

The cross-flow profile, however, changes its shape, if the external inviscid streamline changes its curvature in the plane tangential to the body surface. Then an s-shaped profile will appear in the region just behind the inflexion point of the external streamline, and finally an orientation in the opposite direction, Fig. 7.2.

[2] We remember that the boundary-layer edge is not a streamline, Sub-Section 6.4.2 or here, a stream surface, nearly parallel to the body surface. Only in first-order boundary-layer theory, where the boundary layer is assumed to be infinitely thin, can the inviscid flow at the body surface considered to be the edge flow of the boundary layer. Here it would be the local projections of the loci, where the streamlines cross the boundary-layer edge.

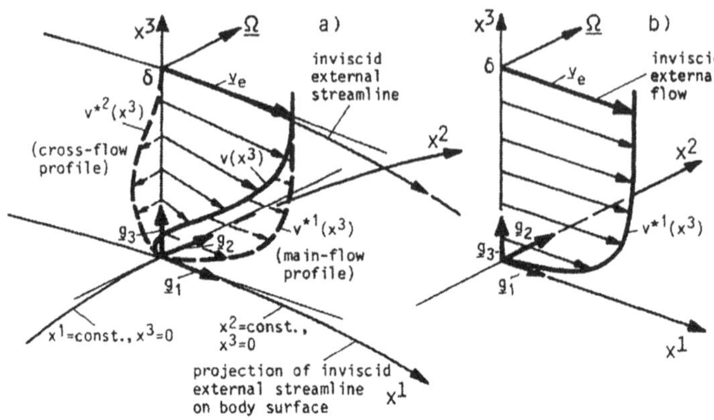

Fig. 7.1. Schematic of boundary-layer profiles [3]: **a**) three-dimensional boundary layer in curvilinear orthogonal external-streamline coordinates, **b**) two-dimensional boundary layer in Cartesian coordinates. The coordinates x^1 and x^2 are tangential to the body surface, x^3 is straight and normal to the body surface. v^{*1} and v^{*2} are the tangential velocity components. The vorticity vector $\underline{\omega}$ in two-dimensional flow points in x^2 direction throughout the boundary layer, while it changes its direction in three-dimensional flow. The vorticity content vector $\underline{\Omega}$ [3] always lies normal to the external inviscid streamline.

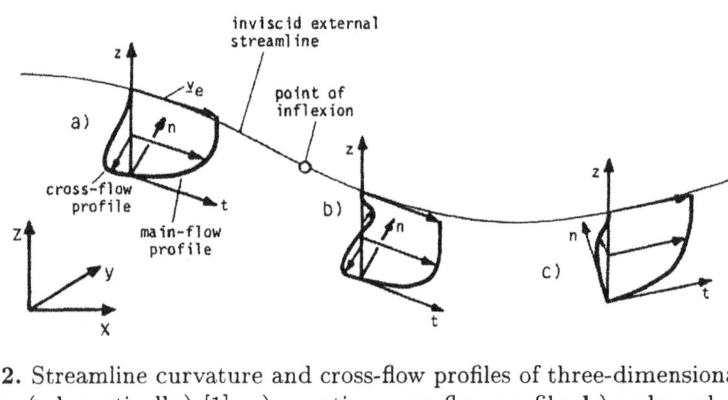

Fig. 7.2. Streamline curvature and cross-flow profiles of three-dimensional boundary layer (schematically) [1]: **a**) negative cross-flow profile, **b**) s-shaped cross-flow profile, **c**) positive cross-flow profile. The coordinates t and n are tangential to the body surface, z is normal to the body surface.

7.1.3 Boundary-Layer Equations

The boundary layer can be considered as the phenomenological model of attached viscous flow. We derive in the following the boundary-layer equations for steady, compressible, three-dimensional flow. We assume laminar flow, but note that the resulting equations also hold for turbulent flows, if we treat them as Reynolds- or Favre-averaged [5] flows.

The derivation is made in Cartesian coordinates. We keep the notation introduced in Fig. 4.1 with x and z (z not indicated there) being the coordinates tangential to the body surface, and y the coordinate normal to it. Accordingly u and w are the tangential velocity components, and v is the component normal to the body surface[3].

We try to keep the derivation as simple as possible in order to concentrate on the basic physical problems (that is also the reason to employ Cartesian coordinates). Therefore, we do not include the description of the typical phenomena connected with hypersonic attached flow like the extra formulations for high-temperature real-gas effects, surface-radiation cooling, and slip-flow effects. Their introduction into the resulting system of equations is straight forward.

The boundary-layer equations are derived from the Navier-Stokes equations, together with the continuity equation and the energy equation, Sub-Section 4.3. They cannot be derived from first principles. One has to introduce the observation - the boundary-layer assumption, originally conceived by Prandtl, [7] - that the extension of the boundary layer in direction normal to the body surface (coordinate y and the involved boundary-layer thicknesses) is very small, like also the velocity component v in the boundary layer normal to the body surface. Actually the observation is that the different boundary-layer thicknesses and v in y-direction are inversely proportional to the square root of the Reynolds number, if the flow is laminar.

We take care of this observation by introducing the so-called boundary-layer stretching, which brings y and v, non-dimensionalized with reference data L_{ref} and v_{ref}, respectively, to O(1):

$$y' = \frac{y\sqrt{Re_{ref}}}{L_{ref}}, \tag{7.1}$$

$$v' = \frac{v\sqrt{Re_{ref}}}{v_{ref}}, \tag{7.2}$$

with the reference Reynolds number defined by

$$Re_{ref} = \frac{\rho_{ref} v_{ref} L_{ref}}{\mu_{ref}}. \tag{7.3}$$

The prime above denotes variables, which were non-dimensionalised and stretched, however, we use it in the following also for variables, which are only non-dimensionalised.

[3] The reader will have noted that we have figures in the text with other notations. We always point this out, because no general nomenclature has been adopted so far in the literature. General surface-oriented "locally monoclinic" coordinates, which make use of the notation of shell theory, denote the tangential coordinates with x^1 and x^2, while the coordinate normal to the body surface is x^3 [4]. Accordingly we have the physical velocity components v^{*1},v^{*2} and v^{*3}.

All other variables are simply made dimensionless with appropriate reference data, and assumed then to be O(1): velocity components u and w with v_{ref}, lengths x and z with L_{ref}, temperature T with T_{ref}, density ρ with ρ_{ref}, pressure[4] p with $\rho_{ref} v^2_{ref}$, the transport coefficients μ and k with μ_{ref} and k_{ref}, respectively, and finally the specific heat at constant pressure c_p with $c_{p_{ref}}$. Each resulting dimensionless variable is marked by a prime, for instance:

$$u' = \frac{u}{v_{ref}}. \tag{7.4}$$

We introduce boundary-layer stretching and non-dimensionalization first into the continuity equation, 4.83. We do this for illustration in full detail. In three dimensions and without the partial time derivative, Section 4.1, we replace u with $u' v_{ref}$, eq. (7.4), v with $v' v_{ref}/\sqrt{Re_{ref}}$, eq. (7.2), et cetera, and find:

$$\frac{\partial \rho' \rho_{ref} u' v_{ref}}{\partial x' L_{ref}} + \frac{\partial \rho' \rho_{ref} v' v_{ref}/\sqrt{Re_{ref}}}{\partial y' L_{ref}/\sqrt{Re_{ref}}} + \frac{\partial \rho' \rho_{ref} w' v_{ref}}{\partial z' L_{ref}} = 0. \tag{7.5}$$

Since all reference parameters, and also Re_{ref} are constants, we find immediately the stretched and dimensionless continuity equation which has the same form - this does not hold for the other equations - as the original equation:

$$\frac{\partial \rho' u'}{\partial x'} + \frac{\partial \rho' v'}{\partial y'} + \frac{\partial \rho' w'}{\partial z'} = 0. \tag{7.6}$$

Consider now the Navier-Stokes equations. Because they were not quoted in full in Sub-Section 4.3.1, we do that now [6]:

$$\rho u \frac{\partial u}{\partial x} + \rho v \frac{\partial u}{\partial y} + \rho w \frac{\partial u}{\partial z} = -\frac{\partial p}{\partial x} - \left(\frac{\partial \tau_{xx}}{\partial x} + \frac{\partial \tau_{yx}}{\partial y} + \frac{\partial \tau_{zx}}{\partial z} \right), \tag{7.7}$$

$$\rho u \frac{\partial v}{\partial x} + \rho v \frac{\partial v}{\partial y} + \rho w \frac{\partial v}{\partial z} = -\frac{\partial p}{\partial y} - \left(\frac{\partial \tau_{xy}}{\partial x} + \frac{\partial \tau_{yy}}{\partial y} + \frac{\partial \tau_{zy}}{\partial z} \right), \tag{7.8}$$

$$\rho u \frac{\partial w}{\partial x} + \rho v \frac{\partial w}{\partial y} + \rho w \frac{\partial w}{\partial z} = -\frac{\partial p}{\partial z} - \left(\frac{\partial \tau_{xz}}{\partial x} + \frac{\partial \tau_{yz}}{\partial y} + \frac{\partial \tau_{zz}}{\partial z} \right). \tag{7.9}$$

4 In classical boundary-layer theory the pressure is made dimensionless with $\rho_{ref} v^2_{ref}$, which has the advantage that the equations describe in this form both compressible and incompressible flows. For general hypersonic viscous flows, however, we choose p_{ref} to make the pressure dimensionless, Sub-Section 7.1.7.

The components of the viscous stress tensor $\underline{\underline{\tau}}$, [8], in eqs. (7.7) to (7.9), with the bulk viscosity κ (see Sub-Section 4.3.1) neglected, because the terms containing it will drop out anyway during the order of magnitude consideration[5], read:

$$\tau_{xx} = -\mu \left[2\frac{\partial u}{\partial x} - \frac{2}{3}\left(\frac{\partial u}{\partial x} + \frac{\partial v}{\partial y} + \frac{\partial w}{\partial z}\right)\right], \tag{7.10}$$

$$\tau_{yy} = -\mu \left[2\frac{\partial v}{\partial y} - \frac{2}{3}\left(\frac{\partial u}{\partial x} + \frac{\partial v}{\partial y} + \frac{\partial w}{\partial z}\right)\right], \tag{7.11}$$

$$\tau_{zz} = -\mu \left[2\frac{\partial w}{\partial z} - \frac{2}{3}\left(\frac{\partial u}{\partial x} + \frac{\partial v}{\partial y} + \frac{\partial w}{\partial z}\right)\right], \tag{7.12}$$

$$\tau_{xy} = \tau_{yx} = -\mu \left(\frac{\partial u}{\partial y} + \frac{\partial v}{\partial x}\right), \tag{7.13}$$

$$\tau_{yz} = \tau_{zy} = -\mu \left(\frac{\partial v}{\partial z} + \frac{\partial w}{\partial y}\right), \tag{7.14}$$

$$\tau_{zx} = \tau_{xz} = -\mu \left(\frac{\partial w}{\partial x} + \frac{\partial u}{\partial z}\right). \tag{7.15}$$

We introduce now non-dimensional and stretched variables, as we did with the continuity equation. We write explicitly all terms of O(1), except for the y-momentum equation, where we write explicitly also terms of $O(1/Re_{ref})$, because we need them for the discussion in Sub-Section 7.1.7, and bundle together all terms, which are of smaller order of magnitude:

$$\rho' u' \frac{\partial u'}{\partial x'} + \rho' v' \frac{\partial u'}{\partial y'} + \rho' w' \frac{\partial u'}{\partial z'} = -\frac{\partial p'}{\partial x'} + \frac{\partial}{\partial y'}\left(\mu' \frac{\partial u'}{\partial y'}\right) + O\left(\frac{1}{Re_{ref}}\right), \tag{7.16}$$

$$\begin{aligned} \frac{1}{Re_{ref}}\left(\rho' u' \frac{\partial v'}{\partial x'} + \rho' v' \frac{\partial v'}{\partial y'} + \rho' w' \frac{\partial v'}{\partial z'}\right) &= -\frac{\partial p'}{\partial y'} + \frac{1}{Re_{ref}}\left\{\frac{\partial}{\partial x'}\left(\mu' \frac{\partial u'}{\partial y'}\right) + \right. \\ &+ \frac{4}{3}\frac{\partial}{\partial y'}\left(\mu' \frac{\partial v'}{\partial y'}\right) - \frac{2}{3}\frac{\partial}{\partial y'}\left[\mu'\left(\frac{\partial u'}{\partial x'} + \frac{\partial w'}{\partial z'}\right)\right] + \left.\frac{\partial}{\partial z'}\left(\mu' \frac{\partial w'}{\partial y'}\right)\right\} + \\ &+ O\left(\frac{1}{Re^2_{ref}}\right), \end{aligned} \tag{7.17}$$

$$\rho' u' \frac{\partial w'}{\partial x'} + \rho' v' \frac{\partial w'}{\partial y'} + \rho' w' \frac{\partial w'}{\partial z'} = -\frac{\partial p'}{\partial z'} + \frac{\partial}{\partial y'}\left(\mu' \frac{\partial w'}{\partial y'}\right) + O\left(\frac{1}{Re_{ref}}\right). \tag{7.18}$$

[5] This is permitted if it can be assumed that $O(\kappa) \leqq O(\mu)$.

Finally we treat the energy equation, eq. (4.63). We write it in full for perfect gas [6]:

$$c_p\left(\rho u\frac{\partial T}{\partial x}+\rho v\frac{\partial T}{\partial y}+\rho w\frac{\partial T}{\partial z}\right)=\frac{\partial}{\partial x}\left(k\frac{\partial T}{\partial x}\right)+\frac{\partial}{\partial y}\left(k\frac{\partial T}{\partial y}\right)+\frac{\partial}{\partial z}\left(k\frac{\partial T}{\partial z}\right)+$$
$$+u\frac{\partial p}{\partial x}+v\frac{\partial p}{\partial y}+w\frac{\partial p}{\partial z}-\left[\tau_{xx}\frac{\partial u}{\partial x}+\tau_{yy}\frac{\partial v}{\partial y}+\tau_{zz}\frac{\partial w}{\partial z}\right]-$$
$$-\left[\tau_{xy}\left(\frac{\partial u}{\partial y}+\frac{\partial v}{\partial x}\right)+\tau_{xz}\left(\frac{\partial u}{\partial z}+\frac{\partial w}{\partial x}\right)+\tau_{yz}\left(\frac{\partial v}{\partial z}+\frac{\partial w}{\partial y}\right)\right]. \tag{7.19}$$

Again we introduce non-dimensional and stretched variables, as we did above. We also write explicitly all terms of O(1), and bundle together all terms, which are of smaller order of magnitude, now except for two of the heat-conduction terms:

$$c_p'\left(\rho' u'\frac{\partial T'}{\partial x'}+\rho' v'\frac{\partial T'}{\partial y'}+\rho' w'\frac{\partial T'}{\partial z'}\right)=\frac{1}{Pr_{ref}}\left\{\frac{\partial}{\partial y'}\left(k'\frac{\partial T'}{\partial y'}\right)+\right.$$
$$+\frac{1}{Re_{ref}}\left[\frac{\partial}{\partial x'}\left(k'\frac{\partial T'}{\partial x'}\right)+\frac{\partial}{\partial z}'\left(k'\frac{\partial T'}{\partial z'}\right)\right]\Bigg\}+$$
$$+E_{ref}\left\{\left[u'\frac{\partial p'}{\partial x'}+v'\frac{\partial p'}{\partial y'}+w'\frac{\partial p'}{\partial z'}\right]+\mu'\left[\left(\frac{\partial u'}{\partial y'}\right)^2+\left(\frac{\partial w'}{\partial y'}\right)^2\right]\right\}+ \tag{7.20}$$
$$+O\left(\frac{E_{ref}}{Re_{ref}}\right)+O\left(\frac{E_{ref}}{Re^2_{ref}}\right).$$

In this equation Pr_{ref} is the reference Prandtl number:

$$Pr_{ref}=\frac{\mu_{ref}c_{p_{ref}}}{k_{ref}}, \tag{7.21}$$

and E_{ref} the reference Eckert number:

$$E_{ref}=(\gamma_{ref}-1)M^2_{ref}. \tag{7.22}$$

We have retained on purpose in this equation two terms, which are nominally of lower order of magnitude. They are the gradients of the heat-conduction terms in x and z direction, which are of $O(1/Re_{ref})$. The reason is that we consider in general radiation-cooled surfaces, where we have to take into account possible strong gradients of T_w in both x and z direction. They appear there on one hand, because the thermal state of the surface changes strongly in the down-stream direction, usually the main-axis direction of a flight vehicle, Chapter 3. On the other hand, strong changes are present in both x and z direction, if laminar-turbulent transition occurs, see, e. g., Sub-Section 7.3.

The question now is, under what conditions can we drop the two terms, regarding the changes of the wall temperature. To answer it, we follow an argumentation given by Chapman and Rubesin [9]. We consider first (in two dimensions) the gradient of the heat-conduction term in direction normal to the surface in dimensional and non-stretched form, eq. (7.19), and introduce finite differences, as we did in Sub-Section 3.2.1:

$$\frac{\partial}{\partial y}\left(k\frac{\partial T}{\partial y}\right) \sim k\frac{T_w - T_\infty}{\delta_T^2}, \tag{7.23}$$

with δ_T being the thickness of the thermal boundary layer.

The gradient of the heat-conduction term in x-direction is written, assuming that $(\partial T/\partial x)_w$ is representative for it:

$$\frac{\partial}{\partial x}\left(k\frac{\partial T}{\partial x}\right) \sim \frac{k}{L}\frac{\partial T}{\partial x}|_w. \tag{7.24}$$

The ratio of the two terms is:

$$\frac{\partial}{\partial x}(k\frac{\partial T}{\partial x})/\frac{\partial}{\partial y}(k\frac{\partial T}{\partial y}) \sim \frac{\delta_T}{L}\,\frac{(\partial T/\partial x)_w}{(T_w - T_\infty)/\delta_T}. \tag{7.25}$$

If the wall-temperature gradient in x-direction is of the same order or smaller than the temperature gradient in direction normal to the surface:

$$\frac{\partial T}{\partial x}|_w \lesseqq \frac{T_w - T_\infty}{\delta_T}, \tag{7.26}$$

we find:

$$\frac{\partial}{\partial x}(k\frac{\partial T}{\partial x})/\frac{\partial}{\partial y}(k\frac{\partial T}{\partial y}) \lesssim \frac{\delta_T}{L}. \tag{7.27}$$

The result is: provided, that eq. (7.26) holds, the gradient term of heat conduction in x-direction can be neglected, because $\delta_T/L \sim 1/(\sqrt{Pr}\sqrt{Re}) \ll 1$ in general:

$$\frac{\partial}{\partial x}(k\frac{\partial T}{\partial x}) \ll \frac{\partial}{\partial y}(k\frac{\partial T}{\partial y}). \tag{7.28}$$

In [9] it is assumed, that the recovery temperature is representative for the wall temperature:

$$T_w = T_r = T_\infty(1 + r\frac{\gamma - 1}{2}M_\infty^2). \tag{7.29}$$

Introducing this into eq. (7.26), together with $\delta_T \approx \delta \approx cL/\sqrt{Re_{ref}}$ for laminar flow, we obtain

$$\frac{\partial T'}{\partial x'}|_w \leqq r\frac{\gamma - 1}{2}M_\infty^2\frac{\sqrt{Re_{ref}}}{c}. \qquad (7.30)$$

With $r = \sqrt{Pr} = \sqrt{0.72}$, $\gamma = 1.4$, $c = 6$ we arrive finally, after having introduced non-dimensional variables, at the Chapman-Rubesin criterion [9]. It says that the term eq. (7.24) can be neglected, if

$$\frac{\partial T'}{\partial x'}|_w \leqq 0.03M_\infty^2\sqrt{Re_{ref}}. \qquad (7.31)$$

This means, that, for instance, for $M_\infty = 1$ and $Re_{ref} = 10^6$, the maximum permissible temperature gradient would be equivalent to a thirty-fold increase of T/T_∞ along a surface of length L_{ref} [9]. From this it can be concluded, that in general for high Mach-number and Reynolds-number flows the Chapman-Rubesin criterion is fulfilled, as long as the surface-temperature distribution is "reasonably smooth and continuous" [9]. The situation can be different for low Mach numbers and Reynolds numbers.

With radiation-cooled surfaces, as we noted above, we do not necessarily have reasonably smooth and continuous surface-temperature distributions in both x and z direction. Moreover, the basic relation eq. (7.23) needs to be adapted, because it does not describe the situation at a radiation-cooled surface, [9]. For that situation we introduce a slightly different formulation for both directions:

$$\left|\frac{\partial T}{\partial x}\right|_w \leqq \left|\frac{T_w - T_r}{\delta_T}\right|, \qquad (7.32)$$

$$\left|\frac{\partial T}{\partial z}\right|_w \leqq \left|\frac{T_w - T_r}{\delta_T}\right|, \qquad (7.33)$$

because, at least for laminar flow, T_r - T_w is the characteristic temperature difference, see Sub-Section 3.2.1. We also introduce the absolute values $|\partial T/\partial x|_w$ and $|\partial T/\partial z|_w$, because the gradients will be negative downstream of the forward stagnation point, Sub-Section 3.2.1, but may be positive or negative in laminar-turbulent transition regimes, Sub-Section 7.3, and in hot-spot and cold-spot situations, Sub-Section 3.2.3.

The modified Chapman-Rubesin criterion is then: if eqs. (7.32) and (7.33) hold, the gradient term of heat conduction in both x and z-direction can be neglected, because again $\delta_T/L \sim 1/(\sqrt{Pr}\sqrt{Re})$:

$$\frac{1}{Re_{ref}}\frac{\partial}{\partial x'}(k'\frac{\partial T'}{\partial x'}) \leqq \frac{\partial}{\partial y'}(k'\frac{\partial T'}{\partial y'}), \qquad (7.34)$$

$$\frac{1}{Re_{ref}}\frac{\partial '}{\partial z}(k'\frac{\partial T'}{\partial z'}) \leqq \frac{\partial}{\partial y'}(k'\frac{\partial T'}{\partial y'}). \qquad (7.35)$$

We refrain to propose detailed criteria, like the original Chapman-Rubesin criterion, eq. (7.31). This could be done for the region downstream of the forward stagnation point, but not in the other regimes. In practice the results of an exploration solution should show, if and where the modified Chapman-Rubesin criterion is violated or not[6].

Provided that the modified Chapman-Rubesin criterion is fulfilled, we arrive at the classical boundary-layer equations by neglecting all terms of $O(1/Re_{ref})$ and $O(1/Re^2_{ref})$ in eqs. (7.6), (7.16) to (7.18), and (7.20). We write the variables without prime, understanding that the equations can be read in either way, non-dimensional, stretched or non-stretched, and dimensional and non-stretched, then without the similarity parameters Pr_{ref} and E_{ref}:

$$\frac{\partial \rho u}{\partial x} + \frac{\partial \rho v}{\partial y} + \frac{\partial \rho w}{\partial z} = 0, \tag{7.36}$$

$$\rho u \frac{\partial u}{\partial x} + \rho v \frac{\partial u}{\partial y} + \rho w \frac{\partial u}{\partial z} = -\frac{\partial p}{\partial x} + \frac{\partial}{\partial y}\left(\mu \frac{\partial u}{\partial y}\right), \tag{7.37}$$

$$0 = -\frac{\partial p}{\partial y}, \tag{7.38}$$

$$\rho u \frac{\partial w}{\partial x} + \rho v \frac{\partial w}{\partial y} + \rho w \frac{\partial w}{\partial z} = -\frac{\partial p}{\partial z} + \frac{\partial}{\partial y}\left(\mu \frac{\partial w}{\partial y}\right), \tag{7.39}$$

$$\begin{aligned} c_p \left(\rho u \frac{\partial T}{\partial x} + \rho v \frac{\partial T}{\partial y} + \rho w \frac{\partial T}{\partial z}\right) &= \frac{1}{Pr_{ref}} \frac{\partial}{\partial y}\left(k \frac{\partial T}{\partial y}\right) + \\ + E_{ref} \left\{ u \frac{\partial p}{\partial x} + w \frac{\partial p}{\partial z} + \mu \left[\left(\frac{\partial u}{\partial y}\right)^2 + \left(\frac{\partial w}{\partial y}\right)^2 \right] \right\}&. \end{aligned} \tag{7.40}$$

These equations are the ordinary boundary-layer equations which describe attached viscous flow fields on hypersonic flight vehicles. If thick boundary layers are present and/or entropy-layer swallowing occurs, they must be employed in second-order formulation, see below. For very large reference Mach numbers[7] M_{ref} the equations become fundamentally changed, see Sub-Section 7.1.7.

With the above equations we can determine the unknowns u, v, w, and T. The unknowns density ρ, viscosity μ, thermal conductivity k, and specific heat

[6] The Chapman-Rubesin criterion in any form is of importance only for boundary-layer methods, viscous shock-layer methods, and for thin-layer formulations of the Navier-Stokes equations, Section 10.2.

[7] Note that locally the boundary-layer edge Mach number M_e is relevant, see Section 4.4. This means that the above equations can be employed, if necessary in second-order formulation, on both CAV-type and RV-type flight vehicles, the latter having at the large angles of attack only small boundary-layer edge Mach numbers, Section 1.2.

at constant pressure c_p are to be found with the equation of state $p = \rho RT$, and the respective relations given in Chapters 4 and 5. If the boundary-layer flow is turbulent, effective transport properties must be introduced, Section 8.3.

Since $\partial p/\partial y$ is zero, the pressure field of the external inviscid flow field, represented by $\partial p/\partial x$ and $\partial p/\partial z$, is imposed on the boundary layer. This means, that in the boundary layer $\partial p/\partial x$ and $\partial p/\partial z$ are constant in y-direction. This holds for first-order boundary layers. If second-order effects are present, $\partial p/\partial x$ and $\partial p/\partial z$ in the boundary layer are implicitly corrected via $\partial p/\partial y \neq 0$ by centrifugal terms, see below.

The equations are first-order boundary-layer equations, based on Cartesian coordinates. In general locally monoclinic surface-oriented coordinates, [4], factors and additional terms are added, which bring in the metric properties of the coordinate system [10], [1]. It should be noted, that the equations for the general coordinates are formulated such that also the velocity components are transformed. This is in contrast to modern Euler and Navier-Stokes/RANS methods formulated for general coordinates. There only the geometry is transformed, Section 12, and not the velocity components.

If locally the boundary-layer thickness is not small compared to the smallest radius of curvature of the surface, the pressure gradient in the boundary layer in direction normal to the surface, $\partial p/\partial y$, is no longer small of higher order, and hence no longer can be neglected[8]. This is a situation found typically in hypersonic flows, where also entropy-layer swallowing can occur, Sub-Section 6.4.2. This situation is taken into account by second-order boundary-layer equations, which basically have the same form as the first-order equations [4], [12]. Information about the curvature properties of the surface is added. The y-momentum equation does not degenerate into $\partial p/\partial y = 0$. $\partial p/\partial y$ is finite because centrifugal forces have to be taken into account. At the outer edge of the boundary layer the boundary conditions are determined by values from within the inviscid flow field, not from the surface as in first-order theory, see the discussion in Sub-Section 6.4.2. Also the first derivatives of the tangential velocity components, of temperature, density and pressure are continuous [12], which is not the case in first-order theory, see, e. g., Fig. 6.21.

7.1.4 Global Characteristic Properties of Attached Viscous Flow

In order to identify the global characteristic properties of attached viscous flow, we need to have a look at the characteristic properties of the boundary-layer equations. We do this for convenience by assuming incompressible flow. Following [13], we introduce characteristic manifolds $\varphi(x, y, z)$, for instance like

[8] There are other higher-order effects, which we do not mention here, see, e. g., [11].

$$\frac{\partial}{\partial x} = \frac{\partial \varphi}{\partial x}\frac{d}{d\varphi} = \varphi_x \frac{d}{d\varphi} \tag{7.41}$$

into eqs. (7.36) to (7.39).

After introduction of the kinematic viscosity $\nu = \mu/\rho$ and some manipulation the characteristic form is found:

$$C = \begin{vmatrix} \varphi_x & \varphi_y & \varphi_z \\ (\Delta - \nu\varphi_y^2) & 0 & 0 \\ 0 & 0 & (\Delta - \nu\varphi_y^2) \end{vmatrix} = \varphi_y(\Delta - \nu\varphi_y^2)^2 = 0, \tag{7.42}$$

with the abbreviation

$$\Delta = u\varphi_x + v\varphi_y + w\varphi_z. \tag{7.43}$$

The pressure gradients $\partial p/\partial x$ and $\partial p/\partial z$ do not enter the characteristic form, because the pressure field is imposed on the boundary layer, i. e. $\partial p/\partial x$ and $\partial p/\partial z$ are forcing functions.

Eq. (7.43) corresponds to the projection of the gradient of the manifold onto the streamline, and represents the boundary-layer streamlines as characteristic manifolds. To prove this, we write the total differential of φ:

$$d\varphi = \varphi_x dx + \varphi_y dy + \varphi_z dz = 0, \tag{7.44}$$

and combine it with the definition of streamlines in three dimensions

$$\frac{dx}{u} = \frac{dy}{v} = \frac{dz}{w}, \tag{7.45}$$

in order to find

$$d\varphi = u\varphi_x + v\varphi_y + w\varphi_z = \Delta = 0. \tag{7.46}$$

Thus it is shown that streamlines are characteristics, too.

The remaining five-fold characteristics φ_y in y-direction in eq. (7.42) are typical for boundary-layer equations. These characteristics are complemented by two-fold characteristics in y-direction coming from the energy equation, eq. (7.40), which we do not demonstrate here. These results are valid for compressible flow, too, and also for second-order boundary-layer equations.

Boundary-layer equations of first or second order, in two or three dimensions, are parabolic, and hence pose a mixed initial condition/boundary condition problem[9]. Where the boundary-layer flow enters the domain under consideration, initial conditions must be prescribed. At the surface of the body, $y = 0$, and at the outer edge of the boundary layer, $y = \delta$, boundary conditions are to be described for u, w, and T, hence the six-fold characteristics in y-direction. For the normal velocity component v only a boundary

[9] The surface boundary conditions for general hypersonic flow have been discussed in detail in Section 4.3.

condition at the body surface, $y = 0$, must be described, which reflects the seventh characteristic.

We have introduced the boundary layer as phenomenological model of attached viscous flow. This model is valid everywhere on the surface of a flight vehicle, where strong interaction phenomena are not present like separation, shock/boundary-layer interaction, hypersonic viscous interaction, et cetera.

In Sub-Section 7.1.1 we have noted the three viscous-flow phenomena, which are directly of interest in vehicle design. If attached hypersonic viscous flow is boundary-layer like, we can now, based on the above analysis, give a summary of its global characteristic properties:

- Attached viscous flow is governed primarily by the external inviscid flow field via its pressure field, and by the surface conditions.
- It has parabolic character, i. e. the boundary conditions in general dominate its properties (seven-fold characteristics in direction normal to the surface), the influence of the initial conditions usually is weak.
- Events in attached viscous flow are felt only downstream, as long as they do not invalidate the boundary-layer assumption. This means, for instance, that a surface disturbance or surface suction or blowing can have a magnitude at most of $O(1/\sqrt{Re_{ref}})$. Otherwise the attached viscous flow loses its boundary-layer properties (strong interaction).
- In attached viscous flow an event is felt upstream only if it influences the pressure field via, e. g., a disturbance of $O(1)$ or if strong temperature gradients in main-flow direction are present ($\partial(k\partial T/\partial x)/\partial x$ and $\partial(k\partial T/\partial z)/\partial z = O(Re_{ref})$. The displacement properties of an attached boundary layer are of $O(1/\sqrt{Re_{ref}})$, and hence influence the pressure field only weakly (weak interaction).
- Separation causes locally strong interaction and may change the onset boundary-layer flow, however only via a global change of the pressure field. Strong interaction phenomena have only small upstream influence, i. e. their influence is felt predominantly downstream (locality principle [3]) and via the global change of the pressure field.
- In two-dimensional attached viscous flow the domain of influence of an event is defined by the convective transport along the - straight - streamlines in downstream direction. Due to lateral molecular or turbulent transport it assumes a wedge-like pattern with small spreading angle.
- In three-dimensional flow, the influence of an event is spread over a domain, which is defined by the strength of the skewing of the stream surface, Fig. 7.3. The skin-friction line alone is not representative. Of course also here lateral molecular or turbulent transport happens. If a boundary-layer method is used for the determination of the flow field, it must take into account the domain of dependence of a point $P(x, y)$ on the body surface, which must be enclosed by the numerical domain of dependence, Fig. 7.3 b).

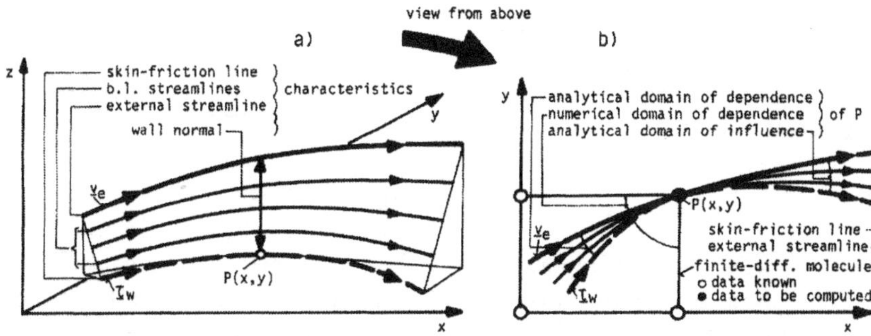

Fig. 7.3. Three-dimensional boundary layer with skewed stream surface (schematically, note that in this figure, [1], x, y are tangential to the surface, and z is normal to it): **a)** the streamlines as characteristics, **b)** domains of dependence and of influence of flow properties in P(x, y).

7.1.5 Wall Compatibility Conditions

In the continuum flow regime the no-slip condition at the surface of a body (tangential velocity components $u_w = w_w = 0$) is the cause for the development of the boundary layer. We assume for the following consideration preliminarily $u_w = w_w = 0$, and also that the normal velocity component at the body surface is zero: $v_w = 0$, although $|v_w/v_{ref}| \leqq O(1/Re_{ref})$ would be permitted. We formulate:

$$u|_{y=0} = 0, \; w|_{y=0} = 0, \; v|_{y=0} = 0. \tag{7.47}$$

In addition, we can make statements about derivatives of u, w and v at the surface. The classical wall compatibility conditions for three-dimensional flow follow from eqs. (7.37) and (7.39). They connect the second derivatives of the tangential flow components u and w at the surface with the respective pressure gradients:

$$\frac{\partial}{\partial y}(\mu \frac{\partial u}{\partial y})|_{y=0} = \frac{\partial p}{\partial x}, \tag{7.48}$$

$$\frac{\partial}{\partial y}(\mu \frac{\partial w}{\partial y})|_{y=0} = \frac{\partial p}{\partial z}. \tag{7.49}$$

The first derivatives of u and w at the surface in attached viscous flow are by definition finite. In external streamline coordinates, Figs. 7.1 a) and 7.2, we obtain for the main-flow direction:

$$\frac{\partial u}{\partial y}|_{y=0} > 0, \tag{7.50}$$

and for the cross-flow direction:

$$\frac{\partial w}{\partial y}|_{y=0} \lesseqgtr 0. \tag{7.51}$$

The first derivative at the surface of the normal velocity component v in y-direction is found from the continuity equation, eq. (7.36):

$$\frac{\partial v}{\partial y}|_{y=0} = 0. \tag{7.52}$$

We generalize now for hypersonic attached flow the classical wall-compatibility conditions by taking into account (also in external streamline coordinates) possible slip flow ($u_w \geqq 0$, $w_w \lesseqgtr 0$), possible suction or blowing ($v_w \lesseqgtr 0$), and temperature gradients in the gas at the wall ($T_{gw} \lesseqgtr 0$), assuming that all obey the boundary-layer assumptions. We find, again from eqs. (7.37) and (7.39):

$$\frac{\partial^2 u}{\partial y^2}|_{y=0} = \left\{\frac{1}{\mu}\left[\rho u\frac{\partial u}{\partial x} + \rho v\frac{\partial u}{\partial y} + \rho w\frac{\partial u}{\partial z} + \frac{\partial p}{\partial x} - \frac{\partial \mu}{\partial T}\frac{\partial T}{\partial y}\frac{\partial u}{\partial y}\right]\right\}_{y=0}, \tag{7.53}$$

$$\frac{\partial^2 w}{\partial y^2}|_{y=0} = \left\{\frac{1}{\mu}\left[\rho u\frac{\partial w}{\partial x} + \rho v\frac{\partial w}{\partial y} + \rho w\frac{\partial w}{\partial z} + \frac{\partial p}{\partial z} - \frac{\partial \mu}{\partial T}\frac{\partial T}{\partial y}\frac{\partial w}{\partial y}\right]\right\}_{y=0}. \tag{7.54}$$

The functions of the tangential velocity components $u(y)$ and $w(y)$ and their derivatives at the outer edge of the boundary layer in first-order boundary-layer theory are found from the asymptotic condition that the boundary-layer equations approach there the (two-dimensional) Euler equations. From eqs. (7.37) and (7.39) we get:

$$u|_{y=\delta} = u_e, \quad \frac{\partial u}{\partial y}|_{y=\delta} = 0, \quad \frac{\partial^2 u}{\partial y^2}|_{y=\delta} = 0, \tag{7.55}$$

$$w|_{y=\delta} = w_e, \quad \frac{\partial w}{\partial y}|_{y=\delta} = 0, \quad \frac{\partial^2 w}{\partial y^2}|_{y=\delta} = 0. \tag{7.56}$$

The normal velocity component $v(y)$ is not defined at the outer edge of the boundary layer, nor its second derivative. From eq. (7.36) we find only the compatibility condition:

$$\frac{\partial \rho v}{\partial y}|_{y=\delta} = -\left(\frac{\partial \rho u}{\partial x} + \frac{\partial \rho w}{\partial z}\right)_{y=\delta} = -\left(\frac{\partial \rho_e u_e}{\partial x} + \frac{\partial \rho_e w_e}{\partial z}\right). \tag{7.57}$$

The compatibility conditions permit to make assertions about the shape of boundary-layer velocity profiles. We demonstrate this with the profile of the tangential velocity component of two-dimensional boundary layers, Fig. 7.1 b), which also holds for the main-flow profile of three-dimensional boundary layers, Fig. 7.1 a). We will come back to the result in Sub-Section 8.1.3.

We consider three possible values of $\partial^2 u/\partial y^2|_w$: < 0 (case 1), $= 0$ (case 2), > 0 (case 3), Fig. 7.4. We see that the second derivative (curvature) is negative above the broken line for all profiles given in Fig. 7.4 a). Hence the second derivative will approach in any case $\partial^2 u/\partial y^2|_{y=\delta} = 0$ with a negative value, Fig. 7.4 a). It can be shown by further differentiation of the x-momentum equation, eq. (7.37), that for incompressible no-slip flow also:

$$\frac{\partial^3 u}{\partial y^3}|_{y=0} = 0. \tag{7.58}$$

With these elements the function $\partial^2 u(y)/\partial y^2$ can be sketched qualitatively, Fig. 7.4 a). Because we consider attached viscous flow, $\partial u/\partial y|_w > 0$ holds in all three cases, Fig. 7.4 b). We obtain finally the result that boundary-layer flow in cases 1 and 2 has profiles $u(y)$ without point of inflexion, and in case 3 has a profile $u(y)$ with point of inflexion.

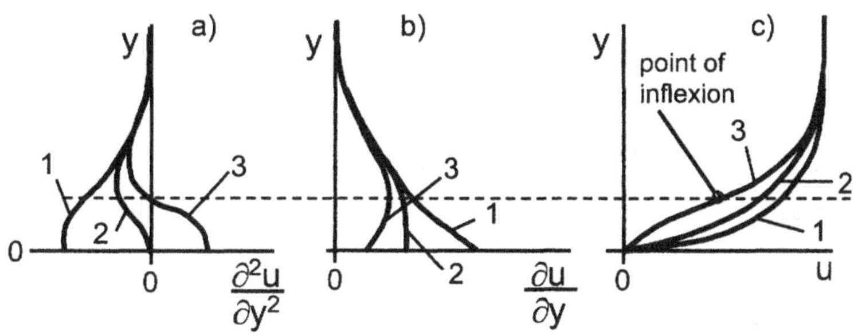

Fig. 7.4. Shape (qualitatively) of **a)** second derivative, **b)** first derivative, and **c)** function of the tangential velocity component $u(y)$ of a two-dimensional boundary layer, or the main-flow profile of a three-dimensional boundary layer. Case 1: $\partial^2 u/\partial y^2|_w < 0$, case 2: $\partial^2 u/\partial y^2|_w = 0$, case 3: $\partial^2 u/\partial y^2|_w > 0$.

The interpretation of this result, Table 7.1, is found through a term by term examination of eq. (7.53). It was assumed, that $\partial u/\partial y|_w$ is always positive, because we consider attached flow only. Also viscosity μ and density ρ are positive. Since we deal with gas or a mixture of gases, the derivative of the viscosity with respect to the temperature is always positive: $\partial\mu/\partial T > 0$ (in liquids, especially in water, it is negative). The derivative $\partial u/\partial x|_w$ in the first term in the bracket of eq. (7.53) in the case of slip flow is assumed to be always negative, i. e. wall-slip is reducing in downstream direction. For convenience the term $\rho w(\partial u/\partial z)|_w$, which can be relevant if slip flow is present, was not taken into account.

We see in Table 7.1 that the single terms in eq. (7.53) may or may not cause a point of inflexion of the tangential velocity profile $u(y)$. In any case an adverse pressure gradient causes it[10], and also heating of the boundary layer, i. e., transfer of heat from the body surface into the flow, or blowing.

In a real flow situation several of the flow features considered in Table 7.1 may be present. Accordingly the sum of the terms in the bracket of eq. (7.53) is the determining factor. The individual terms may weaken or cancel their combined influence, or may enhance it. The factor $1/\mu$ in front of the square bracket is a modifier, which reduces $\partial^2 u/\partial y^2|_w$, if the surface is hot, and enlarges it, if the surface is cold.

Table 7.1. Influence of terms in eq. (7.53) on $\partial^2 u/\partial y^2|_w$.

Term	Flow feature	$\partial^2 u/\partial y^2\|_w$	Point of inflexion
$u_w > 0,\ \partial u/\partial x\|_w < 0$	slip flow	< 0	no
$u_w = 0$	no slip condition	0	no
$v_w > 0$	blowing	> 0	yes
$v_w = 0$	non-permeable surface	0	no
$v_w < 0$	suction	< 0	no
$\partial p/\partial x > 0$	decelerated flow	> 0	yes
$\partial p/\partial x = 0$	Blasius flow, incompressible	0	no
$\partial p/\partial x < 0$	accelerated flow	< 0	no
$\partial T/\partial y\|_{gw} > 0$	cooling of boundary layer	< 0	no
$\partial T/\partial y\|_{gw} = 0$	adiabatic wall	0	no
$\partial T/\partial y\|_{gw} < 0$	heating of boundary layer	> 0	yes

We have mentioned in Sub-Section 7.1.2 that only in a few, singular points streamlines actually impinge on or leave the body surface. This implies that in attached viscous flow close to the surface the flow is parallel to the body surface. In closing this sub-section we use the wall compatibility conditions to determine the flow angle θ, Fig. 4.1, in the limit $y \to 0$. We do this once more only for the profile of the tangential velocity component of two-dimensional boundary layers, Fig. 7.1 b), which also holds for the main-flow profile of three-dimensional boundary layers, Fig. 7.1 a).

With the no-slip condition eq. (7.47), the assumption of attached viscous flow with $\partial u/\partial y|_w > 0$, and condition eq. (7.52), we find by means of a Taylor expansion around a point on the surface for small distances y from the surface:

$$u \sim y, \ v \sim y^2, \tag{7.59}$$

and hence

[10] It is the classical interpretation that an adverse pressure gradient leads to a profile $u(y)$ with point of inflexion, but zero and favorable pressure gradient not. With our generalization we see that also other factors can lead to a point of inflexion of the profile $u(y)$.

$$\tan\theta = \frac{v}{u} \sim y. \tag{7.60}$$

The result is that when the surface is approached in attached viscous flow, the flow in the limit becomes parallel to it:

$$y \rightarrow 0 : \theta \rightarrow 0. \tag{7.61}$$

7.1.6 The Reference Temperature/Enthalpy Method for Compressible Boundary Layers

In [14] it is shown that for laminar high-speed flows boundary-layer skin friction and wall heat transfer can be obtained with good accuracy by employing the relations for incompressible flow, if the fluid property density ρ and the transport property viscosity μ are determined at a suitable reference temperature T^*. The approach is based on the observation that the results of the investigated boundary-layer methods depend strongly on the exponent ω of the employed viscosity relation, see Section 4.2, on the wall temperature T_w, and on the boundary-layer edge Mach number M_e, which is representative of the ratio 'boundary-layer edge temperature' to 'total temperature of the flow', T_e/T_t. The dependence on the Prandtl number Pr is weak.

This is an interesting observation, because the complex interaction of convective and molecular heat transfer, compression and dissipation work, see eq. (4.58) or (7.40), with the boundary conditions at the body surface and at the boundary-layer's outer edge do not suggest it at first glance. We will see later, that theory based on the Lees-Dorodnitsyn transformation to a certain degree supports this observation.

The reference temperature concept was extended to the reference enthalpy concept [15] in order to take high-temperature real-gas effects into account. In [15] it is shown that it can be applied with good results to turbulent boundary layers, too.

Reference temperature (T^*) for perfect gas and reference enthalpy (h^*) for high-temperature real gas, respectively, combine the actual values of the gas at the wall (w), at the boundary layer's outer edge (e), and the recovery value (r) in the following way:

$$T^* = 0.28T_e + 0.5T_w + 0.22T_r, \tag{7.62}$$

$$h^* = 0.28h_e + 0.5h_w + 0.22h_r. \tag{7.63}$$

Eq. (7.63) contains eq. (7.62) for perfect gas.

Because T^*, respectively h^*, depend on boundary-layer edge data and especially on wall data, we note:

$$\textit{on general vehicle surfaces}: \; T^* = T^*(x,z), \; \textit{resp.}\; h^* = h^*(x,z),$$

i. e., reference temperature or enthalpy are not constant on a vehicle surface, especially if this surface is radiation cooled.

The recovery values are found in terms of the boundary-layer edge data:

$$T_r = T_e + r^* \frac{V_e^2}{2c_p}, \tag{7.64}$$

$$h_r = h_e + r_h^* \frac{V_e^2}{2}. \tag{7.65}$$

Eq. (7.64) can also be written as

$$T_r = T_e(1 + r^* \frac{\gamma_e - 1}{2} M_e^2). \tag{7.66}$$

The recovery factor in these relations is

$$r^* = r_h^* = \sqrt{Pr^*}, \tag{7.67}$$

for laminar flow, and with acceptable accuracy

$$r^* = r_h^* = \sqrt[3]{Pr^*}, \tag{7.68}$$

for turbulent flow, with Pr^* being the Prandtl number at reference-temperature conditions:

$$Pr^* = \frac{\mu^* c_p^*}{k^*}. \tag{7.69}$$

With these definitions the reference temperature, eq. (7.62), becomes in terms of the boundary-layer edge Mach number M_e:

$$T^* = 0.5T_e + 0.5T_w + 0.22r^* \frac{\gamma_e - 1}{2} M_e^2 T_e. \tag{7.70}$$

Eqs. (7.62) and (7.63) were found with the help of comparisons of results from solutions of the boundary-layer equations with data from experiments [14], [15]. They are valid for air, and for both laminar and turbulent boundary layers. Other combinations have been proposed, for instance, for boundary layers at swept leading edges, Sub-Section 7.2.4.

We use in the following sub-sections the reference enthalpy or temperature method, although it is an approximate method, to demonstrate thermal surface effects on attached viscous flow. We use '*' to mark reference-temperature data and relate them to overall reference flow parameters, which we mark with '$_{ref}$'. For the Reynolds number Re_x^* we thus find:

$$Re_x^* = \frac{\rho^* v_{ref} x}{\mu^*} = \frac{\rho_{ref} v_{ref} x}{\mu_{ref}} \frac{\rho^*}{\rho_{ref}} \frac{\mu_{ref}}{\mu^*} = Re_{ref,x} \frac{\rho^*}{\rho_{ref}} \frac{\mu_{ref}}{\mu^*}. \tag{7.71}$$

In attached viscous flow the pressure to first order is constant through the boundary layer in direction normal to the surface, and hence we have locally in the boundary layer:

$$\rho^* T^* = \rho_{ref} T_{ref}. \tag{7.72}$$

We introduce this into eq. (7.71) together with the approximate relation $\mu = cT^{\omega}$ for the viscosity, Sub-Section 4.2, and obtain finally:

$$Re_x^* = Re_{ref,x} \left(\frac{T_{ref}}{T^*} \right)^{1+\omega} . \tag{7.73}$$

For flat plates at zero angle of attack, and approximately for slender bodies at small angle of attack, except for the blunt nose region, we can choose '$_{ref}$' = '$_{\infty}$', whereas in general the conditions at the outer edge of the boundary layer are the reference conditions: '$_{ref}$' = '$_{e}$'.

The reference-temperature/enthalpy extension of incompressible boundary-layer relations is not only a simple and effective method to demonstrate thermal-surface effects on attached viscous flow. It is also an effective tool for the actual determination, with sufficient accuracy, of properties of attached compressible laminar or turbulent viscous flows, even for flows with appreciable high-temperature thermo-chemical effects, see, e. g., [16]. In Sub-Section 7.2.1 we give such extended relations for the determination of boundary-layer thicknesses and integral parameters and in Sub-Sections 7.2.3 to 7.2.6 for the determination of skin friction and thermal state of the surface.

The extended relations can be applied on generic surfaces with either inviscid flow data found from impact methods or in combination with inviscid flow field data found by means of Euler solutions. Of course only weak three-dimensionality of the flow is permitted. The stream-wise pressure gradient in principle must be weak, but examples in [16] show that flows with considerable pressure gradients can be treated with good results. Flow separation and re-attachment, see Section 9.1, strong interaction phenomena and hypersonic viscous interaction, see Sections 9.2 and 9.3, must be absent, also slip-flow, Section 9.4. To describe such phenomena Navier-Stokes methods must be employed. Slip flow, however, can also be treated in the frame of boundary-layer theory.

7.1.7 Equations of Motion for Hypersonic Attached Viscous Flow

We study now the equations of motion for hypersonic attached viscous flow. We find these equations in the same way in which we found the boundary-layer equations in Sub-Section 7.1.3, but now making the pressure dimensionless with p_{ref} instead of $\rho_{ref} v_{ref}^2$. We assume perfect gas and keep the bulk viscosity κ because the terms containing it will not drop out, unless we go to the boundary-layer limit. For convenience we consider only the two-dimensional case and keep x as coordinate tangential to the surface and y

normal to it. The equations are written dimensionless, but we leave the prime away. The extension to three dimensions is straight forward.

In analogy to eqs. 7.5, 7.16, 7.17, and 7.20 we find then first the (unchanged) continuity equation:

$$\frac{\partial \rho u}{\partial x} + \frac{\partial \rho v}{\partial y} = 0, \tag{7.74}$$

then the Navier-Stokes equations:

$$\rho u \frac{\partial u}{\partial x} + \rho v \frac{\partial u}{\partial y} = -\frac{1}{\gamma_{ref} M_{ref}^2} \frac{\partial p}{\partial x} + \frac{\partial}{\partial y}\left(\mu \frac{\partial u}{\partial y}\right) + O\left(\frac{1}{Re_{ref}}\right), \tag{7.75}$$

$$\frac{\gamma_{ref} M_{ref}^2}{Re_{ref}}\left(\rho u \frac{\partial v}{\partial x} + \rho v \frac{\partial v}{\partial y}\right) = -\frac{\partial p}{\partial y} + \frac{\gamma_{ref} M_{ref}^2}{Re_{ref}}\left\{\frac{\partial}{\partial y}\left[(\frac{4}{3}\mu + \kappa)\frac{\partial v}{\partial y}\right] - \right.$$
$$\left. - \frac{\partial}{\partial y}\left[\frac{2}{3}(\mu - \kappa)\frac{\partial u}{\partial x}\right] + \frac{\partial}{\partial x}\left(\mu \frac{\partial u}{\partial y}\right)\right\} + O\left(\frac{\gamma_{ref} M_{ref}^2}{Re_{ref}^2}\right), \tag{7.76}$$

and finally the energy equation, where on purpose the Eckert number is not employed:

$$c_p\left(\rho u \frac{\partial T}{\partial x} + \rho v \frac{\partial T}{\partial y} + \right) = \frac{1}{Pr_{ref}} \frac{\partial}{\partial y}\left(k \frac{\partial T}{\partial y}\right) + \frac{\gamma_{ref} - 1}{\gamma_{ref}}\left(u \frac{\partial p}{\partial x} + v \frac{\partial p}{\partial y}\right) +$$
$$+ (\gamma_{ref} - 1) M_{ref}^2 \mu \left(\frac{\partial u}{\partial y}\right)^2 + (\gamma_{ref} - 1)\frac{M_{ref}^2}{Re_{ref}}\left\{(\frac{4}{3}\mu + \kappa)\left[\left(\frac{\partial u}{\partial x}\right)^2 + \right.\right.$$
$$\left.\left. + \left(\frac{\partial v}{\partial y}\right)^2\right] - 2(\frac{2}{3}\mu - \kappa)\frac{\partial u}{\partial x}\frac{\partial v}{\partial y} + 2\mu \frac{\partial u}{\partial y}\frac{\partial v}{\partial x}\right\} + O\left(\frac{1}{Re_{ref}}\right) +$$
$$+ O\left((\gamma_{ref} - 1)\frac{M_{ref}^2}{Re_{ref}^2}\right). \tag{7.77}$$

It is obvious that these equations reduce to the boundary-layer equations, see (7.36) to (7.40), albeit with some different co-factors, if the Reynolds number Re_{ref} is very large and the ratio 'square of the Mach number, M_{ref}^2' to 'Re_{ref}' very small:

$$\frac{M_{ref}^2}{Re_{ref}} \ll 1. \tag{7.78}$$

This means that if eq. (7.78) is true, the flow is of boundary-layer type, because the pressure is constant in the direction normal to the surface.

If on the other hand

$$\frac{M_{ref}^2}{Re_{ref}} \approx O(1), \tag{7.79}$$

we must expect a pressure not constant in direction normal to the surface. If this happens on a flat vehicle configuration element, in the limiting case on a flat plate, we speak of hypersonic viscous interaction, which we treat in Section 9.3, where we will meet again the term M_{ref}^2/Re_{ref}, respectively its square root $M_{ref}/\sqrt{Re_{ref}}$.

We study now in more detail the characteristic properties of eqs. (7.74) to (7.77) in order to get clues regarding the kind of possible numerical computation schemes for such flows.

We introduce again characteristic manifolds φ [13] for derivatives like in eq. (7.41). To make the problem treatable, we simplify the governing equations by keeping only the leading terms in the equations, omitting the cofactors containing M_{ref}, Re_{ref}, γ, and Pr_{ref}, and by assuming constant transport properties μ, k and heat capacity c_p:

$$\rho\frac{\partial u}{\partial x} + u\frac{\partial \rho}{\partial x} + \rho\frac{\partial v}{\partial y} + v\frac{\partial \rho}{\partial y} = 0, \tag{7.80}$$

$$\rho u\frac{\partial u}{\partial x} + \rho v\frac{\partial u}{\partial y} = -\frac{\partial p}{\partial \rho}\frac{\partial \rho}{\partial x} + \mu\frac{\partial^2 u}{\partial y^2}, \tag{7.81}$$

$$\rho u\frac{\partial v}{\partial x} + \rho v\frac{\partial v}{\partial y} = -\frac{\partial p}{\partial \rho}\frac{\partial \rho}{\partial y} + \mu\frac{\partial^2 v}{\partial y^2}, \tag{7.82}$$

$$c_p\left(\rho u\frac{\partial T}{\partial x} + \rho v\frac{\partial T}{\partial y}\right) = k\frac{\partial^2 T}{\partial y^2} + CWT + DWT, \tag{7.83}$$

with CWT and DWT being abbreviations of the compression and dissipation work terms, respectively.

The characteristic matrix then reads, with φ_x and φ_y the characteristic directions of the problem [13]:

$$C = \begin{vmatrix} \rho\varphi_x & \rho\varphi_y & \Delta & 0 \\ (\rho\Delta - \mu\varphi_y^2) & 0 & \frac{\partial p}{\partial \rho}\varphi_x & 0 \\ 0 & (\rho\Delta - \nu\varphi_y^2) & \frac{\partial p}{\partial \rho}\varphi_y & 0 \\ CWT \; + & DWT & 0 & (c_p\rho\Delta - k\varphi_y^2) \end{vmatrix}, \tag{7.84}$$

with $\Delta = u\varphi_x + v\varphi_y$. From this we obtain:

$$\begin{aligned} C =& \rho\left[\rho\Delta - \mu\varphi_y^2\right]\left[c_p\rho\Delta - k\varphi_y^2\right]\Big[(u^2 - \frac{\partial p}{\partial \rho})\varphi_x^2 + 2uv\varphi_x\varphi_y + \\ &+ (v^2 - \frac{\partial p}{\partial \rho})\varphi_y^2 - \nu(u\varphi_x + v\varphi_y)\varphi_y^2\Big], \end{aligned} \tag{7.85}$$

and finally with $\Delta = u\varphi_x + v\varphi_y = 0$, eq. (7.46), and by identifying $\partial p/\partial \rho = a^2$ with the speed of sound:

$$C = \rho\, \mu\, k\, \varphi_y^4 \left[(u^2 - \underline{a^2})\varphi_x^2 + 2uv\varphi_x\varphi_y + (v^2 - a^2)\varphi_y^2\right] = 0. \tag{7.86}$$

The term in this equation coming from the pressure gradient term in eq. (7.75) is underlined. Because we have assumed constant density, no coupling of the continuity equation and the momentum equations with the energy equation exists. Hence from the latter only the convective term and the thermal conduction term in y-direction are reflected in eqs. (7.85) and (7.86), respectively.

We find thus a four-fold characteristic in y-direction and in the angular brackets elliptic characteristics for subsonic flow and hyperbolic ones for supersonic flow. If the underlined pressure-gradient term would be omitted, we would get in the angular brackets for all flow velocities hyperbolic characteristics [17]. The system of equations (7.74) to (7.77) without the terms, which are of smaller order of magnitude, and without the pressure-gradient term in eq. (7.75), thus would constitute a linearized system of equations for the description of a weakly disturbed hypersonic flow with $u^2 \gg a^2$, $v = O(a)$. If the pressure-gradient term in eq. (7.76) would be zero, the whole equation would disappear, and we would get the boundary-layer equations for hypersonic flows.

If we accept the simplifications made in the equations of motion and those made additionally in order to investigate the characteristic properties of the system of equations, we get the result that the equations of motion without the second-order terms in x-direction are essentially of elliptic nature for subsonic flows and of hyperbolic nature for supersonic flows.

If the problem at hand permits it, the pressure-gradient term in eq. (7.75) can be omitted. This is possible for flows where:

$$\frac{1}{M_{ref}^2}\frac{\partial p}{\partial x} \approx O\left(\frac{1}{Re_{ref}}\right). \tag{7.87}$$

In this case the system of equations is of hyperpolic/parabolic type in the whole flow domain and can be solved as initial/boundary value problem with a space-marching numerical scheme, Section 9.3. If the term cannot be omitted, a parabolization scheme, for instance that of Vigneron et al. [18], must be employed, otherwise the solution process will become unstable[11].

[11] Since the unstable behaviour was initially observed in explicit solution schemes, it was attributed to a singularity at the location in the flow where $M(y) = 1$. In [19] it was shown that the singularity is only an apparent one, and then in [17] that the system of equations is of elliptic/parabolic type in the subsonic part of the flow, if the pressure-gradient term in eq. (7.75) is not omitted.

7.2 Basic Properties of Attached Viscous Flow

Basic concepts and results are discussed. Quantitative relations are given for the important properties of attached viscous flow, i. e. boundary-layer thicknesses, wall shear stress, and the thermal state of the surface (either the heat flux in the gas at the wall q_{gw} or the radiation-adiabatic temperature T_{ra}) in two-dimensional flow. The relations for incompressible flow are extended by means of the reference-temperature concept to compressible perfect-gas flow. This suffices to show their basic dependencies on overall flow parameters and on wall temperature. With these relations also fair estimations of the different properties can be made, if the boundary layer under consideration is weakly three-dimensional, and the stream-wise pressure gradient is not too large.

7.2.1 Boundary-Layer Thicknesses and Integral Parameters

We consider exclusively the relevant thicknesses of laminar and turbulent flow boundary layers. Their dependencies on boundary-layer running length x, on the Reynolds number $Re_{ref,x}$, and on the reference temperature T^* are studied. Thermal and mass-concentration boundary layers can be treated likewise.

Boundary-Layer Thicknesses The flow boundary-layer was introduced in Sub-Section 4.3.1. We call it in the following simply "boundary layer". It causes a virtual thickening of a body (via its displacement properties), and it especially prevents a full re-compression of the external inviscid flow at the aft part of the body, where we have a flow-off separation of the boundary layer, and the formation of a wake. This is the cause of the viscosity-induced pressure or form drag of a body. The boundary layer also reduces the aerodynamic effectiveness of lifting, stabilizing, and control surfaces. Its thickness finally governs the height of the boundary-layer diverter in case of airbreathing propulsion, Sub-Section 6.1. In all cases it holds: the thicker the boundary layer, the larger is the adverse effect.

On the other hand we have the effect that the boundary-layer causes and governs the wall-shear stress and the thermal state of the surface. The thicker the boundary layer (with turbulent flow the viscous sub-layer), the smaller the wall-shear stress, the smaller the heat-flux in the gas at the wall, and especially the more effective is surface-radiation cooling.

Attached viscous flows, i. e., boundary layers, have finite thickness everywhere on the body surface, also in stagnation points, and at swept leading edges, i. e. attachment lines in general, see, e. g., [4], [20], [21].

Of interest is the thickness of a boundary layer as such, its displacement thickness, the thickness of the viscous sub-layer, if the flow is turbulent et cetera.

The thickness δ of a laminar boundary layer is not sharply defined. Boundary-layer theory [20] yields the result that the outer edge of the boundary layer lies at $y \to \infty$.

For the incompressible laminar flat-plate boundary layer the theory of Blasius [20] gives for the thickness:

$$\delta_{lam} = c\frac{x}{(Re_{\infty,x})^{0.5}}, \tag{7.88}$$

with[12]

$$Re_{\infty,x} = \frac{\rho_\infty u_\infty x}{\mu_\infty}. \tag{7.89}$$

In terms of the unit Reynolds number $Re^u_\infty = \rho_\infty u_\infty/\mu_\infty$ it reads:

$$\delta_{lam} = c\frac{x^{0.5}}{(Re^u_\infty)^{0.5}}, \tag{7.90}$$

A practical definition of δ for laminar boundary layers is the distance at which locally the tangential velocity component $u(y)$ has approached the inviscid external velocity u_e by ϵu_e (for three-dimensional boundary layers the resultant tangential velocity and the resultant external inviscid velocity are taken[13]):

$$u_e - u(y) \leqq \epsilon u_e. \tag{7.91}$$

Usually, although often not explicitly quoted, the boundary-layer thickness is defined with

$$\epsilon = 0.01. \tag{7.92}$$

For the incompressible flat-plate boundary layer we find with this value from the Blasius solution the constant c in eq. (7.88):

$$\epsilon = 0.01: \; c = 5. \tag{7.93}$$

If $\epsilon = 0.001$ is taken, the constant is $c = 6$. The exact Blasius data are $c = 5$: $u = 0.99155\ u_e$, $c = 6$: $u = 0.99898\ u_e$.

The outer edge of a turbulent boundary layer in reality has a rugged unsteady pattern. In the frame of Reynolds-averaged turbulent boundary-layer theory [5] it is defined as smooth time-averaged edge. The concept of the thermal and the mass-concentration boundary layer, as we use it for laminar flow, is questionable with regard to turbulent flow.

---

[12] This Reynolds number can also be formulated as $Re_{e,x}$ in terms of the boundary-layer edge flow properties, or fully generalized as $Re_{ref,x}$, if a suitable reference state can be established.

[13] In case the inviscid flow is not known, e. g., in Navier-Stokes solutions, but also in experiments, the boundary-layer edge can be defined by vanishing boundary-layer vorticity $|\omega|_{b.l.} \leqq \epsilon$ [4]. However, if shock/boundary-layer interaction has to be taken into account, such a criterion needs to be refined.

For a low-Reynolds number, incompressible turbulent flat-plate boundary layer the boundary-layer thickness is found, by using the $\frac{1}{7}$-th-power velocity distribution law [20], to:

$$\delta_{turb} = 0.37 \frac{x}{(Re_{\infty,x})^{0.2}}, \tag{7.94}$$

and in terms of the unit Reynolds number:

$$\delta_{turb} = 0.37 \frac{x^{0.8}}{(Re^u_{\infty})^{0.2}}. \tag{7.95}$$

Another important thickness of turbulent boundary layers is that of the viscous sub-layer δ_{vs}, which is small compared to the boundary-layer thickness δ_{turb} ($\delta_{vs}/\ \delta_{turb} = O(0.01)$), Fig. 7.5 b).

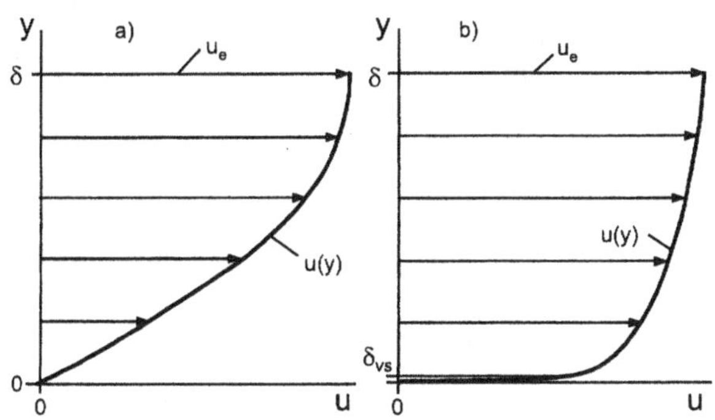

Fig. 7.5. Two-dimensional tangential velocity profile and boundary-layer thickness of **a**) laminar, **b**) turbulent boundary layer (time averaged).

In laminar boundary layers the thickness δ governs shear stress and heat transfer at the wall. In turbulent flow it is δ_{vs}, which is the relevant thickness [22], Fig. 7.5 b). It is found from the definition of the viscous sub-layer:

$$\frac{y \rho u_\tau}{\mu} \lessapprox 5. \tag{7.96}$$

With the friction velocity $u_\tau = \sqrt{\tau_w/\rho}$ and the wall-shear stress of the $\frac{1}{7}$-th-power boundary layer, eq. (7.139) in Sub-Section 7.2.3, we obtain, see also [22], where the different co-factor 72.91 results:

$$\delta_{vs} = 29.06 \frac{x}{(Re_{\infty,x})^{0.9}}. \tag{7.97}$$

The corresponding velocity at δ_{vs} is the viscous sub-layer edge-velocity u_{vs} [22]:

$$u_{vs} = 2.12\frac{u_e}{(Re_{\infty,x})^{0.1}}. \tag{7.98}$$

Simeonides, [23], proposes for scaling purposes, in order to be consistent with the laminar approach, an alternative thickness, δ_{sc}, which is slightly different from the viscous sub-layer thickness δ_{vs}. With the definition of δ_{sc} lying where the non-dimensional velocity u^+ and the wall distance y^+ are equal ($u^+ = u/u_\tau = y^+ = y\rho u_\tau/\mu$) he gets the scaling thickness δ_{sc} for turbulent boundary layers:

$$\delta_{sc} = 33.78\frac{x}{(Re_{\infty,x})^{0.8}}. \tag{7.99}$$

We apply now the reference temperature concept in order to determine the dependence of the boundary-layer thicknesses on the wall temperature T_w and the Mach number, assuming perfect-gas flow. We introduce eq. (7.73) into eq. (7.88) and find, with $c = 5$, the thickness $\delta_{lam,c}$ of the compressible laminar flat-plate boundary layer[14]:

$$\delta_{lam,c} = 5\frac{x}{(Re_{\infty,x})^{0.5}}\left(\frac{T^*}{T_\infty}\right)^{0.5(1+\omega)} = \delta_{lam,ic}\left(\frac{T^*}{T_\infty}\right)^{0.5(1+\omega)}. \tag{7.100}$$

If we assume $T^* \approx T_w$, as well as $\omega \approx 1$, we obtain that the compressible boundary-layer thickness depends approximately directly on the wall temperature:

$$\delta_{lam,c} = 5\frac{x}{(Re_{\infty,x})^{0.5}}\frac{T_w}{T_\infty}. \tag{7.101}$$

At an adiabatic wall the recovery temperature is:

$$T_r = T_w = T_\infty(1 + r\frac{\gamma - 1}{2}M_\infty^2). \tag{7.102}$$

This relation says that for large Mach numbers $T_r \sim M_\infty^2$. Introducing this into eq. (7.101) we get the well known result for the compressible laminar adiabatic flat-plate boundary layer:

$$\frac{\delta_{lam,c,ad}}{x} \sim \frac{M_\infty^2}{(Re_{\infty,x})^{0.5}}. \tag{7.103}$$

For the thickness of the compressible turbulent flat-plate $\frac{1}{7}$-th-power boundary layer we obtain with the reference-temperature extension:

$$\delta_{turb,c} = 0.37\frac{x}{(Re_{\infty,x})^{0.2}}\left(\frac{T^*}{T_\infty}\right)^{0.2(1+\omega)} = \delta_{turb,ic}\left(\frac{T^*}{T_\infty}\right)^{0.2(1+\omega)}, \tag{7.104}$$

[14] In the following we denote thicknesses of compressible boundary layers with '$_c$' and those of incompressible ones with '$_{ic}$'.

and for the thickness of the viscous sub-layer eq. (7.97)

$$\delta_{vs,c} = 29.06 \frac{x}{(Re_{\infty,x})^{0.9}} \left(\frac{T^*}{T_\infty}\right)^{0.9(1+\omega)} = \delta_{vs,ic} \left(\frac{T^*}{T_\infty}\right)^{0.9(1+\omega)} , \quad (7.105)$$

while the turbulent scaling thickness, eq. (7.99), is:

$$\delta_{sc,c} = 33.78 \frac{x}{(Re_{\infty,x})^{0.8}} \left(\frac{T^*}{T_\infty}\right)^{0.8(1+\omega)} = \delta_{sc,ic} \left(\frac{T^*}{T_\infty}\right)^{0.8(1+\omega)} . \quad (7.106)$$

Integral Parameters δ_1 and δ_2 Well defined integral parameters are the boundary-layer displacement thickness δ_1, the momentum thickness δ_2 and others[15]. These parameters appear on the one hand in boundary-layer solution methods ("integral" methods of Kármán-Pohlhausen type [20], [21]), and on the other hand in empirical criteria for laminar-turbulent transition, for separation, et cetera. Often also quotients of them are used, for instance the shape factor $H_{12} = \delta_1/\delta_2$.

In the following we define the more important integral parameters. We do it in all cases for compressible flow. The definitions are valid for both laminar and turbulent flow and contain the definitions for incompressible flow. Because of the importance of the displacement of the external inviscid flow by the boundary layer (weak interaction) we quote δ_1 and the equivalent inviscid source distribution for three-dimensional flow.

In the frame of first-order boundary-layer theory and in Cartesian coordinates the displacement thickness δ_1 of a three-dimensional boundary layer is defined (see, e. g., [4], [1], where it is given in general monoclinic surface-oriented coordinates) by a linear partial differential equation of first order:

$$\frac{\partial}{\partial x}\left[\rho_e u_e(\delta_1 - \delta_{1_x})\right] + \frac{\partial}{\partial z}\left[\rho_e w_e(\delta_1 - \delta_{1_z})\right] = \rho_0 v_0(x,z). \quad (7.107)$$

In this definition surface suction or blowing in the frame of the boundary-layer assumptions is taken into account by $\rho_0 v_0$, which in general can be a function of x and z. The symbols δ_{1_x} and δ_{1_z} denote "components" of δ_1 in x and z-direction, which are determined locally:

$$\delta_{1_x} = \int_{y=0}^{y=\delta} \left(1 - \frac{\rho u}{\rho_e u_e}\right) dy, \quad (7.108)$$

$$\delta_{1_z} = \int_{y=0}^{y=\delta} \left(1 - \frac{\rho w}{\rho_e w_e}\right) dy, \quad (7.109)$$

with ρu and ρw being functions of y.

[15] In literature often δ^* and θ are used instead of δ_1 and δ_2.

The displacement thickness δ_1 of three-dimensional boundary layers can become negative, although the two local components δ_{1_x} and δ_{1_z} are everywhere positive [1]. This is in contrast to two-dimensional boundary layers, where δ_1 is always positive. The effect occurs especially at attachment lines, with a steep negative bulging of the δ_1 surface. Close to the beginning of separation lines a steep positive bulging is observed[16]. These effects are clues for the explanation of the hot-spot and the cold-spot situations along attachment and separation lines on radiation-cooled surfaces, Sub-Section 3.2.3.

For two-dimensional boundary layers eq. (7.107) reduces to

$$\frac{\partial}{\partial x}\left[\rho_e u_e(\delta_1 - \delta_{1_x})\right] = \rho_0 v_0\,(x), \tag{7.110}$$

which is the general definition of the boundary-layer displacement thickness in two dimensions. Only for $\rho_0 v_0 = 0$, and $\rho_e u_e|_{x=0} = 0$, we obtain the classical local formulation:

$$\delta_{1_x} = \delta_1 = \int_{y=0}^{y=\delta}\left(1 - \frac{\rho u}{\rho_e u_e}\right) dy. \tag{7.111}$$

In practice almost always this two-dimensional formulation is employed to determine the displacement properties of a boundary layer, usually without consideration of possible effects of three-dimensionality or, in two dimensions, of initial values, and of variations of the external flow. These can be relevant, for instance, at a wing with round swept leading edge. Here a boundary layer with finite thickness exists already along the leading edge in span-wise direction. If the displacement thickness in chord-wise direction is to be determined, eq. (7.110) must be solved while properly taking into account the displacement thickness of the leading-edge boundary layer as initial value.

If boundary-layer criteria are correlated with the displacement thickness, use of the proper definition is recommended, otherwise the correlation might have deficits. However, criteria correlated with a given displacement thickness should be applied with the same definition, even if the displacement thickness employed in the correlation is not fully representative.

The displacement properties of a boundary layer lead to a virtual thickening of the body. In hypersonic computations by means of coupled Euler/boundary-layer methods this thickening must be taken into account. To change the body contour by a local superposition of the positive or negative displacement thickness would be a cumbersome procedure. A very effective alternative is to employ the equivalent inviscid source distribution $\rho v|_{isd}$ [24]:

$$\rho v|_{isd} = \frac{\partial \rho_e u_e \delta_{1_x}}{\partial x} + \frac{\partial \rho_e w_e \delta_{1_z}}{\partial z} + \rho_0 v_0. \tag{7.112}$$

[16] Navier-Stokes solutions do not exhibit explicitly these properties of attached and separating viscous flow. They can be found by a post-processing of the computed results with eq. (7.107) to (7.109).

This source distribution, which is a function of x and z, is found in a coupled solution method after the first Euler and boundary-layer solution has been performed. It is then employed at the body surface as boundary condition for the next Euler solution. In this way the boundary-layer displacement properties are iteratively taken into account. In practice only one coupling step (perturbation coupling) is sufficient [25].

An integral parameter often used in correlations is the momentum thickness[17] δ_2. In three-dimensional boundary layer flow two such thicknesses appear (δ_{2_x} and δ_{2_z})[4], because the momentum flux is a vector. In practice only the classical two-dimensionalormulation

$$\delta_2 = \int_{y=0}^{y=\delta} \frac{u}{u_e}\left(1 - \frac{\rho u}{\rho_e u_e}\right) dy, \tag{7.113}$$

is employed either for the main-flow profile or assuming two-dimensional flow from the beginning.

So far we have discussed the definitions of integral parameters. We give now displacement and momentum thicknesses for two-dimensional laminar and turbulent boundary layers over flat surfaces, which we also extend by means of the reference temperature to compressible flow. We do this again in order to identify their dependencies on wall temperature and Mach number, and to permit estimations of these thicknesses for practical purposes.

For the laminar Blasius boundary layer we quote [20]:

$$\delta_{1,lam,ic} = 1.7208 \frac{x}{(Re_{\infty,x})^{0.5}}, \tag{7.114}$$

$$\delta_{2,lam,ic} = 0.6641 \frac{x}{(Re_{\infty,x})^{0.5}}, \tag{7.115}$$

and for the turbulent $\frac{1}{7}$-th-power boundary layer [20]:

$$\delta_{1,turb,ic} = 0.0463 \frac{x}{(Re_{\infty,x})^{0.2}}, \tag{7.116}$$

$$\delta_{2,turb,ic} = 0.0360 \frac{x}{(Re_{\infty,x})^{0.2}}. \tag{7.117}$$

The respective shape factors are

$$H_{12,lam,ic} = \frac{\delta_{1,lam,ic}}{\delta_{2,lam,ic}} = 2.591, \tag{7.118}$$

for the Blasius boundary layer, and

[17] The name momentum thickness is used throughout in literature. Actually it is a measure of the loss of momentum in the boundary layer relative to that of the external inviscid flow [20], and hence should more aptly be called momentum-loss thickness.

$$H_{12,turb,ic} = \frac{\delta_{1,turb,ic}}{\delta_{2,turb,ic}} = 1.286, \tag{7.119}$$

for the $\frac{1}{7}$-th-power boundary layer.

Because the tangential velocity profile of a turbulent boundary layer is much fuller than that of a laminar one, Fig. 7.5, its displacement thickness, but also its momentum thickness, is smaller in proportion to the boundary-layer thickness than that of a laminar boundary layer, Table 7.2.

Table 7.2. Ratios of displacement thickness and momentum thickness to boundary-layer thickness for the laminar (Blasius, $\epsilon = 0.01$) and the turbulent ($\frac{1}{7}$-th-power) boundary layer.

Thickness ratio	Laminar boundary layer	Turbulent boundary layer
δ_1/δ	0.3442	0.1328
δ_2/δ	0.1251	0.0973

This points to a very important property of two-dimensional and three-dimensional (main-flow profile) turbulent boundary layers, which in general holds also for hypersonic flow: close to the wall the flow momentum is proportionally larger than that of laminar boundary layers. A turbulent boundary layer can negotiate a stronger adverse pressure gradient than a laminar one, and hence separates later than a laminar one. Although the skin friction exerted by a turbulent boundary layer is larger, a body of finite volume may have a smaller total drag if the flow is turbulent, because with the same pressure field laminar flow will separate earlier. The classical examples are the drag behaviour as function of Reynolds number of the flat plate at zero angle of attack and of the sphere [20]. However, regarding the thermal state of a radiation, or otherwise cooled vehicle surface, the turbulent boundary layer has no such positive secondary effect.

We do not show the derivation of the reference-temperature extensions of the above displacement and the momentum thicknesses to compressible flow, because the effort would be too large. Instead we quote them from [26], taking $Pr = 0.72$, and $r^*_{lam} = \sqrt{Pr} = 0.848$, $r^*_{turb} = \sqrt[3]{Pr} = 0.896$.

For the Blasius boundary layer we obtain:

$$\begin{aligned}\delta_{1,lam,c} =& \delta_{1,lam,ic}\left(-0.122 + 1,122\frac{T_w}{T_\infty} + \right. \\ & \left. + 0.333\frac{\gamma_\infty - 1}{2}M_\infty^2\right)\left(\frac{T^*}{T_\infty}\right)^{0.5(\omega-1)},\end{aligned} \tag{7.120}$$

and

$$\delta_{2,lam,c} = \delta_{2,lam,ic} \left(\frac{T^*}{T_\infty}\right)^{0.5(\omega-1)} . \tag{7.121}$$

The relations for the $\frac{1}{7}$-th-power boundary layer are:

$$\begin{aligned}\delta_{1,turb,c} =&\delta_{1,turb,ic} \left(0.129 + 0.871\frac{T_w}{T_\infty} + \right.\\ &\left. +0.648\frac{\gamma_\infty - 1}{2} M_\infty^2\right) \left(\frac{T^*}{T_\infty}\right)^{0.2(\omega-4)} ,\end{aligned} \tag{7.122}$$

however, with the relation for the incompressible case having a co-factor different to that in eq. (7.116):

$$\delta_{1,turb,ic} = 0.0504 \frac{x}{(Re_{\infty,x})^{0.2}} , \tag{7.123}$$

and

$$\delta_{2,turb,c} = \delta_{2,turb,ic} \left(\frac{T^*}{T_\infty}\right)^{0.2(\omega-4)} . \tag{7.124}$$

The respective shape factor of the Blasius boundary layer with reference-temperature extension is:

$$H_{12,lam,c} = H_{12,lam,ic} \left(-0.122 + 1.122\frac{T_w}{T_\infty} + 0.333\frac{\gamma - 1}{2} M_\infty^2\right) , \tag{7.125}$$

and that of the $\frac{1}{7}$-th-power boundary layer:

$$H_{12,turb,c} = H_{12,turb,ic} \left(0.129 + 0.871\frac{T_w}{T_\infty} + 0.648\frac{\gamma_\infty - 1}{2} M_\infty^2\right) . \tag{7.126}$$

The shape factor for incompressible turbulent flow $H_{12,turb,ic}$ is here, according to eqs. (7.123) and (7.117):

$$H_{12,turb,ic} = 1.4. \tag{7.127}$$

Summary We discuss the relations for the boundary-layer thicknesses and the integral parameters together. First however we consider the different factors, which we quoted for the displacement thickness of incompressible $\frac{1}{7}$-th-power turbulent boundary layers, eqs. (7.116) and (7.123), and the different values of the shape factor in eqs. (7.119) and (7.127).

Discrepancies can be found in literature regarding all simple relations for two-dimensional incompressible and compressible flat-plate turbulent boundary layers. Possible reasons in such cases are that not enough information is given:

- The relations in general are valid only in a certain Reynolds-number range. The thicknesses for the $\frac{1}{7}$-th-power turbulent boundary layer, which we gave above, $\delta_{turb,ic}$, eq. (7.94), δ_{vs}, eq. (7.97), as well as the integral parameters $\delta_{1,turb,ic}$, eq. (7.116), and $\delta_{2,turb,ic}$, eq. (7.117), are valid only for moderate Reynolds-numbers of $O(10^6)$ [20]. For larger Reynolds numbers other factors in these relations or completely other relations are required, which can be found in the literature.
- The above relations assume that the boundary layer is turbulent from the leading edge of the flat plate onwards [20]. Relations or factors derived from measurement data may not take this into account and hence differ from these relations. The classical shape-parameter data of Schubauer and Klebanoff [27] (see also [20]) vary in the transition region of a flat plate in incompressible flow from $H_{12} = 2.6$ in the laminar part to $H_{12} = 1.4$ in the turbulent part. They are of course locally determined data. The value for the laminar part compares well with the value given in eq. (7.118), whereas the value for the turbulent part is larger than that given in eq. (7.119). Obviously it was employed in eq. (7.123).
  We mention that the location of laminar-turbulent transition, if desired, can easily be taken into account in the determination of the boundary-layer thicknesses with simple relations, Fig. 7.6. At the transition location x_{tr}, the boundary-layer thickness δ_{turb} is taken equal to δ_{lam}. The laminar boundary layer has its origin at $x_{lam,0}$. The apparent origin of the turbulent boundary layer is then $x_{turb,0}$, which is found with the help of eq. (7.95) for incompressible flow or eq. (7.104) for compressible flow.

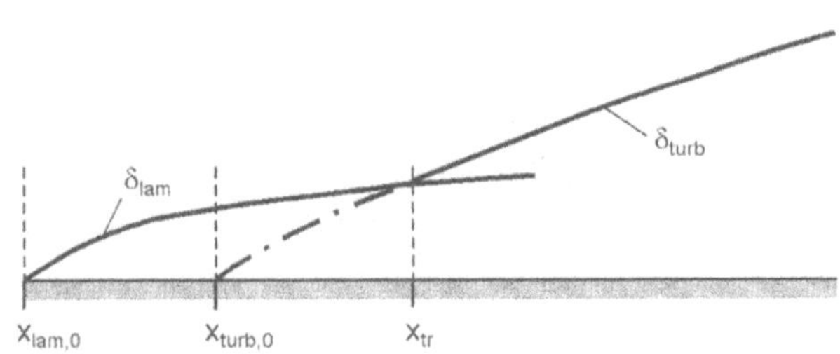

Fig. 7.6. Schematic to take into account the laminar-turbulent transition location in the determination of boundary-layer thicknesses by means of simple relations.

- Wall temperature and Mach number influence may have not be taken properly into account.

It is not attempted here to rate the available relations or to recommend one or the other. Basically it is intended to show the dependencies of thicknesses and integral parameters on typical flow and other parameters, which the engineer or researcher should know. Besides that the relations should permit to obtain an estimation of the magnitude of these entities with reasonable accuracy.

We combine now the above results in Table 7.3. We substitute '∞' conditions by general '$_{ref}$' conditions[18] and chose $\omega = \omega_\mu = 0.65$ in the viscosity law, Section 4.2. We break up all Reynolds numbers Re into the unit Reynolds number Re^u and the running length x in order to show the explicit dependencies on these parameters[19].

An inspection of the reference-temperature relation eq. (7.70) and the reference-temperature extensions eqs. (7.100), (7.104), (7.105) and (7.106), (7.120) to (7.124) shows, that we can write all temperature-extension terms symbolically as:

$$\left(\frac{T_*}{T_{ref}}\right)^c = \left((1-a) + a\frac{T_w}{T_{ref}} + bM_{ref}^2\right)^c . \qquad (7.128)$$

This form shows the desired limit behaviour, since for incompressible flow with $M_{ref} = 0$ the last term in the large brackets vanishes, and with $T_w = T_{ref}$ the two first terms in sum have the value '1'.

The reference-temperature extension relations of the integral parameters $\delta_{1,lam,c}$, eq.(7.120), and $\delta_{1,turb,c}$, eq. (7.122), have different factors in the large brackets, but these factors are quite close to those in the original reference temperature relation, eq. (7.70). Therefore we use eq. (7.128) in order to contract these relations for an easier discussion. In Table 7.3 they are marked accordingly with '$_m$'.

The third and the fourth column in Table 7.3 give the basic dependencies of the thicknesses of both incompressible and compressible boundary layers on running length x and unit Reynolds number Re^u_{ref}, the fifth column the dependence of the thicknesses of compressible boundary layers on the reference-temperature ratio T^*/T_{ref}, i. e., on wall temperature T_w and Mach number M_{ref}, eq. (7.128).

The dependencies summarized in Table 7.3 give us insight into the basic behaviour of thicknesses and integral parameters of two-dimensional flat-plate boundary layers.

- Dependence on the boundary-layer running length x.
  All boundary-layer thicknesses increase with increasing x. The thicknesses of turbulent boundary-layers grow stronger ($\sim x^{0.8}$) with x than those of laminar boundary layers ($\sim x^{0.5}$). The thickness of the viscous sub-layer

18 Also here it holds that for flat plates at zero angle of attack, and hence also at CAV-type flight vehicles at small angle of attack, except for the blunt nose region, we can choose '$_{ref}$' = '$_\infty$', whereas at RV-type vehicles the conditions at the outer edge of the boundary layer are the reference conditions: '$_{ref}$' = '$_e$'.

19 The implicit dependence of the thicknesses, and later of skin friction and thermal state of the surface, on x via the reference temperature is not given, because it is, in general, rather weak.

Table 7.3. Dependence of boundary-layer thicknesses on running length x, unit Reynolds number Re^u_{ref}, and reference-temperature ratio T^*/T_{ref} ($\omega = \omega_\mu = 0.65$). Manipulated relations (see above) are marked with '$_m$'.

Thickness	eq.	x	Re^u_{ref}	T^*/T_{ref}
δ_{lam}	(7.100)	$\sim x^{0.5}$	$\sim (Re^u_{ref})^{-0.5}$	$\sim \left(\frac{T^*}{T_{ref}}\right)^{0.825}$
δ_{turb}	(7.104)	$\sim x^{0.8}$	$\sim (Re^u_{ref})^{-0.2}$	$\sim \left(\frac{T^*}{T_{ref}}\right)^{0.33}$
$\delta_{1,lam}$	(7.120)	$\sim x^{0.5}$	$\sim (Re^u_{ref})^{-0.5}$	$\sim \left(\frac{T^*}{T_{ref}}\right)_m^{0.825}$
$\delta_{1,turb}$	(7.122)	$\sim x^{0.8}$	$\sim (Re^u_{ref})^{-0.2}$	$\sim \left(\frac{T^*}{T_{ref}}\right)_m^{0.33}$
$\delta_{2,lam}$	(7.121)	$\sim x^{0.5}$	$\sim (Re^u_{ref})^{-0.5}$	$\sim \left(\frac{T^*}{T_{ref}}\right)^{-0.175}$
$\delta_{2,turb}$	(7.124)	$\sim x^{0.8}$	$\sim (Re^u_{ref})^{-0.2}$	$\sim \left(\frac{T^*}{T_{ref}}\right)^{-0.67}$
δ_{vs}	(7.105)	$\sim x^{0.1}$	$\sim (Re^u_{ref})^{-0.9}$	$\sim \left(\frac{T^*}{T_{ref}}\right)^{1.485}$
δ_{sc}	(7.106)	$\sim x^{0.2}$	$\sim (Re^u_{ref})^{-0.8}$	$\sim \left(\frac{T^*}{T_{ref}}\right)^{1.32}$

grows only very weakly, $\sim x^{0.1}$, and also that of the turbulent scaling thickness, $\sim x^{0.2}$, with x.

- Dependence on the unit Reynolds number Re^u_{ref}.
  All thicknesses of boundary-layers depend on the inverse of some power of the unit Reynolds number. The larger Re^u_∞, the smaller are boundary-layer thicknesses. Thicknesses of laminar boundary layers react stronger on changes of the unit Reynolds number ($\sim (Re^u_{ref})^{-0.5}$) than those of turbulent boundary layers ($\sim (Re^u_{ref})^{-0.2}$). Strongest reacts the thickness of the viscous sub-layer, $\sim (Re^u_{ref})^{-0.9}$, and that of the turbulent scaling thickness, $\sim (Re^u_{ref})^{-0.8}$.
- Dependence on T^*/T_{ref}.
  The larger T^*/T_{ref}, the larger are the boundary-layer thicknesses, except for the momentum thicknesses $\delta_{2,lam}$ and $\delta_{2,turb}$. For a given M_{ref} and a given T_{ref} an increasing wall temperature T_w leads to an increase of the laminar boundary-layer thicknesses δ_{lam} and $\delta_{1,lam}$ ($\sim (T^*/T_{ref})^{0.825}$), which is stronger than the increase of the turbulent boundary-layer thicknesses δ_{turb} and $\delta_{1,turb}$ ($\sim (T^*/T_{ref})^{0.33}$). Strongest is the increase of the thickness of the viscous sub-layer δ_{vs}, $\sim (T^*/T_{ref})^{1.485}$, and that of the turbulent scaling thickness, $\sim (T^*/T_{ref})^{1.32}$. Concerning the momentum thicknesses that of the turbulent boundary layer decreases stronger with increasing wall temperature than that of the laminar boundary layer.

These are the basic dependencies of flat-plate boundary-layer thicknesses on flow parameters and wall temperature. On actual configurations other dependencies exist, which are mentioned in the following. In general no explicit relations are available to prescribe them. We note only the dependencies of the boundary-layer thickness δ, which holds for laminar and turbulent flow.

- Pressure gradient $\partial p/\partial x$ in main-stream direction.
  In general a negative pressure gradient (accelerated flow) reduces δ, a positive (decelerated flow) increases it. The cross-flow pressure gradient $\partial p/\partial z$ governs boundary-layer profile skewing and the form of the cross-flow boundary-layer profile (see Sub-Sections 7.1.2 and 7.1.3).
- Change of the body cross-section dA/dx in main-stream direction.
  If the cross-section grows in the main-stream direction ($dA/dx > 0$), see, e. g., Fig. 6.20 a), the boundary layer gets "stretched" in circumferential direction, because the wetted surface grows in the main-stream direction. δ becomes with the same pressure field and the same boundary conditions smaller at such a body than over a flat plate. This effect is the Mangler effect [28], [20]. Both wall-shear stress and heat flux in the gas at the wall are correspondingly larger. The inverse Mangler effect enlarges δ. It occurs where the body cross-section in the main-stream direction is reduced ($dA/dx < 0$). A typical example is the boat-tailed after-body of a flight vehicle (Fig. 6.20 a), if the flow would come from the right-hand side).
- 3-D effects.
  We mention here only the most obvious 3-D effects, which we find at attachment and separation lines. We have discussed them in the context of surface-radiation cooling already in Sub-Section 3.2.3. At an attachment line, due to the diverging flow pattern, the boundary layer is effectively thinned, compared to that in the vicinity, Fig. 3.5. At a separation line, the flow has a converging pattern and hence the tendency is the other way around, Fig. 3.6. Consequently the boundary-layer thickness δ is reduced at attachment lines, and enlarged at separation lines.
- Surface parameter k.
  We take k here as a representative permissible surface property, see Section 1.1, like surface roughness, waviness, et cetera. It influences boundary-layer properties, if it is not sub-critical, see Section 8.1. A roughness k is critical, if its height is larger than a characteristic boundary-layer thickness, for instance the displacement thickness δ_1 (see Sub-Section 8.1.5). It must be remembered, that boundary layers are thin at the front part of flight vehicles, and become thicker in the main-stream direction. The influence of surface roughness and the like on laminar-turbulent transition, turbulent wall-shear stress and the thermal state of the surface is known, at least empirically. Its influence on δ, however, which probably exists, at least in turbulent boundary layers, is not known.

7.2.2 Boundary-Layer Thickness at Stagnation Point and Attachment Lines

The attached viscous flow, i. e. the boundary layer, has its origin[20] at the forward stagnation point of a flight vehicle configuration.

The primary attachment point, depending on the angle of attack, can lie away from the nose-point on the lower side of the configuration, as we have seen for the Blunt Delta Wing, Fig. 3.16. Also the primary attachment lines on a hypersonic flight vehicle with the typical large leading-edge sweep of the wing, can lie away from the blunt-wing leading edge on the lower side of the wing. Secondary and tertiary attachment lines can be present. We have shown this for the BDW in Figs. 3.16 and 3.17 (see also Figs. 3.19 and 3.20). In such cases we can have an infinite swept wing flow situation, i. e., zero or only weak changes of flow parameters along the attachment line, like on the (two) primary and the tertiary attachment line of the BDW. The same situation can be present at separation lines. This all is typical for RV-type flight vehicles.

The boundary layer at the forward stagnation point has finite thickness despite the fact that there both the external inviscid velocity u_e and the tangential boundary-layer velocity $u(y)$ are zero. The situation is similar at a three-dimensional attachment line, however there a finite velocity along it exists.

In the following and also for the consideration of wall shear stress and thermal state of the surface we idealize the flow situation. We study the situation at the stagnation point of a sphere, and of a circular cylinder (2-D case), and at the attachment line of an infinite swept circular cylinder. The reason is that the velocity gradient $du_e/dx|_{x=0}$, which governs the flow there (at the swept cylinder it is the gradient across the attachment line), can be introduced explicitly as function of the radius R, Sub-Section 6.7.2. This may be a rather crude approximation of the situation found in reality on a RV-type flight vehicle, but it fits the situation more or less exactly for a CAV-type vehicle. Nevertheless, it permits us to gain insight into the basic dependencies of the boundary-layer thickness, and later also on the wall-shear stress, Sub-Section 7.2.4, and on the thermal state of the surface, Sub-Section 7.2.6, at an attachment point and at a primary attachment line.

The classical approach to describe the boundary layer at a stagnation point is to replace in the boundary-layer equation, explicitly or implicitly, for instance the velocity $u|_{x=0}$ by $(\partial u/\partial x)|_{x=0}$, which, like $(\partial u_e/\partial x)|_{x=0}$, is finite. This operation can be made, for instance, by differentiating the momentum equation in question with respect to the corresponding tangential coordinate [4].

---

[20] Although it is actually only one boundary layer, which develops on the vehicle surface, usually the boundary layers on the lower and the upper side of the configuration are distinguished.

In [29], for instance, the similarity transformation normal to the surface reads:

$$\eta = y\sqrt{\frac{\rho_e}{\mu_e}\frac{du_e}{dx}|_{x=0}}, \tag{7.129}$$

with ρ_e and μ_e being the values at the edge of the stagnation-point boundary layer.

We are interested here only in the basic dependencies of the boundary-layer thickness and find:

$$\delta|_{x=0} \sim \frac{1}{\sqrt{\frac{\rho_e}{\mu_e}\frac{du_e}{dx}|_{x=0}}}. \tag{7.130}$$

With eq. (6.163) the boundary-layer thickness becomes:

$$\delta|_{x=0} \sim \frac{\sqrt{\frac{R}{k}}}{\sqrt{\frac{\rho_e}{\mu_e}\left[\frac{2(p_s - p_\infty)}{\rho_s}\right]^{0.5}}}. \tag{7.131}$$

Here $k = 1$ for the sphere and $k = 1.33$ for the circular cylinder (2-D case).

We interpret the data ρ_e, μ_e as reference data ρ_{ref}, μ_{ref} which can later be replaced by reference-temperature data ρ^*, μ^* in order to obtain a dependence on the wall temperature, too[21]. We find finally, after replacing the subscript 'e' by 's' for 'stagnation value', like in eq. (6.163):

$$\delta|_{x=0} \sim \frac{\sqrt{\frac{R}{k}}}{\sqrt{\frac{\rho_{ref}}{\mu_{ref}}\left[\frac{2(p_s - p_\infty)}{\rho_s}\right]^{0.5}}}\left(\frac{T^*}{T_{ref}}\right)^{0.5(1+\omega)}, \tag{7.132}$$

where ρ_{ref}, μ_{ref} and T_{ref} are equal to ρ_s, μ_s and T_s.

We consider now the flow along an attachment line. At the attachment line we have a finite inviscid velocity component along it, and hence a boundary layer of finite thickness. Both may not or only weakly change in the direction of the attachment line.

In order to obtain the basic dependencies of the boundary-layer thickness, we assume that the attachment-line flow can locally be represented by the flow at the attachment line of an infinite swept circular cylinder, Fig. 6.34 b). There we have constant flow properties at the attachment line in the z-direction.

We only replace the velocity gradient for the stagnation point in eq. (7.132) by that for the swept infinite cylinder, eq. (6.164), and find for the laminar boundary layer:

---

[21] In [29] these data are finally interpreted as wall data only.

$$\delta|_{x=0} \sim \frac{\sqrt{\frac{R}{1.33\cos\varphi}}}{\sqrt{\frac{\rho_{ref}}{\mu_{ref}}\left[\frac{2(p_s-p_\infty)}{\rho_s}\right]^{0.5}_{\varphi=0}}}\left(\frac{T^*}{T_{ref}}\right)^{0.5(1+\omega)} . \qquad (7.133)$$

For a subsonic leading edge, i. e., $M_\infty \cos\varphi < 1$, exact theory shows that the dependence of δ is indeed $\sim 1/\sqrt{\cos\varphi}$, but for a supersonic leading edge it is somewhat thicker [30].

Summary We summarize the dependencies in Table 7.4. We chose, like in Sub-Section 7.2.1, $\omega = \omega_\mu = 0.65$ in the viscosity law, Section 4.2. In all cases δ is proportional $\sqrt{R}$, i. e., the larger the nose radius or the leading-edge radius, the thicker is the boundary layer. Also in all cases δ increases with increasing reference-temperature ratio, this means especially also with wall temperature.

In the case of the infinite swept circular cylinder δ increases $\sim (\cos\varphi)^{-0.5}$, at least for small sweep angles φ. This result will hold also for turbulent flow. For $\varphi \to 90°$ we get $\delta \to \infty$. This is consistent with the situation on an infinitely long cylinder aligned with the free-stream direction.

Table 7.4. Sphere/circular cylinder and infinite swept circular cylinder: dependence of boundary-layer thicknesses on radius R, sweep angle φ, and reference-temperature ratio T^*/T_{ref} ($\omega = \omega_\mu = 0.65$).

Body	Thickness	eq.	R	φ	T^*/T_{ref}
Sphere	δ_{sp}	(7.132)	$\sim \sqrt{R}$	-	$\sim \left(\frac{T^*}{T_{ref}}\right)^{0.825}$
Cylinder (2-D case)	δ_{cy}	(7.132)	$\sim \sqrt{R}$	-	$\sim \left(\frac{T^*}{T_{ref}}\right)^{0.825}$
Infinite swept cylinder, laminar	δ_{scy}	(7.133)	$\sim \sqrt{R}$	$\sim \frac{1}{\sqrt{\cos\varphi}}$	$\sim \left(\frac{T^*}{T_{ref}}\right)^{0.825}$

7.2.3 Wall Shear Stress at Flat Surface Portions

Wall shear stress (skin friction) τ_w is the cause of the viscous drag, which is exerted by the flow on the flight vehicle. For a CAV-type flight vehicle this drag can be of the order of 40 per cent of total drag. The wall-shear stress can also considerably influence lift forces, especially during flight at high angles of attack, and moments around the pitch and the yaw axis. Finally it is a deciding factor in erosion phenomena of surface coatings of thermal protection systems of RV-type flight vehicles.

The wall shear stress in the continuum regime and in Cartesian coordinates is defined as, see also eq. (4.36):

$$\tau_w = -\tau_{xy}|_w = \mu_w \frac{\partial u}{\partial y}|_w. \tag{7.134}$$

We consider here wall shear stress on flat surfaces, idealized as flat plates at zero angle of attack. We apply the reference-enthalpy/temperature extension of incompressible-flow relations to the compressible case.

For the incompressible laminar flat-plate boundary layer the theory of Blasius [20] gives for the wall-shear stress:

$$\tau_{w,lam} = 0.332 \frac{\rho_\infty u_\infty^2}{(Re_{\infty,x})^{0.5}}. \tag{7.135}$$

In terms of the unit Reynolds number $Re_\infty^u = \rho_\infty u_\infty / \mu_\infty$ this reads:

$$\tau_{w,lam} = 0.332 \frac{\rho_\infty u_\infty^2}{(Re_\infty^u)^{0.5} x^{0.5}}. \tag{7.136}$$

Introducing the skin-friction coefficient with $q_\infty = \rho_\infty u_\infty^2 / 2$:

$$c_f = \frac{\tau_w}{q_\infty}, \tag{7.137}$$

we obtain:

$$c_{f,lam} = \frac{0.664}{(Re_{\infty,x})^{0.5}}. \tag{7.138}$$

For a low-Reynolds number, incompressible turbulent flat-plate boundary layer the wall shear stress is, by using the $\frac{1}{7}$-th-power velocity distribution law, [20]:

$$\tau_{w,turb} = 0.0296 \frac{\rho_\infty u_\infty^2}{(Re_{\infty,x})^{0.2}}, \tag{7.139}$$

and in terms of the unit Reynolds number:

$$\tau_{w,turb} = 0.0296 \frac{\rho_\infty u_\infty^2}{(Re_\infty^u)^{0.2} x^{0.2}}. \tag{7.140}$$

The skin-friction coefficient is:

$$c_{f,turb} = \frac{0.0592}{(Re_{\infty,x})^{0.2}}. \tag{7.141}$$

For the reference-temperature extension, density and viscosity in eqs. (7.135) and (7.139) are all to be interpreted as data at reference temperature conditions.

The equation for the laminar flat-plate boundary layer then reads:

$$\tau_{w,lam,c} = 0.332 \frac{\rho_\infty u_\infty^2}{(Re_{\infty,x})^{0.5}} \left(\frac{\rho^*}{\rho_\infty}\right)^{0.5} \left(\frac{\mu^*}{\mu_\infty}\right)^{0.5}. \tag{7.142}$$

We introduce $\rho^* T^* = \rho_\infty T_\infty$ together with the power-law formulation of viscosity. With the nomenclature used for the thicknesses of compressible boundary layers, where the subscript 'c' stands for the compressible and 'ic' for the incompressible case, we obtain:

$$\tau_{w_{lam,c}} = 0.332 \frac{\rho_\infty u_\infty^2}{(Re_{\infty,x})^{0.5}} \left(\frac{T^*}{T_\infty}\right)^{0.5(\omega-1)} = \tau_{w_{lam,ic}} \left(\frac{T^*}{T_\infty}\right)^{0.5(\omega-1)} . \tag{7.143}$$

Similarly we find the skin friction with reference-temperature extension of the turbulent flat-plate boundary layer:

$$\tau_{w_{turb,c}} = 0.0296 \frac{\rho_\infty u_\infty^2}{(Re_{\infty,x})^{0.2}} \left(\frac{T^*}{T_\infty}\right)^{0.2\omega-0.8} = \tau_{w_{turb,ic}} \left(\frac{T^*}{T_\infty}\right)^{0.2(\omega-4)} . \tag{7.144}$$

The two expressions can be written in generalized form for perfect gas [31]:

$$\tau_{w,c} = C \frac{\rho_\infty u_\infty^2}{(Re_{\infty,x})^n} \left(\frac{T^*}{T_\infty}\right)^{n(1+\omega)-1} , \tag{7.145}$$

with $C = 0.332$ and $n = 0.5$ for laminar flow, and $C = 0.0296$ and $n = 0.2$ for turbulent flow.

An alternative formulation is:

$$\tau_{w,c} = C \frac{\mu_\infty u_\infty}{L} \left(\frac{L}{x}\right)^n (Re_{\infty,x})^{1-n} \left(\frac{T^*}{T_\infty}\right)^{n(1+\omega)-1} , \tag{7.146}$$

Before we summarize these results, we have a look at the result, which exact theory yields with use of the Lees-Dorodnitsyn transformation for the wall shear stress of a self-similar compressible laminar boundary layer [32]. We quote the result in a form given by Anderson [33]:

$$\tau_w = \frac{1}{2} \rho_e u_2^2 \sqrt{2} \left(\frac{T_w}{T_\infty}\right)^{\omega-1} \frac{f''(0)}{(Re_{\infty,x})^{0.5}} . \tag{7.147}$$

The dependence of τ_w on the Reynolds number is like in eq. (7.143). The function $f''(0)$ is the derivative of the velocity function $f' = u/u_e$ at the wall. It implicitly is a function of the boundary-layer edge Mach number M_e, the Prandtl number Pr, and the ratio of specific heats γ. Hence we have a dependence like in eq. (7.135) on these parameters, however implicitly and in different form.

Summary Like for the boundary-layer thicknesses discrepancies can be found in literature regarding the simple relations for the wall shear stress of two-dimensional incompressible and compressible flat-plate turbulent boundary layers. Again we do not pursue this problem further.

We summarize the results, eqs. (7.143), and (7.144), respectively eq. (7.145), in Table 7.5. We substitute also '∞' conditions by general '$_{ref}$' conditions and[22] choose $\omega = \omega_\mu = 0.65$ in the viscosity law, Section 4.2. We introduce the dynamic pressure q_{ref} and break up all Reynolds numbers Re into the unit Reynolds number Re^u and the running length x in order to show explicitly the dependencies on these parameters.

Table 7.5. Flat surface portions: dependence of boundary-layer wall shear stress on running length x, dynamic pressure q_{ref}, unit Reynolds number Re^u_{ref} and reference-temperature ratio T^*/T_{ref} ($\omega = \omega_\mu = 0.65$).

Wall shear stress	eq.	x	q_{ref}	Re^u_{ref}	T^*/T_{ref}
τ_{lam}	(7.143)	$\sim x^{-0.5}$	$\sim q_{ref}$	$\sim (Re^u_{ref})^{-0.5}$	$\sim \left(\frac{T^*}{T_{ref}}\right)^{-0.175}$
τ_{turb}	(7.144)	$\sim x^{-0.2}$	$\sim q_{ref}$	$\sim (Re^u_{ref})^{-0.2}$	$\sim \left(\frac{T^*}{T_{ref}}\right)^{-0.670}$

The third, fourth and the fifth column in Table 7.5 give the basic dependencies of the wall shear stress of both incompressible and compressible boundary layers on running length x, unit Reynolds number Re^u_{ref} and dynamic pressure q_{ref}, the sixth column the dependence of the wall shear stress of compressible boundary layers on the reference-temperature ratio T^*/T_{ref}, i. e., on wall temperature T_w and Mach number M_{ref}, eq. (7.128).

These dependencies give us insight into the basic behaviour of the wall shear stress of two-dimensional flat-plate boundary layers.

- Dependence on the boundary-layer running length x.
  The wall shear stress decreases with increasing x. The wall shear stress of laminar boundary-layers reduces stronger ($\sim x^{-0.5}$) with x than that of turbulent boundary layers ($\sim x^{-0.2}$). Remember that the thickness of the viscous sub-layer grows only very weakly with x compared to the thickness of the laminar boundary layer.
- Dependence on the dynamic pressure q_{ref}.
  The wall shear stress of laminar as well as of turbulent boundary layers increases linearly with increasing dynamic pressure.

[22] We remember that for flat plates at zero angle of attack, and hence also at CAV-type flight vehicles at small angle of attack, except for the blunt nose region, we can choose '$_{ref}$' = '$_\infty$', whereas at RV-type vehicles we must choose the conditions at the outer edge of the boundary layer: '$_{ref}$' = '$_e$'.

- Dependence on the unit Reynolds number Re^u_{ref}.
  The wall shear stress depends on the inverse of some power of the unit Reynolds number. The larger Re^u_∞, the smaller is the wall shear stress. The wall shear stress of laminar boundary layers reacts stronger on changes of the unit Reynolds number ($\sim (Re^u_{ref})^{-0.5}$) than that of turbulent boundary layers ($\sim (Re^u_{ref})^{-0.2}$). This result is not surprising, because the dynamic pressure q_∞ was isolated, see, e. g., eq. (7.145). In the alternate formulation of eq. (7.146) we have $\sim (Re^u_{ref})^{0.5}$ for the laminar, and $\sim (Re^u_{ref})^{0.8}$ for the turbulent case.
- Dependence on T^*/T_{ref}.
  The smaller T^*/T_{ref}, the larger is the wall shear stress. For a given M_{ref} and a given T_{ref} an increase in wall temperature T_w leads to a decrease of the wall shear stress of a turbulent boundary-layer ($\tau_{w_{1,turb}} \sim (T^*/T_{ref})^{-0.67}$), which is stronger than the decrease of the wall shear stress of a laminar boundary-layer ($\tau_{w_{1,lam}} \sim (T^*/T_{ref})^{-0.175}$). A drag-sensitive hypersonic of CAV-type flight vehicle with predominantly turbulent boundary layer therefore should be flown with a surface as hot as possible.

These are the basic dependencies of the wall shear stress for flat-plate boundary-layers on flow parameters and wall temperature. On actual configurations dependencies exist, which are similar to the dependencies of the boundary-layer thickness δ mentioned in Sub-Section 7.2.1. In general it holds that a larger boundary-layer thickness leads to a smaller wall shear stress. Important is to note, that a super-critical roughness or waviness of the surface will increase the wall shear stress of turbulent boundary layers [20]. The laminar boundary layer in such a case might be forced to become turbulent (unintentional turbulence tripping). A drag-sensitive CAV-type flight vehicle of must have a surface with sub-critical roughness or waviness everywhere.

7.2.4 Wall Shear Stress at Attachment Lines

The forward stagnation point is a singular point in which the wall shear stress is zero. In three-dimensional flow along an attachment line, like along a separation line, we have finite wall shear stress, $\tau_w > 0$. The direction of the wall shear stress vector $\underline{\tau}_w$ is coincident with the direction of the attachment line, which also holds for a separation line [1]. If we have an infinite swept situation, the wall shear stress is constant along the respective line.

We remember the discussion at the begin of Sub-Section 7.2.2 and consider the wall shear stress along the attachment line of an infinite swept circular cylinder, '$_{scy}$', with the reference-temperature extension in generalized form [31], assuming perfect-gas flow. This wall shear stress is constant in the z-direction, Fig. 6.34 b):

$$\tau_{w,scy} = \rho_\infty u_\infty^2 f_{scy}, \tag{7.148}$$

with

$$f_{scy} = \\ = C\left(\frac{w_e}{u_\infty}\right)^{2(1-n)} \left(\frac{\rho^*}{\rho_\infty}\right)^{1-n} \left(\frac{\mu^*}{\mu_\infty}\right)^n \left(\frac{R}{u_\infty}\frac{du_e}{dx}|_{x=0}\right)^n \frac{1}{(Re_{\infty,R})^n}. \tag{7.149}$$

R is the radius of the cylinder, $du_e/dx|_{x=0}$ the gradient of the inviscid external velocity normal to the attachment line, eq. (6.164), and $w_e = u_\infty \sin\varphi$ the inviscid external velocity along it, Fig. 6.34 b).

For laminar flow $C = 0.57$, $n = 0.5$. As reference-temperature values are recommended, [31], $\rho^* = \rho_e^{0.8}\rho_w^{0.2}$ and $\mu^* = \mu_e^{0.8}\mu_w^{0.2}$.

For turbulent flow $C = 0.0345$, $n = 0.21$. The reference-temperature values in this case, following a proposal from [36] to put more weight on the recovery temperature and less on the wall temperature, are taken at:

$$T^* = 0.30T_e + 0.10T_w + 0.60T_r. \tag{7.150}$$

The density at reference temperature conditions, ρ^*, is found with T^* and the external pressure p_e.

An alternate formulation is:

$$\tau_{w,scy} = \frac{\mu_\infty u_\infty}{R} f_{scy}^*, \tag{7.151}$$

with

$$f_{scy}^* = \\ = C\left(\frac{w_e}{u_\infty}\right)^{2(1-n)} \left(\frac{p_e}{p_\infty}\right)^{1-n} \left(\frac{T^*}{T_\infty}\right)^{n(1+\omega)-1} \left(\frac{R}{u_\infty}\frac{du_e}{dx}|_{x=0}\right)^n \frac{1}{(Re_{\infty,R})^n}. \tag{7.152}$$

Summary In Table 7.6 the general dependencies of the wall shear stress τ_w in eq. (7.148) are summarized for laminar and turbulent flow. Again we choose $\omega = \omega_\mu = 0.65$ in the viscosity law, Section 4.2. We introduce, like before, the dynamic pressure q_{ref} and break up the Reynolds number Re into the unit Reynolds number Re^u and the radius R. Regarding the reference-temperature dependencies the differentiation recommended in [31] for the laminar and the turbulent case (see above) is not made. Instead we assume:

$$\left(\frac{p_e}{p_\infty}\right)^{1-n} \left(\frac{T^*}{T_\infty}\right)^{n(1+\omega)-1} \sim \left(\frac{T^*}{T_\infty}\right)^{n(1+\omega)-1}. \tag{7.153}$$

The results are:

Table 7.6. Attachment line at the infinite swept circular cylinder: dependence of wall shear stress τ_w, eq. (7.148), of the compressible laminar and turbulent boundary layer on dynamic pressure q_∞, radius R, sweep angle φ, unit Reynolds number Re^u_∞, and reference-temperature ratio T^*/T_∞ ($\omega = \omega_\mu = 0.65$).

$\tau_{w,scy}$	q_∞	R	φ	Re^u_∞	T^*/T_∞
$\tau_{w,lam}$	$\sim q_\infty$	$\sim R^{-0.5}$	$\sim \sin\varphi(\cos\varphi)^{0.5}$	$\sim (Re^u_\infty)^{-0.5}$	$\sim \left(\frac{T^*}{T_\infty}\right)^{-0.175}$
$\tau_{w,turb}$	$\sim q_\infty$	$\sim R^{-0.21}$	$\sim (\sin\varphi)^{1.58}(\cos\varphi)^{0.21}$	$\sim (Re^u_\infty)^{-0.21}$	$\sim \left(\frac{T^*}{T_\infty}\right)^{-0.653}$

- Dependence on the dynamic pressure q_∞.
  The wall shear stress of the laminar as well as of the turbulent boundary layer increases linearly with increasing dynamic pressure.
- Dependence on the radius R.
  The wall shear stress decreases with increasing R stronger for laminar, $\sim R^{-0.5}$, than for turbulent flow, $\sim R^{-0.21}$.
- Dependence on the sweep angle φ.
  For $\varphi = 0$ we have the case of the non-swept circular cylinder (2-D case), where $\tau_{w,scy}$ is zero in the stagnation point. For $\varphi \to 90°$ also $\tau_{w,scy} \to 0$. This means that the attachment line ceases to exist for $\to 90°$ and we get the situation on an infinitely long cylinder aligned with the free-stream direction, where finally $\tau_{w,scy}$ becomes zero. In between we observe that $\tau_{w,scy}$ first increases with increasing φ and that stronger for laminar than for turbulent flow. This reflects the behaviour of the component w_e of the external inviscid flow, Fig. 6.34 b), which grows stronger with increasing φ than the velocity gradient du_e/dx across the attachment line declines with it. At large φ finally the effect reverses, first for laminar then for turbulent flow, and $\tau_{w,scy}$ drops to zero.
- Dependence on the unit Reynolds number Re^u_∞.
  The wall shear stress depends on the inverse of some power of the unit Reynolds number in the same way as on flat surface portions with $\sim (Re^u_\infty)^{-0.5}$ for the laminar and $\sim (Re^u_\infty)^{-0.21}$ for the turbulent boundary case. The larger Re^u_∞, the smaller is the wall shear stress. Again one has to keep in mind the dependence on the dynamic pressure, which has been isolated in this consideration.
- Dependence on T^*/T_∞.
  The smaller T^*/T_{ref}, the larger is the wall shear stress, like in the case of the boundary layer on flat surface portions. For the wall shear stress of the laminar boundary-layer holds $\tau_{w_{1,lam}} \sim (T^*/T_\infty)^{-0.175}$ and for that of the turbulent boundary-layer $\tau_{w_{1,turb}} \sim (T^*/T_\infty)^{-0.653}$.

For the influence of other factors see the summary for the wall shear stress on flat surface portions at the end of Sub-Section 7.2.3.

At a general three-dimensional attachment line the velocity gradient $du_e/dx|_{x=0}$ and φ are not connected explicitly to a geometrical property of the configuration. In any case it can be stated that the larger du_e/dx, the smaller the relevant boundary-layer thickness, and the larger the wall shear stress. Of course also here the magnitude of w_e plays a role. τ_w depends also here inversely on the wall temperature and that, like in general, stronger for turbulent than for laminar flow.

7.2.5 Thermal State of Flat Surface Portions

The thermal state of the surface governs thermal-surface effects on wall and near-wall viscous and thermo-chemical phenomena, as well as thermal loads on the structure, Section 1.4. Important is the fact that external surfaces of hypersonic flight vehicles basically are radiation cooled. Above we have seen, how the thermal state of the surface influences via T_w the boundary layer thicknesses, and hence also the wall-shear stress of laminar or turbulent flow. In view of viscous thermal-surface effects, the thermal state of the surface is of large importance for CAV-type flight vehicles. Thermo-chemical thermal-surface effects are especially important for RV-type vehicles. Thermal loads finally are of large importance for all vehicle classes.

We have defined the thermal state of a surface by two entities, the temperature of the gas at the wall which in the continuum-flow regime is the wall temperature $T_{gw} = T_w$, and the temperature gradient $\partial T/\partial y|_{gw}$ in the gas at the wall. For perfect gas or a mixture of thermally perfect gases in equilibrium the latter can be replaced by the heat flux in the gas at the wall q_{gw}.

In Section 3.1 several cases regarding the thermal state of the surface were distinguished. In the following we consider the first two cases.

If the vehicle surface is radiation cooled and the heat flux into the wall, q_w, is small, the radiation-adiabatic temperature $T_w = T_{ra}$ is to be determined. This is case 1:

$$q_{gw}(x,z) \approx \sigma\epsilon T_{ra}^4(x,z) \;\rightarrow\; T_w = T_{ra}(x,z) = \,?$$

Case 2 is the case with T_w prescribed directly, because, for instance, of design considerations, or because the situation at a cold-wall wind-tunnel model is studied. Hence the heat flux in the gas at the wall, q_{gw}, is to be determined. This is case 2:

$$T_w = T_w(x,z) \;\rightarrow\; q_{gw}(x,z) = \,?$$

To describe the thermal state of the surface we remain with the reference temperature extension, following [34], [31], and perfect-gas flow. We begin with the discussion of the situation at flat surface portions, '$_{fp}$'. The situation at the stagnation-point region and at attachment lines is considered in Sub-Section 7.2.6.

The basis of the following relations is the Reynolds analogy[23], respectively the Chilton-Colburn analogy [31]:

$$St = \frac{q_{gw}}{\rho_\infty u_\infty (h_r - h_w)} = Pr^{-2/3} \frac{\tau_w}{\rho_\infty u_\infty^2}. \tag{7.154}$$

The heat flux in the gas at the wall in the continuum regime and for a perfect gas or a mixture of thermally perfect gases in thermo-chemical equilibrium is defined, see also eq. (4.62), as:

$$q_{gw} = -k_w \frac{\partial T}{\partial y}|_w. \tag{7.155}$$

We find for case 1 for perfect gas [31]:

$$T^4_{ra,fp} = \frac{q_{rad}}{\sigma\epsilon} = \frac{q_{gw}}{\sigma\epsilon} = \frac{k_\infty}{\sigma\epsilon} Pr^{1/3} g_{fp} \frac{1}{L} T_r (1 - \frac{T_{ra}}{T_r}), \tag{7.156}$$

with g_{fp} being:

$$g_{fp} = C \left(\frac{T^*}{T_\infty}\right)^{n(1+\omega)-1} \left(\frac{L}{x}\right)^n (Re_{\infty,L})^{1-n}. \tag{7.157}$$

Here again $C = 0.332$ and $n = 0.5$ for laminar flow, and $C = 0.0296$ and $n = 0.2$ for turbulent flow.

The heat flux in the gas at the wall q_{gw} with given T_w, Case 2, reads:

$$q_{gw,fp} = k_\infty Pr^{1/3} g_{fp} \frac{1}{L} T_r (1 - \frac{T_w}{T_r}). \tag{7.158}$$

Summary We summarize the dependencies in Table 7.7. We substitute also here '∞' conditions by general '$_{ref}$' conditions[24] and chose $\omega = \omega_\mu = 0.65$ in the viscosity law, Section 4.2. We break up the Reynolds numbers Re into the unit Reynolds number Re^u and the running length x in order to show the explicit dependencies on these parameters. They reflect inversely the behaviour of the boundary-layer thickness of laminar flow and the viscous sub-layer thickness of turbulent flow, Sub-Section 7.2.1. As was to be expected, the qualitative results from Sub-Section 3.2.1 are supported. We see also that indeed in case 2 with given T_w the heat flux in the gas at the wall q_{gw} has the same dependencies as T^4_{ra}.

[23] In [9] it is shown that the Reynolds analogy is valid strictly only for constant wall temperature. For radiation-cooled surface it hence holds only approximately.

[24] Again we remember that for flat plates at zero angle of attack, and hence also at CAV-type flight vehicles at small angle of attack, except for the blunt nose region, we can choose '$_{ref}$' = '$_\infty$', whereas at RV-type vehicles we must choose the conditions at the outer edge of the boundary layer: '$_{ref}$' = '$_e$'.

Table 7.7. Flat surface portions: dependence of the thermal state of the surface, case 1 and case 2, on boundary-layer running length x, unit Reynolds number Re^u_{ref}, reference-temperature ratio T^*/T_{ref}, and temperature difference $T_r - T_w$ ($\omega = \omega_\mu = 0.65$).

Item	eq.	x	Re^u_{ref}	T^*/T_{ref}	$T_r - T_w$
Case 1: $T^4_{ra,lam}$	(7.156)	$\sim x^{-0.5}$	$\sim (Re^u_{ref})^{0.5}$	$\sim \left(\frac{T^*}{T_{ref}}\right)^{-0.175}$	$\sim (T_r - T_{ra})$
Case 1: $T^4_{ra,turb}$	(7.156)	$\sim x^{-0.2}$	$\sim (Re^u_{ref})^{0.8}$	$\sim \left(\frac{T^*}{T_{ref}}\right)^{-0.67}$	$\sim (T_r - T_{ra})$
Case 2: $q_{gw,lam}$	(7.158)	$\sim x^{-0.5}$	$\sim (Re^u_{ref})^{0.5}$	$\sim \left(\frac{T^*}{T_{ref}}\right)^{-0.175}$	$\sim (T_r - T_w)$
Case 2: $q_{gw,turb}$	(7.158)	$\sim x^{-0.2}$	$\sim (Re^u_{ref})^{0.8}$	$\sim \left(\frac{T^*}{T_{ref}}\right)^{-0.67}$	$\sim (T_r - T_w)$

- Dependence on the boundary-layer running length x.
  Both T^4_{ra}, case 1, and q_{gw}, case 2, decrease with increasing x. This is stronger in the laminar, ($\sim x^{-0.5}$), than in the turbulent cases ($\sim x^{-0.2}$).
- Dependence on the unit Reynolds number Re^u_{ref}.
  The larger Re^u_∞, the larger T^4_{ra}, case 1, and q_{gw}, case 2, because the respective boundary-layer thicknesses become smaller with increasing Re^u_∞. The thermal state of the surface with a laminar boundary layer reacts less strongly on changes of the unit Reynolds number, $\sim (Re^u_{ref})^{0.5}$, than that of the turbulent boundary layer, $\sim (Re^u_{ref})^{0.8}$.
- Dependence on T^*/T_{ref}.
  The larger T^*/T_{ref}, the smaller are T^4_{ra}, case 1, and q_{gw}, case 2. For a given M_{ref} and a given T_{ref} an increase in wall temperature T_w would lead to a decrease of them. The effect is stronger for turbulent flow, $\sim (T^*/T_{ref})^{-0.67}$, than for laminar flow, $\sim (T^*/T_{ref})^{-0.175}$. However, the major influence is that of the $T_r - T_w$, next item.
- Dependence on $T_r - T_w$.
  For both case 1 and case 2, laminar and turbulent flow, we have a linear dependence on the temperature difference $T_r - T_w$. The larger this difference, the larger is T^4_{ra} or q_{gw}. However, note that in case 2 T_w is given, whereas in case 1 $T_w = T_{ra}$ is the unknown.

These are the basic dependencies of the thermal state of the surface. On actual configurations dependencies exist, which are inversely similar to the dependencies of the respective boundary-layer thicknesses, Sub-Section 7.2.1. Important is that super-critical wall roughness, waviness et cetera increase either T^4_{ra} or q_{gw} if the flow is turbulent.

7.2.6 Thermal State of Stagnation Point and Attachment Lines

Remembering the discussion at the beginning of Sub-Section 7.2.2 we study the situation at a sphere, respectively the circular cylinder (2-D case), and at the attachment line of an infinite swept circular cylinder. We are aware that this is a more or less good approximation of the situation at stagnation points and (primary) attachment lines at a hypersonic flight vehicle in reality.

We find at the sphere, '$_{sp}$', respectively the circular cylinder (2-D case), for case 1 for perfect gas with the generalized reference-temperature formulation [31] like before:

$$T_{ra}^4 = \frac{k_\infty}{\sigma\epsilon} Pr^{1/3} g_{sp} \frac{1}{R} T_r (1 - \frac{T_{ra}}{T_r}), \tag{7.159}$$

where:

$$g_{sp} = C \left(\frac{p_e}{p_\infty}\right)^{0.5} \left(\frac{T^*}{T_\infty}\right)^{0.5(\omega-1)} \left(\frac{R}{u_\infty}\frac{du_e}{dx}|_{x=0}\right)^{0.5} (Re_{\infty,R})^{0.5}, \tag{7.160}$$

with $C = 0.763$ for the sphere and $C = 0.57$ for the circular cylinder. The velocity gradient $du_e/dx|_{x=0}$ is found with eq. (6.163). This is a laminar case.

The meanwhile classical formulation for q_{gw} at the stagnation point of a sphere with given wall enthalpy h_w, generalized case 2, in the presence of high-temperature real-gas effects is that of Fay and Riddell [35]. This result of an exact similar solution ansatz reads:

$$q_{gw,sp} = kPr_w^{-0.6}(\rho_w\mu_w)^{0.1}(\rho_e\mu_e)^{0.4} \cdot \\ \cdot \left[1 + (Le^m - 1)(\frac{h_D}{h_e})\right](h_e - h_w)\left(\frac{du_e}{dx}|_{x=0}\right)^{0.5}. \tag{7.161}$$

The stagnation point values are denoted here as boundary-layer edge 'e' values. For the sphere $k = 0.763$ and for the circular cylinder $k = 0.57$ [34]. The term in square brackets contains the Lewis number Le, eq. (4.93). Its exponent is $m = 0.52$ for the equilibrium and $m = 0.63$ for the frozen case over a catalytic wall.

The term h_D is the average atomic dissociation energy times the atomic mass fraction in the boundary-layer edge flow [35]. For perfect-gas flow the value in the square brackets reduces to one. Note that in eq. (7.161) the external flow properties $\rho_e\mu_e$ have a stronger influence on q_{gw} than the properties at the wall $\rho_w\mu_w$. This reflects the dependence of the boundary-layer thickness in the stagnation point on the boundary-layer edge parameters, eq. (7.132).

Because we wish to show the basic dependencies also here, we use the equivalent generalized reference-temperature formulation for case 2 in the following. The heat flux in the gas at the wall, q_{gw}, for the sphere or the circular cylinder (2-D case) with given T_w, reads, [34]:

$$q_{gw,sp} = k_\infty Pr^{1/3} g_{sp} \frac{1}{R} T_r \left(1 - \frac{T_w}{T_r}\right). \tag{7.162}$$

The radiation-adiabatic temperature, case 1, at the attachment line of an infinite swept circular cylinder, 'scy', is found from:

$$T^4_{ra,scy} = \frac{Pr^{\frac{1}{3}}}{\sigma\epsilon} k_\infty g^*_{scy} \frac{1}{R} T_r \left(1 - \frac{T_{ra}}{T_r}\right), \tag{7.163}$$

where g^*_{scy} is:

$$g^*_{scy} = C \left(\frac{w_e}{u_\infty}\right)^{1-2n} \left(\frac{p_e}{p_\infty}\right)^{1-n} \left(\frac{T^*}{T_\infty}\right)^{n(1+\omega)-1} \left(\frac{R}{u_\infty} \frac{du_e}{dx}|_{x=0}\right)^n Re^{1-n}_{\infty,R}. \tag{7.164}$$

The heat flux in the gas at the wall (case 2) finally is:

$$q_{gw,scy} = Pr^{\frac{1}{3}} k_\infty g^*_{scy} \frac{1}{R} T_r \left(1 - \frac{T_w}{T_r}\right). \tag{7.165}$$

Like in Sub-Section 7.2.4, R is the radius of the cylinder, $du_e/dx|_{x=0}$ the gradient of the inviscid external velocity normal to the attachment line, eq. (6.164), and $w_e = u_\infty \sin\varphi$ the inviscid external velocity along it, Fig. 6.34 b). For laminar flow $C = 0.57$, $n = 0.5$ and $\rho^* = \rho_e^{0.8}\rho_w^{0.2}$, $\mu^* = \mu_e^{0.8}\mu_w^{0.2}$.

For turbulent flow $C = 0.0345$, $n = 0.21$. The reference-temperature values are found with eq. (7.150), and ρ^* again with T^* and the external pressure p_e.

Summary In Table 7.8 the general dependencies of the radiation-adiabatic temperature, (case 1), are summarized for laminar and turbulent flow. Again we choose $\omega = \omega_\mu = 0.65$ in the viscosity law, Section 4.2. We break up, like before, the Reynolds numbers Re into the unit Reynolds number Re^u and the radius R. The reference-temperature dependencies are also taken in simplified form. The dependencies of q_{gw} (case 2) are the same, see eqs. (7.162) and (7.165) and are therefore not shown.

The results are:

- Dependence on the radius R.
  The fourth power of the radiation adiabatic temperature T^4_{ra} is the smaller, the larger R is. It depends (infinite swept circular cylinder) on R stronger for laminar, $\sim R^{-0.5}$, than for turbulent flow, $\sim R^{-0.21}$.
- Dependence on the sweep angle φ.
  This dependence holds only for the infinite swept circular cylinder. For $\varphi = 0.0°$ we have the case of the non-swept circular cylinder (2-D case). For $\varphi \to 90°$ $T^4_{ra} \to 0$. This means that we get the situation on an infinitely

Table 7.8. The radiation-adiabatic temperature, (case 1), at the sphere, respectively the circular cylinder (2-D case), and at the attachment line of the infinite swept circular cylinder: dependence on radius R, sweep angle φ, unit Reynolds number Re^u_∞, and reference-temperature ratio T^*/T_∞ ($\omega = \omega_\mu = 0.65$). The temperature difference T_r - $T_w = T_r - T_{ra}$ is not included.

Item	eq.	R	φ	Re^u_∞	T^*/T_∞
$T^4_{ra,sp}$	(7.159)	$\sim R^{-0.5}$	-	$\sim (Re^u_\infty)^{0.5}$	$\sim \left(\frac{T^*}{T_\infty}\right)^{-0.175}$
$T^4_{ra,scy,lam}$	(7.163)	$\sim R^{-0.5}$	$\sim (\cos\varphi)^{0.5}$	$\sim (Re^u_\infty)^{0.5}$	$\sim \left(\frac{T^*}{T_\infty}\right)^{-0.175}$
$T^4_{ra,scy,turb}$	(7.163)	$\sim R^{-0.21}$	$\sim (\sin\varphi)^{0.58}(\cos\varphi)^{0.21}$	$\sim (Re^u_\infty)^{0.79}$	$\sim \left(\frac{T^*}{T_\infty}\right)^{-0.653}$

long cylinder aligned with the free-stream direction, where finally δ and $\delta_{vs} \to \infty$, and hence T^4_{ra} becomes zero.

- Dependence on the unit Reynolds number Re^u_∞.
  T^4_{ra} depends on some power of the unit Reynolds number in the same way as on flat surface portions with $\sim (Re^u_\infty)^{0.5}$ for the laminar and $\sim (Re^u_\infty)^{0.79}$ for the turbulent case. The larger Re^u_∞, the larger is T^4_{ra}, because δ and δ_{vs} become smaller with increasing Re^u_∞.
- Dependence on T^*/T_{ref}.
  The larger T^*/T_{ref}, the smaller is T^4_{ra}. For a given M_{ref} and a given T_{ref} an increase in wall temperature T_w would lead to a decrease of them. The effect is stronger for turbulent flow, $\sim (T^*/T_{ref})^{-0.67}$, than for laminar flow, $\sim (T^*/T_{ref})^{-0.175}$. However, the major influence is that of the $T_r - T_w$, next item.
- Dependence on $T_r - T_w$ (not in Table 7.8).
  For laminar and turbulent flow, we have a linear dependence on the temperature difference $T_r - T_{ra}$. The larger this difference, the larger is T^4_{ra}. Note that in case 2 , which is not included in Table 7.8, $T_w$is given, whereas in Case 1 $T_w = T_{ra}$ is the unknown.

These are the basic dependencies of the thermal state of the surface. On actual configurations dependencies exist especially for the infinite swept circular cylinder, which are inversely similar to the dependencies of the respective boundary-layer thicknesses, Sub-Section 7.2.1. Important is that in the case of the infinite swept circular cylinder super-critical wall roughness, waviness et cetera also increase either T^4_{ra} or q_{gw} if the flow is turbulent.

7.3 Case Study: Wall Temperature and Skin Friction at the SÄNGER Forebody

We discuss the results of Navier-Stokes solutions ([37], based on [38], [39]) for the forebody of the lower stage of SÄNGER, Section 1.1, which is a CAV-type flight vehicle. The numerical simulations were made for a flight situation and a wind-tunnel situation, Table 7.9. We check and interpret the computed data with the help of the approximate, reference-temperature extended boundary-layer relations for flat surface portions which we have presented in the preceding sub-sections.

Table 7.9. Parameters of the SÄNGER forebody computation cases. L is the length of the forebody, see Fig. 7.10.

Case	M_∞	H $[km]$	T_∞ $[K]$	T_w [K]	Re^u_∞ $[m^{-1}]$	L $[m]$	α [°]
Flight situation	6.8	33.0	231.5	*variable*	$1.5{\cdot}10^6$	55.0	6.0
H2K situation	6.8	-	61.0	300.0	$8.7{\cdot}10^6$	0.344	6.0

In Figs. 7.7, [40], and 7.8, [41], the configuration is shown with computed skin-friction line patterns for the case under consideration. The first ramp of the inlet lies at approximately 67 per cent length on the lower side. Along the lower symmetry line the pattern shows streamlines which are with good approximation parallel. On the upper side this is the case to a lesser extent.

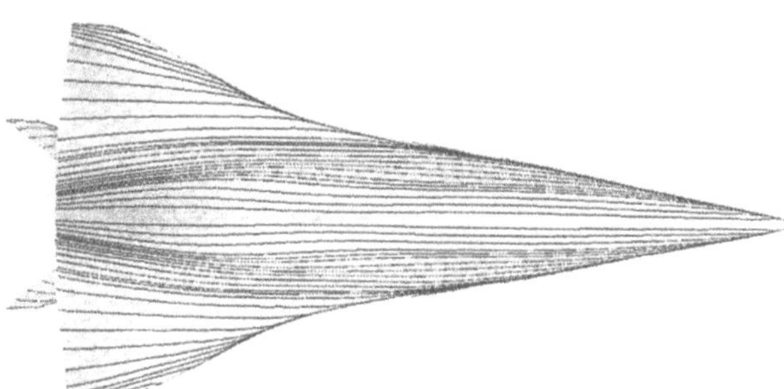

Fig. 7.7. Lower stage of SÄNGER without propulsion system: computed pattern of skin-friction lines on the lower side without propulsion system [40].

The presented and discussed results are wall temperatures and skin-friction coefficients on the lower, and partly on the upper symmetry line of the forebody. For the flight situation several assumptions were made regarding surface-radiation cooling, gas model, and state of the boundary layer (laminar/turbulent).

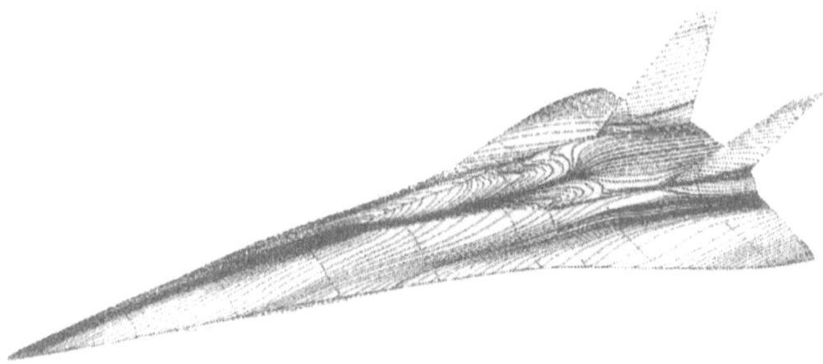

Fig. 7.8. Lower stage of SÄNGER: computed pattern of skin-friction lines on the upper side without upper stage [41].

We consider first the results on the lower (windward) symmetry line of the forebody. We see in Fig. 7.9 that the state of the boundary layer, laminar or turbulent, does not affect strongly the adiabatic (recovery) wall temperature $T_w = T_r$ (cases $\epsilon = 0$). This can be understood by looking at the definition of the wall-heat flux in eq. (7.155). If the temperature gradient at the wall is zero by definition, the thermal conductivity k at the wall (always the laminar value), and in its vicinity (laminar or turbulent) can have any value without strong influence on the balance of thermal convection and conduction, compression and dissipation work, Sub-Section 4.3.2. The zero temperature gradient is also the reason why T_r does not depend on the inverse of some power of the boundary-layer running length x, like the radiation-adiabatic temperature T_{ra} or q_{gw}.

Another explanation for the small influence of the state of the boundary layer can be obtained from a look at the relation for the estimation of the recovery temperature T_r at flat plates, eq. (3.7). If the recovery factor r is equal to $\sqrt{Pr}$ for laminar flow, and equal to $\sqrt[3]{Pr}$ for turbulent flow, T_r will not be much different for laminar and turbulent flow, because for air $Pr = O(1)$. The recovery temperature estimated with eq. (3.7) is constant. Note, however, that the computed T_w drops from the nose region, perfect gas $T_w \approx 2{,}300.0\ K$ (total temperature: $T_t = 2{,}372.4\ K$, recovery temperature laminar with eq. (3.7): $T_r = 2{,}069.4\ K$), by about $150.0\ K$, and only then is approximately constant[25].

High-temperature real-gas effects cannot be neglected at $M_\infty = 6.8$ in the flight situation, at least regarding T_r. The switch from the perfect-gas model

[25] The flow on the lower side of the forebody is not exactly two-dimensional, Fig. 7.7. Hence all results are only more or less monotonic in x/L.

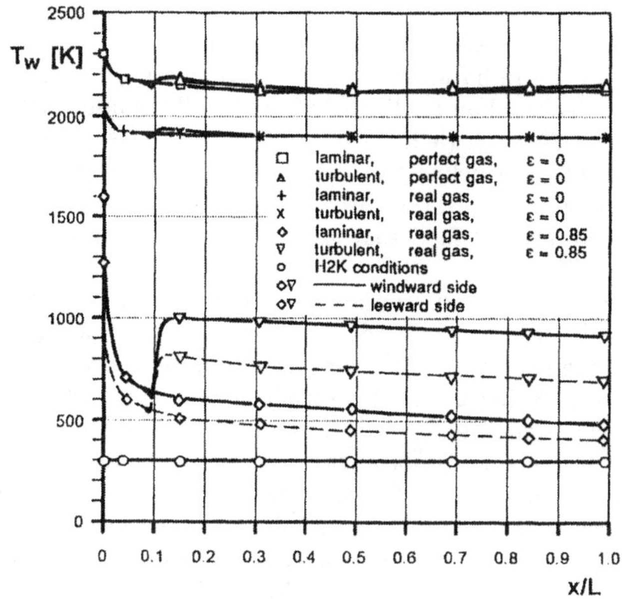

Fig. 7.9. Wall temperatures on the lower and the upper symmetry line of the forebody of the lower stage of SÄNGER in the flight and the wind-tunnel (H2K) situation [37], [38] (configuration see Fig. 7.10). Influence of the state of the boundary layer (laminar/turbulent), gas model, radiation cooling, location (windward side/lee side) on the wall temperature. H2K-data: Navier-Stokes solution [39].

to the equilibrium real-gas model results in a drop of the recovery temperature by approximately 250.0 to 300.0 K for both laminar and turbulent flow.

If we switch on radiation cooling ($\epsilon = 0.85$), the picture changes drastically. The wall temperature is now the radiation-adiabatic temperature. In the case of laminar flow it drops from the stagnation-point temperature, now $T_w = T_{ra} \approx 1{,}600.0\ K$, very fast to temperatures between 600.0 K and 500.0 K. This drop follows quite well the proportionality $T_w \sim x^{-0.125}$ on flat surface portions, see eq. (3.25), see also Table 7.7. It is due to the inverse of the growth of the laminar boundary-layer thickness ($\sim x^{0.5}$), Table 7.3, and to the fact, that the surface-radiation flux is proportional to T_w^4.

The case of turbulent flow reveals at once that the radiation-adiabatic temperature reacts much stronger on the state of the boundary layer, laminar or turbulent, than the adiabatic temperature. In our case we have an increase of $T_w = T_{ra}$ behind the location of laminar-turbulent transition by about 400.0 K.

The transition location was chosen arbitrarily[26] to lie between approximately $x/L = 0.08$ and 0.12. In reality a transition zone of larger extension in

[26] No non-empirical transition criterion is available today, see Section 8.2.

the x-direction may be expected, and also different positions at the windward and the lee side, with possible "tongues" extending upstream or downstream.

The drop of the radiation-adiabatic wall temperature in the case of turbulent flow is weaker than for laminar flow. It follows quite well the proportionality $T_w \sim x^{-0.05}$, eq. (3.25), see also Table 7.7. This is due to the inverse of the growth of the thickness of the viscous sub-layer, respectively of the turbulent scaling length, of the turbulent boundary-layer, $\sim x^{0.2}$, Table 7.3, and of course also to the fact, that the surface-radiation flux is proportional to T_w^4.

In Fig. 7.9 also the radiation-adiabatic temperatures on the upper symmetry line (lee- side of the forebody) are indicated. Qualitatively they behave like those on the windward side, however, the data for laminar flow are approximately 80.0 K lower, and for turbulent flow even 200.0 K lower than on the windward side. We observe that here the increase behind the laminar-turbulent transition zone is not as severe as on the windward side.

The explanation for the different temperature levels on the windward and the lee side lies with the different unit Reynolds numbers on the two sides, and hence with the different boundary-layer thicknesses. We show this with data found with the help of the RHPM-flyer, Chapter 11. The ratio of the unit Reynolds numbers on the leeward (Re_l^u) and on the windward (Re_w^u) symmetry lines at $x/L = 0.5$ for the flight Mach number $M_\infty = 6.8$ and the angle of attack $\alpha = 6.0°$, as well as the computed and the approximatively scaled, eq. (3.34), temperatures are given in Table 7.10.

Table 7.10. Computed and approximatively scaled radiation-adiabatic temperature ratios (leeward ('l') side to windward ('w') side) at $x/L = 0.5$ on the symmetry lines of the SÄNGER forebody at $M_\infty = 6.8$ and $\alpha = 6.0°$.

	Re_l^u/Re_w^u	$T_{ra,l}/T_{ra,w}$ laminar	$T_{ra,l}/T_{ra,w}$ turbulent
Computed data, Fig. 7.9	-	≈ 0.82	≈ 0.76
Scaled data, eq. (3.34)	≈ 0.368 (Chapter 11)	≈ 0.88	≈ 0.84

The skin-friction coefficient on the symmetry line of the windward side of the forebody is given in Fig. 7.10. The influence of the wall temperature on the skin friction is very small, if the boundary layer is laminar, Table 7.5. The skin-friction coefficient drop follows well the primary proportionality $\tau_w \sim x^{-0.5}$, eq. (7.145), see also Table 7.5. This is due to the inverse of the growth of the laminar boundary-layer thickness, $\sim x^{0.5}$, Table 7.3.

The influence of the wall temperature on the skin friction in the cases of turbulent flow is very large. We note first for all three cases a non-monotonic behaviour of c_f behind the transition location in the downstream direction. The cause of this behaviour is not clear. It can be the reaction on the enforced

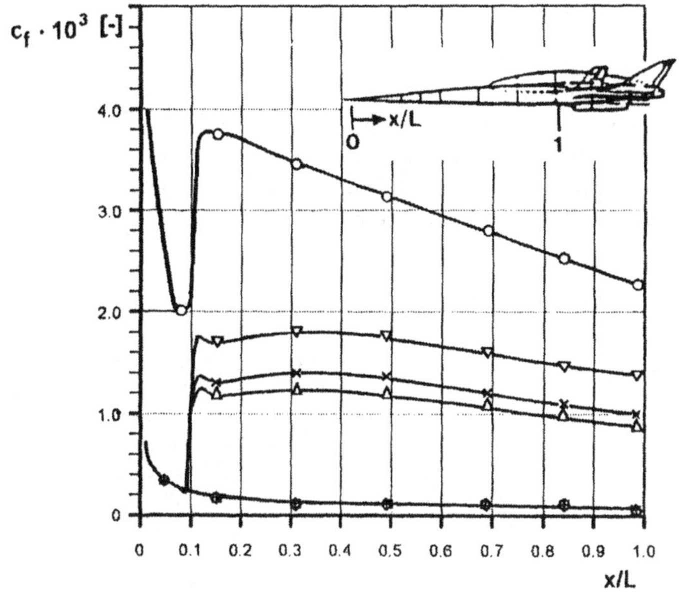

Fig. 7.10. Skin-friction coefficients on the lower symmetry line of the forebody of the lower stage of SÄNGER in the flight and the wind-tunnel (H2K) situation [37], [38] (symbols see Fig. 7.9). Influence of the state of the boundary layer (laminar/turbulent), gas model, radiation cooling on the skin-friction coefficient (windward side only). H2K-data: Navier-Stokes solution [39].

transition of the simple algebraic turbulence model employed in [38], as well as three-dimensional effects. Only for $x/L \gtrapprox 0.5$ the skin-friction coefficient drops approximately proportional to $x^{-0.2}$, which is the primary proportionality for turbulent flow, Table 7.5.

Important is the observation, that the turbulent skin friction is lowest for the perfect gas, adiabatic ($\epsilon = 0$) wall case, and largest (approximately 45 per cent larger) for the radiation-cooling ($\epsilon = 0.85$) case. For the equilibrium gas, adiabatic wall case c_f is somewhat larger than for the perfect gas, adiabatic wall case, i. e., high-temperature real-gas effects can play a non-negligible role regarding turbulent skin friction.

The general result is: the smaller the wall temperature, the larger is the skin-friction of a turbulent boundary layer. This behaviour follows with good approximation the dependence $\tau_w \sim (T^*/T_{ref})^{n(1+\omega)-1}$, eq. (7.145), see also Table 7.5.

We give the data computed at $x/L = 0.5$ in Table 7.11. We compare them with data found with the simple relation for flat surface portions with reference-temperature extension, eq. (7.145). This is done with a scaling by means of T^*/T_{ref}. The ratio T^*/T_{ref}, eq. (7.70), is computed with T_w taken from Fig. 7.9, with $\gamma_e = 1.4$, $M_e = M_\infty = 6.8$, and $r^* = r^*(T_w)$. We find for turbulent flow that the skin-friction coefficient at the radiation-adiabatic surface is approximately 1.45 times larger than at the adiabatic surface with perfect gas assumption, and approximately 1.3 times at the adiabatic surface with equilibrium real-gas assumption. The scaled data are in fair agreement with the computed data. The scaling of the skin-friction coefficient for laminar flow gives an influence of the wall temperature of less than 10 per cent, which is barely discernible in the data plotted in Fig. 7.10.

Table 7.11. Computed and scaled skin-friction ratios on the windward side at $x/L = 0.5$ of the SÄNGER forebody at $M_\infty = 6.8$ and $\alpha = 6.0°$. 'a' denotes the case with perfect gas, $\epsilon = 0$ (Δ), 'b' the case with real gas, $\epsilon = 0$ (x), and 'c' the case with real gas, $\epsilon = 0.85$ (∇), Fig. 7.9.

	$c_{f,c}/c_{f,a}$	$c_{f,c}/c_{f,b}$
Computed data, Fig. 7.10	≈ 1.477	≈ 1.286
Scaled data, eq. (7.145)	≈ 1.356	≈ 1.296

The results in general show that for the estimation of the viscous drag of CAV-type flight vehicles the surface temperature, which is governed predominantly by radiation cooling, must be properly taken into account, at least for turbulent flow. In view of the fact that the viscous drag may account for 30 to 40 per cent of the total drag, the vehicle surface should be flown as hot as possible.

For a typical cold hypersonic wind-tunnel situation, that of the H2K at DLR Köln-Porz, Germany, a computation has been performed for the SÄNGER forebody, too [39]. The Mach number was the same as the flight Mach number, but the flow parameters were different, and especially the surface temperature T_w was the typical ambient model temperature, Table 7.9. The resulting skin friction is quantitatively and qualitatively vastly different from that computed for the flight situation. In the laminar regime just ahead of the location of enforced transition it is about eight times larger than that computed for the flight situation. The data do not scale properly. In the turbulent regime the skin friction initially is about 2.2 times larger than for the flight situation. The scaling yields 2.65, which is a reasonable result. The slope of c_f in the turbulent regime is much steeper in the wind-tunnel situation than in the flight situation. Both the unit Reynolds number, and the small body length may play a role in so far, as transition was enforced at a local Reynolds number which is too small to sustain turbulence. In the wind tunnel transition was not enforced and the state of the boundary layer was not clearly established.

The large influence of the state of the boundary layer, laminar or turbulent, on wall temperature and on skin friction, if the surface is radiation cooled, poses big problems in hypersonic flight vehicle design. The transition location is very important in view of thermal-surface effects, as well as of thermal loads. The prediction and verification of viscous drag is very problematic, if the flight vehicle is drag critical, which in general holds for CAV-type flight vehicles. Fig. 7.10 indicates that in such cases with present-day wind-tunnel techniques, especially with cold model surfaces, the skin friction cannot be found with the needed degree of accuracy and reliability.

7.4 Problems

Problem 7.1 The flow past a flat plate has a unit Reynolds number $Re^u_\infty = 10^6\ m^{-1}$. Assume incompressible flow and determine on the plate at $x = 1.0\ m$ for a) laminar and b) turbulent flow the boundary-layer thicknesses δ, δ_1, δ_2, the shape factors H_{12}, and for turbulent flow in addition δ_{vs}, and δ_{sc}.

Solution: a) $\delta = 0.005\ m$, $\delta_1 = 0.00172\ m$, $\delta_2 = 0.00066\ m$, $H_{12} = 2.591$. b) $\delta = 0.0233\ m$, $\delta_1 = 0.00292\ m$, $\delta_2 = 0.00271\ m$, $H_{12} = 1.286$, $\delta_{vs} = 0.000116\ m$, $\delta_{sc} = 0.000535\ m$.

Problem 7.2 The flow past a flat plate is that of Problem 7.1. Assume incompressible flow and determine on the plate, but now at $x = 10.0\ m$ for a) laminar and b) turbulent flow the boundary-layer thicknesses δ, δ_1, δ_2, the shape factors H_{12}, and for turbulent flow in addition δ_{vs}, and δ_{sc}.

Problem 7.3 Compare and discuss the results from Problem 7.1 and 7.2 in view of Table 7.3. Program the equations for the thicknesses and plot them as functions of x in the interval $0.0\ m \leqq x \leqq 20.0\ m$.

Problem 7.4 Derive the formula (perfect gas) for the friction drag D_f (with reference-temperature extension) of one side of a flat plate with the length L_{ref} and the reference area A_{ref} (the width of the plate is $b_{ref} = A_{ref}/L_{ref}$). Write the formula for both fully laminar and fully turbulent flow.

Solution: Integrate the general expression for the skin friction, eq. (7.145), along the flat plate and find with $Re_{ref} = \rho_{ref}\, v_{ref}\, L_{ref}/\mu_{ref}$, and the exponent of the power-law relation for the viscosity, eq. (4.15), $\omega_{\mu 2} = 0.65$:

$$C_{D,f} = \frac{2}{1-n} C \frac{1}{(Re_{ref})^n} \left(\frac{T^*}{T_{ref}}\right)^{n(1+\omega_{\mu 2})-1}, \qquad (7.166)$$

$$C_{D,f,lam} = 1.328 \frac{1}{(Re_{ref})^{0.5}} \left(\frac{T^*}{T_{ref}}\right)^{-0.175}, \qquad (7.167)$$

$$C_{D,f,turb} = 0.074 \frac{1}{(Re_{ref})^{0.2}} \left(\frac{T^*}{T_{ref}}\right)^{-0.67}, \qquad (7.168)$$

and

$$D_f = C_{D,f}\, q_{ref}\, A_{ref}. \qquad (7.169)$$

Remember that eq. (7.145), and hence also eqs. (7.166) to (7.168), hold only in a certain Reynolds number range. See in this regard the discussion in the summary of Sub-Section 7.2.1.

Problem 7.5 Compute the friction drag D_f of a flat plate at zero angle of attack. The parameters are similar to those in Problem 6.8: $M_\infty = 6$, $H = 30.0\ km$, $A_{ref} = 1{,}860.0\ m^2$, $L_{ref} = 80.0\ m$. Disregard possible hypersonic viscous interaction, Section 9.3. Use the power-law relation, eq. (4.15), for the viscosity, take in the relation for the reference temperature, eq. (7.70), $\gamma = 1.4$, $Pr^* = 0.74$. Assume laminar flow and two wall temperatures: a) $T_w = 1{,}000.0\ K$, $T_w = 2{,}000.0\ K$.

Solution: $Re_{\infty,L} = 1.692 \cdot 10^8$, $q_\infty = 30{,}181.62\ Pa$. $D_{f,lam}$ = a) 8,962.96 N, b) 8,310.64 N.

Problem 7.6 Compute the friction drag D_f of a flat plate at zero angle of attack like in Problem 7.5, but now for turbulent flow.

Solution: $D_{f,turb}$ = a) 72,507.6 N, b) 54,453.68 N.

Problem 7.7 Discuss the results from Problem 7.5 and 7.6, especially also the influence of the wall temperature on the friction drag.

Problem 7.8 Compute the friction forces L_f and D_f for the RHPM-CAV-flyer of Problem 6.8. On the windward (w) side we have: $M_w = 5.182$, $Re^u_w = 3.123 \cdot 10^6\ m^{-1}$, $u_w = 1{,}773.63\ m/s$, $\rho_w = 0.0327\ kg/m^3$, $T_w = 291.52\ K$, and on the lee (l) side: $M_l = 6.997$, $Re^u_l = 1.294 \cdot 10^6\ m^{-1}$, $u_l = 1{,}840.01\ m/s$, $\rho_l = 0.00927\ kg/m^3$, $T_l = 172.12\ K$. Assume fully turbulent flow, wall temperatures $T_{w,w} = 1{,}000.0\ K$, $T_{w,l} = 800.0\ K$, and the other parameters like in Problem 7.6.

Solution: Compute the friction forced on both sides, add them to find the total friction force $D_{f,t}$ and decompose it to find L_f = - 8,950.6 N and D_f = 85,180.3 N.

Problem 7.9 Add the inviscid parts, Problem 6.8, and the viscous parts, Problem 7.8, of the lift and the drag of the RHPM-CAV-flyer, compute the lift to drag ratio and discuss the results.

References

1. E. H. HIRSCHEL. "Evaluation of Results of Boundary-Layer Calculations with Regard to Design Aerodynamics". AGARD R-741, 1986, pp. 5-1 to 5-29.
2. J. COUSTEIX, D. ARNAL, B. AUPOIX, J. PH. BRAZIER, A. LAFON. "Shock Layers and Boundary Layers in Hypersonic Flows". Progress in Aerospace Sciences, Pergamon Press, Vol. 30, 1994, pp. 95 - 212.

3. E. H. Hirschel. "Vortex Flows: Some General Properties, and Modelling, Configurational and Manipulation Aspects". AIAA-Paper 96-2514, 1996.

4. E. H. Hirschel, W. Kordulla. "Shear Flow in Surface-Oriented Coordinates". *Notes on Numerical Fluid Mechanics, Vol. 4*. Vieweg, Braunschweig/Wiesbaden, 1981.

5. D. C. Wilcox. "Turbulence Modelling for CFD". DCW Industries, La Cañada, CAL., USA, 1998.

6. R. B. Bird, W. E. Stewart, E. N. Lightfoot. "Transport Phenomena". John Wiley, New York and London/Sydney, 2nd edition, 2002.

7. L. Prandtl. "Über Flüssigkeitsbewegung bei sehr kleiner Reibung". Proc. Third International Mathematical Congress, Heidelberg, Germany, 1904.

8. D. Frederick, T. S. Chang. "Continuum Mechanics". Allyn and Bacon, Boston, Mass., 1965.

9. D. R. Chapman, M. W. Rubesin. "Temperature and Velocity Profiles in the Compressible Laminar Boundary Layer with Arbitrary Distribution of Surface Temperature". Journal of the Aeronautical Sciences, Vol. 16, 1949, pp. 547 - 565.

10. E. H. Hirschel. "Boundary-Layer Coordinates on General Wings and Fuselages". Zeitschrift für Flugwissenschaften und Weltraumforschung (ZFW), Vol. 6, No. 3, 1982, pp. 194 - 202.

11. M. van Dyke. "Perturbation Methods in Fluid Mechanics". The Parabolic Press, Stanford, 1975.

12. B. Aupoix, J. Ph. Brazier, J. Cousteix, F. Monnoyer. "Second-Order Effects in Hypersonic Boundary Layers". *J. J. Bertin, J. Periaux, J. Ballmann (eds.), Advances in Hypersonics, Vol. 3, Computing Hypersonic Flows*. Birkhäuser, Boston, 1992, pp. 21 - 61.

13. R. Courant, D. Hilbert. "Methods of Mathematical Physics". Vol. II. John Wiley-Interscience, New York, 1962.

14. M. W. Rubesin, H. A. Johnson. "A Critical Review of Skin Friction and Heat Transfer Solutions of the Laminar Boundary Layer of a Flat Plate". Trans. ASME, Vol. 71, 1949, pp. 385 - 388.

15. E. R. G. Eckert. "Engineering Relations of Friction and Heat Transfer to Surfaces in High-Velocity Flow". J. Aeronautical Sciences, Vol. 22, No. 8, 1955, pp. 585 - 587.

16. G. Simeonides, L. Walpot, M. Netterfield, G. Tumino. "Evaluation of Engineering Heat Transfer Prediction Methods in High Enthalpy Flow Conditions". AIAA-Paper 96-1860, 1996.

17. E. H. Hirschel. "Untersuchung grenzschichtähnlicher Strömungen mit Druckgradienten bei grossen Anström-Machzahlen (Investigation of Boundary Layer Like Flows at Large Free-Stream Mach Numbers)". Inaugural Thesis, RWTH Aachen, Germany, 1975, also DLR FB 76-22, 1976.

18. Y. C. Vigneron, J. V. Rakich , J. C. Tannehill. "Calculation of Supersonic Viscous Flows over Delta Wings with Sharp Subsonic Leading Edges". AIAA-Paper 78-1137, 1978.

19. E. H. Hirschel. "Untersuchungen zum Problem Normaler Druckgradienten in Hyperschallgrenzschichten". *Laminare und turbulente Grenzschichten.* DLR Mitt. 71-13, 1971, pp. 51 - 73.
20. H. Schlichting. "Boundary Layer Theory". 7th edition, McGraw-Hill, New York, 1979.
21. F. M. White. "Viscous Fluid Flow". McGraw-Hill, New York, 2nd edition, 1991.
22. E. R. G. Eckert, R. M. Drake. "Heat and Mass Transfer". MacGraw-Hill, New York, 1950.
23. G. Simeonides. "On the Scaling of Wall Temperature Viscous Effects". ESA/ESTEC EWP - 1880, 1996.
24. M. J. Lighthill. "On Displacement Thickness". J. Fluid Mechanics, Vol. 4, 1958, pp. 383 - 392.
25. F. Monnoyer, Ch. Mundt, M. Pfitzner. "Calculation of the Hypersonic Viscous Flow past Reentry Vehicles with an Euler-Boundary Layer Coupling Method". AIAA-Paper 90-0417, 1990.
26. G. Simeonides. "Hypersonic Shock Wave Boundary Layer Interactions over Compression Corners". Doctoral Thesis, University of Bristol, U.K., 1992.
27. G. B. Schubauer, P. S. Klebanoff. "Contributions on the Mechanics of Boundary Layer Transition". NACA TN 3489, 1955, and NACA R 1289, 1956.
28. W. Mangler. "Zusammenhang zwischen ebenen und rotatiossymmetrischen Grenzschichten in kompressiblen Flüssigkeiten". ZAMM, Vol. 28, 1948, pp. 97 - 103.
29. E. Reshotko. "Heat Transfer to a General Three-Dimensional Stagnation Point". Jet Propulsion, Vol. 28, 1958, pp. 58 - 60.
30. E. Reshotko, I. E. Beckwith. "Compressible Laminar Boundary Layer over a Yawed Infinite Cylinder with Heat Transfer and Arbitrary Prandtl Number". NACA R 1379, 1958.
31. G. Simeonides. "Generalized Reference-Enthalpy Formulation and Simulation of Viscous Effects in Hypersonic Flow". Shock Waves, Vol. 8, No. 3, 1998, pp. 161 - 172.
32. L. Lees. "Laminar Heat Transfer over Blunt-Nosed Bodies at Hypersonic Speeds". Jet Propulsion, Vol. 26, 1956, pp. 259 - 269.
33. J. D. Anderson. "Hypersonic and High Temperature Gas Dynamics". McGraw-Hill, New York, 1989.
34. G. Simeonides. "Simple Formulations for Convective Heat Transfer Prediction over Generic Aerodynamic Configurations and Scaling of Radiation-Equilibrium Wall Temperature". ESA/ESTEC EWP 1860, 1995.
35. J. A. Fay, F. R. Riddell. "Theory of Stagnation Point Heat Transfer in Dissociated Gas". Journal of Aeronautical Science, Vol. 25, No. 2, 1958, pp. 73 - 85.
36. D. I. A. Poll. "Transition Description and Prediction in Three-Dimensional Flow". AGARD-R-709, 1984, pp. 5-1 to 5-23.

37. E. H. HIRSCHEL. "Thermal Surface Effects in Aerothermodynamics". *Proc. Third European Symposium on Aerothermodynamics for Space Vehicles, Noordwijk, The Netherlands, November 24 - 26, 1998.* ESA SP-426, 1999, pp. 17 - 31.

38. M. A. SCHMATZ, R. K. HÖLD, F. MONNOYER, CH. MUNDT, H. RIEGER, K. M. WANIE. "Numerical Methods for Aerodynamic Design II". Space Course 1991, RWTH Aachen, 1991, pp. 62-1 to 62-40.

39. R. RADESPIEL. Personal communication. 1994.

40. I. OYE, H. NORSTRUD. "Personal communication". 2003.

41. W. SCHRÖDER, R. BEHR, S. MENNE. "Analysis of Hypersonic Flows Around Space Transportation Systems via CFD Methods". AIAA-Paper 93-5067, 1993.

8 Laminar-Turbulent Transition and Turbulence in High-Speed Viscous Flow

The state of the boundary layer, laminar or turbulent, influences strongly the thermal state of the surface, in particular if the surface is radiation cooled. The thermal state governs thermal surface effects and thermal loads. Regarding the former, a strong back-coupling to the state of the boundary layer exists. Especially the behaviour of hydrodynamic stability, and hence laminar-turbulent transition, are affected by the thermal state of the surface.

We have seen that laminar-turbulent transition strongly rises the radiation-adiabatic temperature, Figs. 3.3 and 7.9. This is important on the one hand for the structure and materials layout of a hypersonic flight vehicle. On the other hand, transition rises also the wall shear stress, Fig. 7.10, to a large extent. Both the temperature and the shear stress rise are due to the fact that the characteristic boundary-layer thickness of the ensuing turbulent boundary layer, the thickness of the viscous sub-layer, δ_{vs}, is much smaller than the characteristic thickness of the - without transition - laminar boundary layer. The latter is the thermal (δ_T), respectively the flow (δ_{flow}) boundary-layer thickness, with $\delta_T \approx \delta_{flow}$ in general.

Boundary-layer turbulence and its modeling is a wide-spread topic also in aerothermodynamics, but perhaps not so much - at least presently - the origin of turbulence, i. e., the phenomenon of laminar-turbulent transition and its modeling.

RV-type flight vehicles in general are not very sensitive to laminar-turbulent transition. Transition there concerns mainly thermal loads. Above approximately 60.0 to 40.0 km altitude the attached viscous flow is laminar, below that altitude it becomes turbulent, beginning usually at the rear part of the vehicle. Because the largest thermal loads occur at approximately 70.0 km altitude, the trajectory part with laminar flow is the governing one regarding thermal loads. Of course other trajectory patterns than the present baseline pattern, especially also contingency abort trajectories, can change the picture. Transition phenomena, however, may also appear at high altitudes locally on RV-type flight vehicles, for instance, on deflected trim or control surfaces, due to shock/boundary-layer interaction and local separation.

Important is the observation that due to the large angles of attack, at least down to approximately 40.0 km altitude, Fig. 1.3 in Section 1.2, the boundary-layer edge Mach numbers are rather small on the windward side of RV-type flight vehicles[1]. On the windward side of the Space Shuttle, the transition location lies at approximately 90 per cent vehicle length at approximately 50.0 km altitude while $\alpha \approx 35°$, and has moved forward to approximately 10 per cent vehicle length at approximately 40.0 km altitude while $\alpha \approx 30°$ [1]. This means that on RV-type flight vehicles in general laminar-turbulent transition happens on the windward side actually not in a hypersonic boundary layer, but in an at most low supersonic boundary layer[2].

On (airbreathing) CAV-type flight vehicles, laminar-turbulent transition is not only a matter of thermal loads, but also, since such vehicles are drag-sensitive in general, also a matter of viscous drag and of airframe/propulsion integration, see, e. g., [2].

For the US National Aerospace Plane (NASP/X-30) it was reported that the uncertainty of the location of laminar-turbulent transition affects the take-off mass of the vehicle by a factor of two or more [3]. NASP/X-30, [4], was an extremely ambitious project. Strongly influenced by laminar-turbulent transition were mainly thermal loads, viscous drag, and the engine inlet onset flow (height of the boundary-layer diverter).

In the background of such a case looms a vicious snow-ball effect, see, e. g., [5]. Uncertainties in vehicle mass and total drag prediction lead to design margins, e. g. [6], which make, for instance, more engine thrust necessary. As a consequence bigger engines and a larger fuel tank volume are needed, hence a larger engine and tank mass, a larger airframe volume and a larger wetted vehicle surface, consequently a larger total drag, and finally a larger take-off mass.

A flight vehicle is weight-critical, or mass-sensitive, if the take-off mass grows strongly with the ratio 'empty-vehicle mass' to 'take-off mass', see, e. g., [7]. Large mass-growth factors together with small payload fractions are typical for CAV-type flight vehicles. Such vehicles usually are viscous-effects dominated and especially transition sensitive. Laminar-turbulent transition definitely is the key problem in the design of CAV-type and ARV-type flight vehicles.

At CAV-type flight vehicles transition occurs indeed in hypersonic boundary layers. These vehicles typically fly at angles of attack which are rather small, see Fig. 1.3 for the SÄNGER space transportation system up to separation of the upper stage at about 35.0 km altitude. Because of the small

[1] In [1] it is reported that during a Space Shuttle re-entry these are typically at most $M_e \approx 2.5$, and mostly below $M_e \approx 2$.

[2] The windward side boundary layer on a RV-type flight vehicle at large angle of attack is initially a subsonic, then a transonic, and finally a low supersonic, however, not ordinary boundary layer. It is characterized by large temperatures and hence high-temperature real-gas effects (although in the respective speed/altitude regime non-equilibrium effects don't necessarily play a role), by strong wall-normal temperature gradients due to surface-radiation cooling, and is, depending on the type of the TPS, influenced by surface roughness.

angles of attack, the boundary-layer edge Mach numbers will be of the order of magnitude of the flight Mach number. On the windward side, with pre-compression in order to reduce the necessary inlet capturing area, the boundary-layer edge Mach number will be somewhat smaller, but in any case the boundary layer also here is a hypersonic boundary layer.

The problems with laminar-turbulent transition and with turbulence are the insufficient understanding of the involved phenomena on the one hand, and the deficits of the ground-simulation means on the other hand. This holds for both ground-facility and computational simulation.

However, once hypersonic attached viscous flow can be considered as turbulent, i. e., if shape and location of the transition zone have been somehow established, it usually is possible to compute the properties of such flow to a fair degree of accuracy, see, e. g. [8], [9], and also Section 7.2. The situation changes negatively if turbulent strong interaction phenomena and flow separation are present.

In hypersonic ground-simulation facilities basically the low attainable Reynolds numbers, the (in general wrong) disturbance[3] environment, which the tunnel poses for the boundary layer on the model, and the thermal state of the model surface are the problems, Section 10.3. Either the Reynolds number (though lower than in flight) is large enough for laminar-turbulent transition to occur in a ground-simulation facility, although in general with wrong shape and location of the transition zone due to the wrong disturbance environment and the wrong thermal state of the surface, or artificial turbulence triggering must be employed (where, again, shape and location of the transition zone must have been somehow guessed). In the latter case, too, the Reynolds number still must be large enough to sustain the artificially created turbulence. If that is the case, turbulent attached flow and strong interaction phenomena and separation can be simulated, however, in general without taking into account the proper thermal state of the surface. In general the model surface is cold, which is in contrast to the actual flight situation, see, e. g., Figs. 3.3 and 7.9.

With laminar-turbulent transition the situation is different to that of turbulence. *Boundary-layer transition is a problem that has plagued several generations of aerodynamicists. There are very few things about transition that are known with certainty, other than the fact that it happens if the Reynolds number is large enough* (K. F. Stetson, 1992 [10]). Certainly we know much more about laminar-turbulent transition today than in 1992, but an empirical or semi-empirical transition prediction with the needed accuracy and reliability, if a hypersonic vehicle design is viscous-effects sensitive, or even a non-empirical transition prediction, is not yet possible. Unfortunately, an experimental determination of the transition location is also not possible in ground-simulation facilities (see above).

[3] The (unstable) boundary layer responds to disturbances which are present in flight or, in general wrongly, in the ground-simulation facility, Sub-Section 8.1.6.

In the following Section 8.1 we try to draw a picture of the different instability and transition phenomena and their dependencies on flow-field parameters and vehicle surface properties, including the thermal state of the surface[4]. In Section 8.2 state-of-the-art transition criteria for hypersonic flight-vehicle design purposes are given with due reservations regarding their applicability and accuracy.

Turbulence in hypersonic flows and its modeling will be treated rather briefly and with emphasis on computational simulation in Section 8.3.

8.1 Laminar-Turbulent Transition as Hypersonic Flow Phenomenon

Laminar-turbulent transition in high-speed flow is a phenomenon with a multitude of possible instability and receptivity mechanisms, which depend on a multitude of flow, surface and environment parameters. In the frame of this book only an overview over the most important issues can be given. Detailed introductions to the topic are found in, e. g., [11], [12], [13], [10].

Possibly two basic transition scenarios can be distinguished, which, however, may overlap to a certain degree[5]:

1. regular transition,
2. forced or by-pass transition.

These two scenarios can be characterized as follows:

- Regular transition occurs if, once a boundary layer is unstable, low-intensity level disturbances, which fit the receptivity properties of the unstable boundary layer, undergo first linear, then non-linear amplification(s), until turbulent spots appear and actual transition to self-sustained turbulence happens. In [14] this is called "transition emanating from exponential instabilities".
  This scenario has been discussed in detail in the classical paper by Morkovin, [11], see also [12], who considers the (two-dimensional) laminar boundary layer as "linear and non-linear operator" which acts on small disturbances like free-stream vorticity, sound, entropy spots, but also high-frequency vibrations. This begins with linear amplification of Tollmien-Schlichting type disturbance waves, which can be modified by boundary-layer and surface properties like those which occur on real flight-vehicle

[4] The thermal state of the surface is seen especially important in view of research activities in ground-simulation facilities and in view of hypersonic flight experiments. In both so far the thermal state of the surface usually is either uncontrolled or not recorded.

[5] For a detailed discussion in a recent publication see [14].

configurations: pressure gradients, thermal state of the surface, three-dimensionality, small roughness, waviness et cetera. It follow non-linear and three-dimensional effects, secondary instability and scale changes and finally turbulent spots and transition.
Probably this is the major transition scenario which can be expected to exist on CAV-type flight vehicles. An open question is how propulsion-system noise and airframe vibrations fit into this scenario.

- Forced transition is present, if large amplitude disturbances, caused, e. g., by surface irregularities, lead to turbulence without the boundary layer acting as convective exponential amplifier, like in the first scenario. Morkovin calls this "high-intensity bypass" transition.
  Probably this is the major scenario on RV-type flight vehicles with a thermal protection system (TPS) consisting of tiles or shingles which pose a surface of large roughness. Forced transition can also happen at the junction, of, for instance, a ceramic nose cone and the regular TPS, if under thermal and mechanical loads a surface step of sufficient size appears. Similar surface disturbances, of course, also can be present on CAV-type flight vehicles, but must be avoided if possible. Attachment-line contamination, Sub-Section 8.1.5, falls also under this scenario.
  Boundary-layer tripping on wind-tunnel models, if the Reynolds number is too small for regular transition to occur, is forced transition on purpose. However, forced transition can also be an - unwanted - issue in ground-simulation facilities, if a large disturbance level is present in the test section. Indeed Tollmien-Schlichting instability originally could only be studied and verified in (a low-speed) experiment after such a wind-tunnel disturbance level was discovered and systematically reduced in the classical work of Schubauer and Skramstadt [15]. In supersonic/hypersonic wind tunnels the sound field radiated from the turbulent boundary layers of the tunnel wall was shown by Kendall, [16], to govern transition at $M = 4.5$, see the discussion of Mack in [17].

In the following sub-sections we sketch basic issues of stability and transition, and kind and influence of the major involved phenomena. We put emphasis on regular transition.

8.1.1 Some Basic Observations

We consider laminar-turbulent transition of the two-dimensional boundary layer over a flat plate as prototype of regular transition, and ask what can be observed macroscopically at its surface. We study the qualitative behaviour of wall shear stress along the surface[6], Fig. 8.1. We distinguish three branches of τ_w. The laminar branch (I) is sketched in accordance with $\sim x^{-0.5}$, and

[6] The radiation-adiabatic temperature and the heat flux in the gas at the wall with fixed cold wall temperature show a similar qualitative behaviour.

the turbulent branch (III) with $\sim x^{-0.2}$, Table 7.5. We call the distance between x_{cr} and $x_{tr,u}$ the transitional branch (II). In consists of the instability sub-branch (IIa) between x_{cr} and $x_{tr,l}$, and the transition sub-branch (IIb) between $x_{tr,l}$ and $x_{tr,u}$. The instability sub-branch overlaps with the laminar branch (see below).

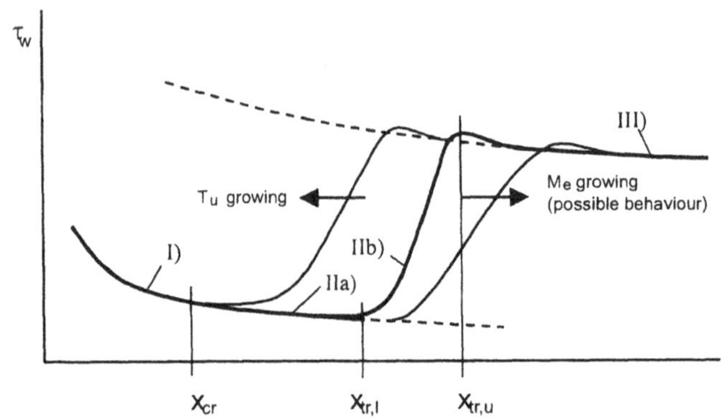

Fig. 8.1. Schematic of behaviour of wall shear stress τ_w in flat-plate boundary-layer flow undergoing laminar-turbulent transition: **I)** laminar branch, **II)** transitional branch with **IIa)** instability sub-branch and **IIb)** transition sub-branch, and **III)** turbulent branch of the boundary layer. x_{cr} is the location of primary instability, $x_{tr,l}$ the "lower" and $x_{tr,u}$ the "upper" location of transition.

In Fig. 8.1 x_{cr} denotes the point of primary instability (critical point). Upstream of x_{cr} the laminar boundary layer is stable, i. e., a small disturbance introduced into it will be damped out. At x_{cr} the boundary layer is neutrally stable, and downstream of it is unstable. Disturbances there trigger Tollmien-Schlichting waves (normal modes of the boundary layer) whose amplitudes grow rather slowly[7]. Secondary instability sets in after the Tollmien-Schlichting amplitudes have reached approximately 1 per cent of u_e, i. e. at amplitudes where non-linear effects are still rather small regarding the (primary) Tollmien-Schlichting waves. Finally turbulent spots appear and the net-production of turbulence begins (begin of sub-branch IIb). This location is the "lower" location of transition, $x_{tr,l}$. At the "upper" location of transition, $x_{tr,u}$, the boundary layer is fully turbulent[8]. This means that now

[7] Tollmien-Schlichting waves can propagate with the wave vector aligned with the main-flow direction ("normal" wave as a two-dimensional disturbance, wave angle $\psi = 0$) or lying at a finite angle to it ("oblique" wave as three-dimensional disturbance, wave angle $\psi \neq 0$). The most amplified Tollmien-Schlichting waves are usually in two-dimensional low-speed flows the two-dimensional waves, and in two-dimensional supersonic and hypersonic flows the oblique waves.

[8] In [14] the branches I and II are, less idealized, divided into five stages : 1) disturbance reception (branch I ahead of x_{cr} and part of sub-branch IIa), 2) linear

the turbulent fluctuations transport fluid and momentum towards the surface such that the full time-averaged velocity profile shown in Fig. 7.5 b) develops.

The length of the transition region, related to the location of primary instability, can be defined either as:

$$\Delta x_{tr} = x_{tr,l} - x_{cr}, \tag{8.1}$$

or as

$$\Delta x_{tr} = x_{tr,u} - x_{cr}. \tag{8.2}$$

In Fig. 8.1 some important features of the transition region are indicated:

- The time-averaged ("mean") flow properties practically do not deviate in the instability sub-branch (IIa) between x_{cr} and $x_{tr,l}$ from those of laminar flow (branch I). This feature permits the formulation of stability and especially transition criteria and models based on the properties of the laminar flow branch. It is of very large importance for practical instability and transition predictions, the latter still based on empirical or semi-empirical models and criteria.
- The transition sub-branch (IIb), i. e. $\Delta x'_{tr} = x_{tr,u}$ - $x_{tr,l}$, usually is very narrow. It is characterized by the departure of τ_w from that of the laminar branch and by its joining with that of branch III. For boundary-layer edge flow Mach numbers $M_e \gtrapprox 4$ to 5 the (temporal) amplification rates of disturbances can decrease with increasing Mach number, therefore a growth of $\Delta x'_{tr}$ is possible. In such cases transition criteria based on the properties of the laminar flow branch would become questionable. The picture in reality, however, is very complicated, as was shown first by Mack in 1965, [18], [17]. We will come back to that later.
- At the end of the transitional branch (II), $x_{tr,u}$, the wall shear stress overshoots shortly[9] that of the turbulent branch (III). This overshoot occurs also for the heat flux in the gas at the wall. At a radiation-adiabatic surface this overshoot can lead to a hot-spot situation relative to the nominal branch III situation.

growth of (unstable) disturbances (largest part of sub-branch IIa), 3) non-linear saturation (towards the end of sub-branch IIa), 4) secondary instability (towards the very end of sub-branch IIa), 5) breakdown (begin of sub-branch IIb). The term "breakdown" has found entry into the literature. It is a somewhat misleading term in so far as it suggests a sudden change of the (secondary unstable) flow into the turbulent state. Actually it means the "breakdown" of identifiable structures in the disturbance flow. In sub-branch IIb a true "transition" into turbulence occurs. Sub-branch IIb, i. e., the length $x_{tr,u}$ - $x_{tr,l}$, can be rather large, especially in high-speed flows.

[9] This overshoot occurs, because the (viscous sub-layer) thickness of the turbulent boundary layer is initially smaller than what it would have been, if the boundary layer grew turbulent from the plate's leading edge on.

- With increasing disturbance level, for example increasing free-stream turbulence T_u in a ground-simulation facility, the transition sub-branch (IIa) will move upstream while becoming less narrow, see, e. g., [19]. When disturbances grow excessively and transition becomes forced transition, transition criteria based on the properties of the laminar flow branch (I) become questionable.
- In general it can be observed, that boundary-layer mean flow properties, which destabilize the boundary layer, see Sub-Section 8.1.3, shorten the length of sub-branch IIa ($x_{tr,l}$ - x_{cr}) as well as that of sub-branch IIb ($x_{tr,u}$ - $x_{tr,l}$). The influence of an adverse stream-wise pressure gradient is most pronounced in this regard. If the mean flow properties have a stabilizing effect, the transition sub-branches IIa and IIb become longer.

8.1.2 Outline of Stability Theory

We sketch now some features of linear stability theory. This will give us insight into the basic dependencies of instability but also of transition phenomena, [20], [14]. Of course, linear stability theory does not explain all of the many phenomena of regular transition which can be observed. It seems, however, that at a sufficiently low external disturbance level linear instability is the primary cause of regular transition, [17], [11].

For the sake of simplicity we consider only the two-dimensional incompressible flat plate case (Tollmien-Schlichting instability) with due detours to our topic, instability and transition in high-speed attached viscous flow. The basic approach and many of the formulations, e. g., concerning temporally and spatially amplified disturbances, however are the same for both incompressible and compressible flow [21], [22].

Tollmien-Schlichting theory begins with the introduction of splitted flow parameters $\widehat{q} = q + q'$ into the Navier-Stokes equations and their linearization (q denotes mean flow, and q' disturbance flow parameters). It follows the assumption of parallel boundary-layer mean flow[10], i. e., $v \equiv 0$. The consequence is $\partial u/\partial x \equiv 0$. Hence in this theory only a mean flow $u(y)$ is considered, without dependence on x. Therefore we speak about linear and local stability theory. The latter means that only locally, i. e., in locations x on the surface under consideration, which can be arbitrarily chosen, stability properties of the boundary layer are investigated[11].

The disturbances q' are then formulated as sinusoidal disturbances:

---

[10] There are situations in which this is not allowed, because information of the mean flow is lost, which is of importance for the stability/instability behaviour of the boundary layer ("non-parallel" effects, [14]).

[11] Non-local stability theory and methods, see Sub-Section 8.2.1, are based on parabolized stability equations (PSE). With them the stability properties are investigated taking into account the whole boundary-layer domain of interest.

$$q'(x,y,t) = q'_A(y)e^{i(\alpha x - \omega t)}. \tag{8.3}$$

Here $q'_A(y)$ is the complex disturbance amplitude as function of y, and α and ω are parameters regarding the disturbance behaviour in space and time.

The complex wave number α is:

$$\alpha = \alpha_r + i\alpha_i, \tag{8.4}$$

with

$$\alpha_r = \frac{2\pi}{\lambda_x}, \tag{8.5}$$

λ_x being the complex length of the in x-direction propagating disturbance wave, and ω the complex circular frequency:

$$\omega = \omega_r + i\omega_i, \tag{8.6}$$

with

$$\omega_r = 2\pi f, \tag{8.7}$$

f being the complex frequency of the wave.

The phase velocity is:

$$c = c_r + ic_i = \frac{\omega}{\alpha}. \tag{8.8}$$

Temporal amplification of an amplitude A is found, with α real-valued, by:

$$\frac{1}{A}\frac{dA}{dt} = \frac{d}{dt}(lnA) = \omega_i = \alpha_r c_i, \tag{8.9}$$

and spatial amplification by:

$$\frac{1}{A}\frac{dA}{dx} = \frac{d}{dx}(lnA) = -\alpha_i. \tag{8.10}$$

We see that a disturbance is amplified, if $\omega_i > 0$, or $\alpha_i < 0$. It is damped, if $\omega_i < 0$, or $\alpha_i > 0$, and neutral, if $\omega_i = 0$, or $\alpha_i = 0$.

The total amplification rate in the case of temporal amplification follows from eq. (8.9) with

$$\frac{A(t)}{A_0} = e^{\int_{t_0}^{t} \omega_i dt}, \tag{8.11}$$

and in the case of spatial amplification from eq. (8.10) with

$$\frac{A(x)}{A_0} = e^{\int_{x_0}^{x} (-\alpha_i) dx}. \tag{8.12}$$

If we assume ω_i or $-\alpha_i$ to be constant in the respective integration intervals, we observe for the amplified cases from these equations the unlimited exponential growth of the amplitude A which is typical for linear stability theory with

$$\frac{A(t)}{A_0} = e^{\omega_i t}, \tag{8.13}$$

and

$$\frac{A(x)}{A_0} = e^{-\alpha_i x} \tag{8.14}$$

respectively.

With the introduction of a disturbance stream function $\Psi'(x,y,t)$, where $\Phi(y)$ is the complex amplitude:

$$\Psi'(x,y,t) = \Phi(y)e^{i(\alpha x - \omega t)} \tag{8.15}$$

into the linearized and parallelized Navier-Stokes equations finally the Orr-Sommerfeld equation is found[12]:

$$(u-c)(\Phi_{yy} - \alpha^2\Phi) - u_{yy}\Phi = -\frac{1}{\alpha Re_\delta}(\Phi_{yyyy} - 2\alpha^2\Phi_{yy} + \alpha^4\Phi). \tag{8.16}$$

The properties of the mean flow, i. e., the tangential boundary-layer velocity profile, appear as $u(y)$ and its second derivative $u_{yy} = d^2u/dy^2(y)$. The Reynolds number Re_δ on the right-hand side is defined locally with boundary-layer edge data[13] and the boundary-layer thickness[14] δ:

$$Re_\delta = \frac{\rho_e u_e \delta}{\mu_e}. \tag{8.17}$$

Obviously stability or instability of a boundary layer depend locally on the mean-flow properties u, u_{yy}, and the Reynolds number Re_δ. A typical stability chart is sketched in Fig. 8.2. The boundary layer is temporally unstable in the shaded area ($c_i > 0$, see eq. (8.9)) for $0 < \alpha \leqq \alpha_{max}$ and $Re \geqq Re_{cr}$, Re_{cr} being the critical Reynolds number. For $Re \gg Re_{cr}$ we see that the domain of instability shrinks asymptotically to zero, the boundary layer becomes stable again.

For large Re_δ the right-hand side of eq.(8.16) can be neglected which leads to the Rayleigh equation:

$$(u-c)(\Phi_{yy} - \alpha^2\Phi) - u_{yy}\Phi = 0. \tag{8.18}$$

---

[12] $()_{yy}$ et cetera stands for twofold differentiation with respect to y et cetera.

[13] The boundary-layer thickness δ, found, for instance, with eq. (7.91), Sub-Section 7.2.1, defines the boundary-layer edge.

[14] For practical purposes usually the displacement thickness δ_1 or the momentum thickness δ_2 is employed.

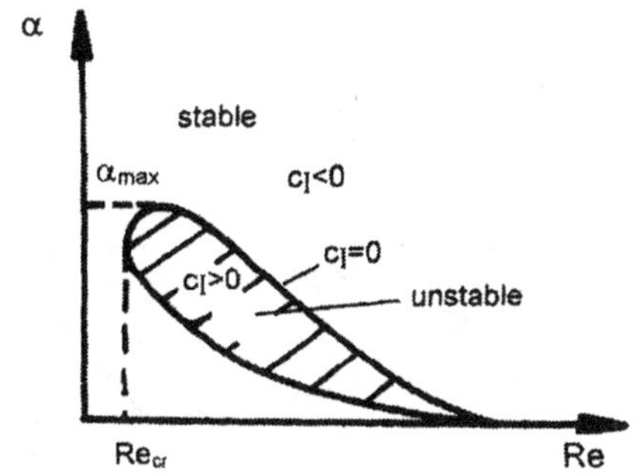

Fig. 8.2. Schematic of a temporal stability chart of a boundary layer at a flat plate ($c_I \equiv c_i$).

Stability theory based on this equation is called "inviscid" stability theory. This is sometimes wrongly understood. Of course, only the viscous terms in the Orr-Sommerfeld equation are neglected, but stability properties of viscous flow can properly be investigated with it, except, of course, for the plain flat-plate flow.

8.1.3 Inviscid Stability Theory and the Point-of-Inflexion Criterion

Inviscid stability theory gives insight into instability mechanisms with the point-of-inflexion criterion, which follows from the Rayleigh equation. It says basically, [20], [14], that the presence of a point of inflexion is a sufficient condition for the existence of amplified disturbances with a phase speed $0 \leqq c_r \leqq u_e$. In other words, the considered boundary-layer profile $u(y)$ is unstable, if it has a point of inflexion[15]:

$$\frac{d^2u}{dy^2} = 0 \tag{8.19}$$

lying in the boundary layer at y_{ip}:

$$0 < y_{ip} \leqq \delta. \tag{8.20}$$

The stability chart of a boundary layer with a point-of-inflexion deviates in a typical way from that without a point of inflexion. We show such a stability chart in Fig. 8.3. For small *Re* the domain of instability has the same form as for a boundary layer without point of inflexion, Fig. 8.2. For large *Re* its upper boundary reaches an asymptotic inviscid limit at finite wave number α. For large *Re* the boundary layer thus remains unstable.

[15] The reader is warned that this is a highly simplified discussion. The objective is only to arrive at insights into the basic instability behaviour, not to present detailed theory.

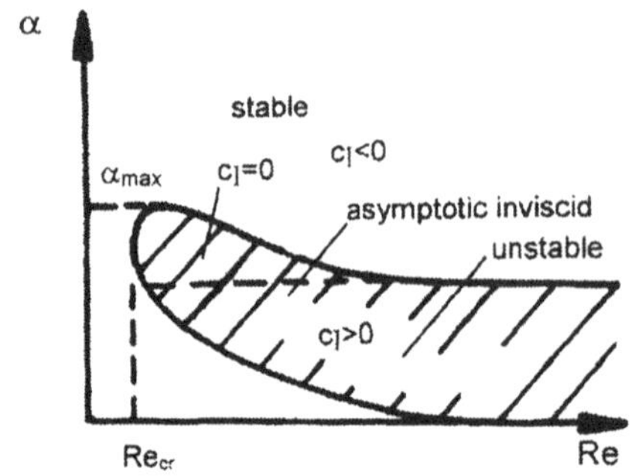

Fig. 8.3. Schematic of a temporal stability chart of a boundary-layer with inviscid instability ($c_I \equiv c_i$).

When does a boundary-layer profile have a point of inflexion? We remember Fig. 7.4 and Table 7.1 in Sub-Section 7.1.5. There the results of the discussion of the generalized wall-compatibility conditions, eqs. (7.53) and (7.54), are given. We treat here only the two-dimensional case, and recall that a point of inflexion exists away from the surface at $y_{poi} > 0$, if the second derivative of $u(y)$ at the wall is positive: $d^2u/dy^2|_w > 0$. In a Blasius boundary layer, the point of inflexion lies at $y = 0$.

With the help of Table 7.1 we find that an air boundary layer is:

- destabilized by an adverse pressure gradient[16,17] ($\partial p/\partial x > 0$), by heating, i. e., a heat flux from the surface into the boundary layer ($\partial T_w/\partial y|_{gw} < 0$) and by blowing through the surface $v_w > 0$),
- stabilized by a favorable pressure gradient ($\partial p/\partial x < 0$), by cooling, i. e., a heat flux from the boundary layer into the surface ($\partial T_w/\partial y|_{gw} > 0$), by suction through the surface $v_w < 0$, and by slip flow $u_w > 0$.

---

[16] This is the classical interpretation of the point-of-inflexion instability. Regarding turbulent boundary layers on bodies of finite thickness, it can be viewed in the following way: downstream of the location of largest thickness of the body $\partial p/\partial x > 0$, hence a tendency of separation of the laminar boundary layer. Point-of-inflexion instability signals the boundary layer to become turbulent, i. e., to increase the lateral transport of momentum (also mass and energy) towards the body surface. The ensuing time-averaged turbulent boundary-layer profile is fuller than the laminar one, Fig. 7.5, which reduces the tendency of separation. Although the wall shear stress goes up, total drag remains small. The flat-plate boundary layer is a special case, where this does not apply.

[17] If the adverse pressure gradient is too strong, the ordinary transition sequence will not happen. Instead the (unstable) boundary layer separates and forms a usually very small and flat separation bubble. At the end of the bubble the flow re-attaches then turbulent (separation-bubble transition, see, e. g. [14] and [23] with further references).

These results basically hold also for hypersonic boundary layers. Stability theory there deals with a generalized point of inflexion [24]:

$$\frac{d}{dy}\left(\rho\frac{du}{dy}\right)_{y_s} = \frac{d}{dy}\left(\frac{1}{T}\frac{du}{dy}\right)_{y_s} = 0, \; y_s > y_0, \tag{8.21}$$

and with even more generalized (doubly generalized) forms, which take into account the metric properties of the body surface, see, e. g., [25].

A sufficient condition for the existence of unstable disturbances is the presence of this point of inflexion at $y_s > y_o$. Here y_0 is the point at which:

$$\frac{u}{u_e} = 1 - \frac{1}{M_e}. \tag{8.22}$$

Further it is required that the "relative Mach number" [21]

$$M_{rel} = M - \frac{c}{a} \tag{8.23}$$

is subsonic throughout the boundary layer. In fact, the condition is $M_{rel}^2 < 1$. Here M is the local Mach number ($M = u/a$), c is the phase velocity of the respective disturbance wave which is constant across the boundary layer ($c \neq c(y)$), and $a = a(y)$ is the local speed of sound.

The physical interpretation of the generalized point of inflexion is, [24]:

- $\frac{d}{dy}\left(\frac{1}{T}\frac{du}{dy}\right) > 0$: energy is transferred from the mean flow to the disturbance, the boundary-layer flow is unstable (sufficient condition),
- $\frac{d}{dy}\left(\frac{1}{T}\frac{du}{dy}\right) < 0$: energy is transferred from the disturbance to the mean flow, the boundary-layer flow is stable,
- $\frac{d}{dy}\left(\frac{1}{T}\frac{du}{dy}\right) = 0$: no energy is transferred between disturbance and mean flow, the boundary-layer flow is neutrally stable.

A systematic connection of the generalized point of inflexion to the surface flow parameters, as we established above for incompressible flow, is possible in principle, but not attempted here. Important is the observation that the distribution of the relative Mach number $M_{rel}(y)$ depends also on the wall temperature. Thus the thermal state of the surface is an important parameter regarding stability or instability of a compressible boundary layer.

8.1.4 Influence of the Thermal State of the Surface and the Mach Number

The original formulation of stability theory for compressible flow of Lees and Lin, [24], with the generalized point of inflexion, led to the result that sufficient cooling can stabilize the boundary layer in the whole Reynolds and Mach number regime of flight [26].

This is an interesting finding, which could help to reduce the thermal load and drag problems of CAV-type flight vehicles with radiation-cooled surfaces, because such vehicles fly with cryogenic fuel. An appropriate layout of the airframe surface as heat exchanger would combine both cooling of the surface and stabilization of the attached laminar viscous flow past it. This would be possible even for flow portions with adverse stream-wise pressure gradient, because, as we can see from eq. (7.53), the influence of the pressure-gradient term can be compensated by sufficiently strong cooling. If then the flow past the flight vehicle would not become turbulent, the heating and drag increments due to the occurrence of transition, see Figs. 7.9 and 7.10, could be avoided.

Unfortunately this conclusion is not true. It was shown almost two decades later by Mack, [18], see also [17], [21], and the discussion in [12], [13], that for $M^2_{rel} > 1$ higher modes (the so-called "Mack modes") appear, which cannot be stabilized by cooling, in contrary, they are amplified by it. The first of these higher modes, the "second mode", if the low-speed mode is called first mode, in general is of largest importance at high boundary-layer edge Mach numbers, because it is most amplified. For an adiabatic, flat-plate boundary-layer higher modes appear at edge Mach numbers larger than $M_e \approx 2.2$. At $M_e \gtrapprox 4$ the second mode has a frequency low enough to definitely influence the instability behaviour. This result is illustrated qualitatively in Fig. 8.4 for a boundary layer on an adiabatic flat plate.

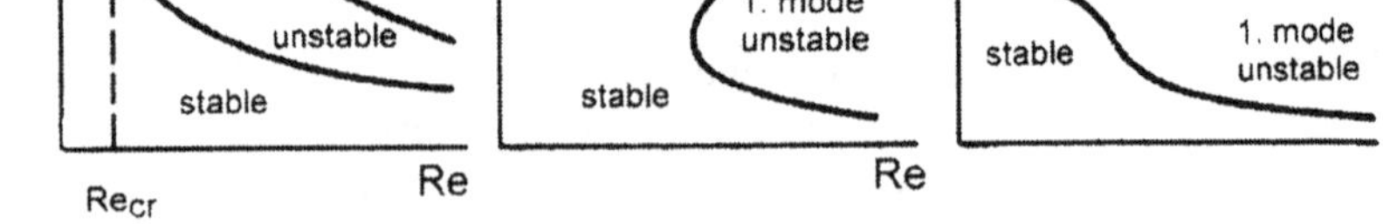

Fig. 8.4. Temporal stability charts of the boundary layer at an adiabatic flat plate for different boundary-layer edge Mach numbers ($c_I \equiv c_i$). Qualitative presentation of the results of Mack, following Reshotko [12]: **a)** $M_e = 0$, **b)** $M_e = 4.5$, **c)** $M_e = 5.8$.

At $M_e = 0$, case a), only the classical (first) instability mode of incompressible flow exists. At $M_e = 4.5$, case b), a second instability mode has appeared, which at $M_e = 5.8$ has merged with the first mode.

If the wall is cooled, the second mode can become important already at low supersonic Mach numbers. The significant finding of Mack is, that the first mode is damped by cooling, whereas the second mode is amplified by it. We show this with numerical results of Kufner, [27], in Fig. 8.5. There the dependence of the spatial amplification rates α_i of the first and the second instability mode on the thermal state of the surface, in this case the wall temperature T_w, and hence implicitly also on the heat flux in the gas at the wall, q_{gw}, is illustrated for a blunt cone (Stetson cone).

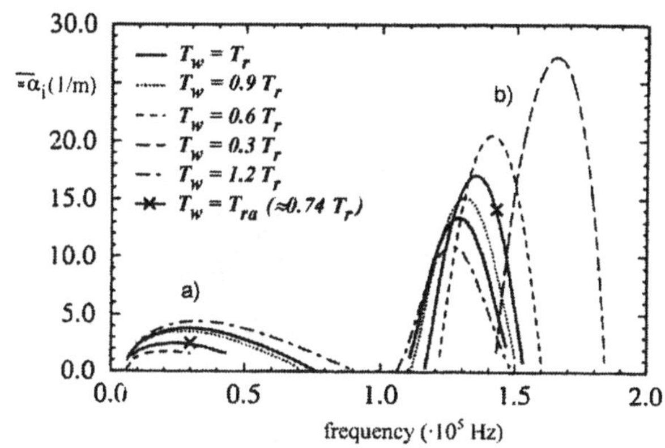

Fig. 8.5. Influence of wall temperature T_w on the spatial amplification rates of **a)** first, and **b)** second instability mode in the boundary layer at a blunt cone at $M_\infty = 8$, $Re^u_\infty = 3.28 \cdot 10^6\ m^{-1}$, $\alpha = 0°$ [27].

The first instability mode, a), on the left-hand side (always the result for the wave angle with maximum amplification is given), exhibits the classical result, that with cooling, i. e., wall temperature lower than the recovery temperature, $T_w < T_r$, the amplification rate is reduced and the maxima are shifted to smaller frequencies. This alone would lead to the transition pattern indicated for increasing Mach number in Fig. 8.1, which is characterized by a widened and flattened transitional branch. In contrast to this, the second instability mode, b), on the right-hand side, is strongly amplified by cooling, with a shift of the maxima towards larger frequencies[18].

It was noted above that for sufficiently small external disturbance levels transition is governed directly by linear instability. Regarding the influence of the thermal state of the surface on the instability and transition behaviour of the boundary layer, this appears to be corroborated by data of Vignau, see [13]. Results of the application of the e^n- method, a semi-empirical transition-prediction method, see Sub-Section 8.2, to flat-plate flow are in fairly good agreement with experimental trends. Fig. 8.6 shows the ratio of 'transition Reynolds number at given wall temperature' to 'transition Reynolds number

[18] Of course, this plot does not indicate how finally the disturbance amplification will behave.

at the adiabatic wall', $Re_{x,t}/Re_{x,t_0}$, as function of the ratio 'wall temperature' to 'adiabatic wall temperature', T_w/T_r, for different boundary-layer edge Mach numbers M_e.

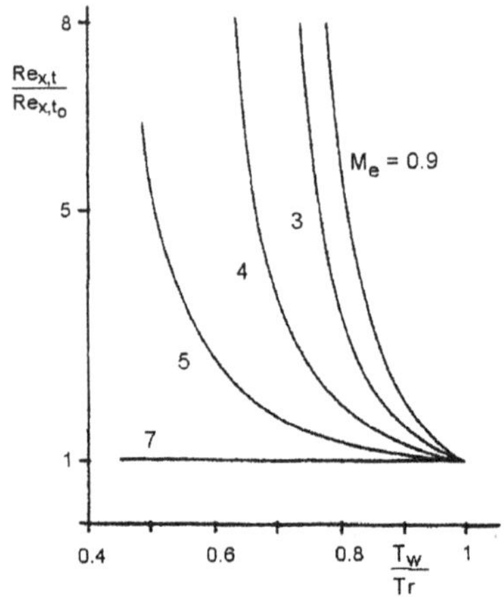

Fig. 8.6. Results of a numerical study by Vignau, see [13], of the influence of wall cooling on the flat-plate transition Reynolds number at different boundary-layer edge Mach numbers M_e. $Re_{x,t}$ is the transition Reynolds number at given wall temperature, Re_{x,t_0} that at the adiabatic wall, T_w the wall temperature, and T_r the adiabatic wall temperature.

The transition Reynolds number $Re_{x,t}$ for $M_e \lessapprox 3$ becomes progressively larger with smaller T_w/T_r. This is due to the reduction of the growth rate of the first instability mode, which is the only mode present at $M_e \lessapprox 3$. The boundary layer with surface cooling appears to be almost fully stabilized. For $M_e \gtrapprox 3$ the transition Reynolds number $Re_{x,t}$ reacts less strongly on the decreasing wall temperature, because now the second instability mode exists which becomes amplified with wall cooling. For $M_e = 7$ wall cooling has no more an effect. Although the use of a constant $n = 9$ in this study can be questioned, it illustrates well the influence of wall cooling on the transition behaviour of a flat-plate boundary layer with hypersonic edge flow Mach numbers.

8.1.5 Real Flight-Vehicle Effects

The infinitely thin flat plate is the canonical configuration of boundary-layer theory and also of stability and transition research. Basic concepts and fundamental results are gained with and for the boundary-layer flow past it. However, we have seen in Section 7.2 that on real configurations, which first of all have a finite volume, boundary-layer flow is influenced by a number of effects, which are not present in planar boundary layers on the flat plate. Regarding stability and transition, the situation is similar.

On CAV- and RV-type flight-vehicle configurations large flow portions exist, which are only weakly three-dimensional, see, e. g., the computed skin-friction line patterns in Figs. 7.7, 7.8, 9.4, and 9.5. At such configurations appreciable three-dimensionality of the boundary layer is found usually only at blunt noses and leading edges, and at attachment and separation lines.

This holds also for possible CAV-type vehicle configurations with conical shape and for configurations, where the upper and the side faces are aligned with the free-stream flow, and the lower side is a fully integrated ramp-like lift and propulsion surface, see, e. g., [28]. The lower sides of such, typically slender configurations exhibit more or less parallel flow between the primary attachment lines, for airbreathing vehicles especially necessary to obtain an optimum inlet onset flow. Of course on axisymmetric configurations at angle of attack and spinning configurations the attached viscous flow is fully three-dimensional.

In the following we discuss shortly the influence of the most important real-vehicle effects on stability and transition. Other possible real-vehicle effects like noise of the propulsion system transmitted through the airframe and dynamic aeroelastic surface deformations (vibrations, panel flutter) are difficult to assess quantitatively. To comment on them is not possible in the frame of this book.

Cone versus Planar Boundary-Layer Transition The simplest way to take into account the shape of a non-axisymmetric configuration for stability and transition considerations is to approximate it by a conical configuration.

Considerable confusion arose around this approach [10]. Older experimental data had shown consistently that cone flow exhibits higher transition Reynolds numbers than planar flow, in any case for Mach numbers between 3 and 8, [10]. But stability analyses, first by Mack [29], and then by Malik [30], did show that disturbances grow slower on flat plates than on cones. We illustrate this in Fig. 8.7 with results obtained with linear stability theory by Fezer and Kloker [31]. From these results, with small initial disturbance levels, one should expect that the transition Reynolds numbers on sharp cones are smaller than on flat plates.

This was indeed found with experiments in the "quiet tunnel" of NASA Langley, Sub-Section 8.1.6, at $M = 3.5$ [10]. However, at $M = 8$ Stetson found later (with cooled model surfaces) transition on the cone to occur again somewhat more downstream than on the flat plate. Different disturbance levels and different instability phenomena on cone and on flat plate are the cause of the diverse observations. In [31] also spatial direct numerical simulation (DNS) was applied. For $M_\infty = 6.8$ and radiation-cooled surfaces the result was obtained that in principle the transition mechanisms work in the same manner on cone and flat plate. But on the cone fundamental breakdown is accelerated by the decreasing propagation angle of the secondary 3-D wave,

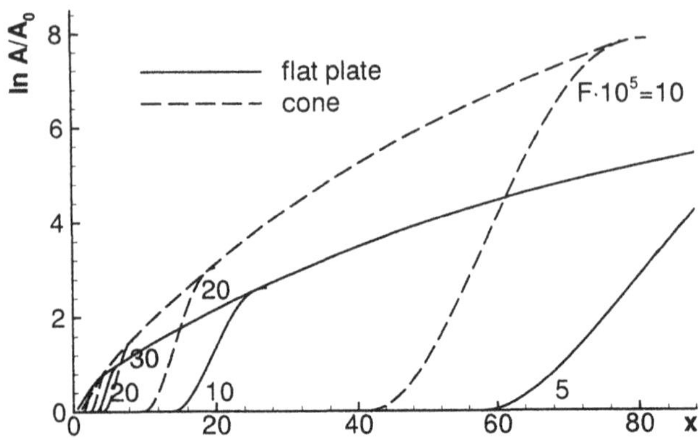

Fig. 8.7. Transition on a flat plate and on a sharp 7° half-angle cone: results from linear stability theory show the different amplitude growths of individual disturbance amplification curves and envelopes. In both cases $M_e = 6.8$, $Re_e^u = 5.72{\cdot}10^5$ m^{-1}, $T_w = 975.0\ K$, hypersonic viscous interaction neglected, 2-d disturbances, various frequency parameters F [31].

while oblique breakdown probably is the dominant transition mechanism on the flat plate[19].

A direct first-principle connection of these results and the results of linear stability theory to the different mean-flow patterns on a cone and a flat plate has not been established so far, although second-mode disturbances are tuned to the boundary-layer thickness [10]. On the cone the axisymmetric boundary layer is thinner than on the flat plate due to the Mangler effect, Section 7.2, and the streamlines in the boundary layer show a divergent (conical) pattern compared to the parallel flow on the flat plate. In addition we have, with the same free-stream conditions in both cases, a smaller boundary-layer edge Mach number and a larger edge unit Reynolds number on the cone than on the flat plate.

Effect of Nose Bluntness - Entropy-Layer Instability Configurations of hypersonic flight vehicles have blunt noses in order to cope with the large thermal loads there. We have seen in Sections 3.2 and 7.2 that the nose radius is important regarding the efficiency of radiation cooling. Nose bluntness on the other hand is the cause of the entropy, Sub-Section 6.4.2. This entropy layer is a shear layer, which is or is not swallowed by the boundary layer, depending on the nose radius and the Reynolds number of the flow case.

[19] Three transition scenarios can be distinguished [14]: "fundamental breakdown", i. e., transition due to secondary instability of Tollmien-Schlichting waves, "streak breakdown" due to secondary instability of stream-wise vortices, "oblique breakdown", i. e., transition governed by the growth of oblique waves, which in general happens in two-dimensional supersonic and hypersonic flows.

It is justified to surmise that the entropy layer can play a role in laminar-turbulent transition of the boundary layer over blunt bodies. Experimental studies on slender cones have shown that this is the case, see, e. g., the overview and discussion in [10]. A small nose-tip bluntness increases the transition Reynolds number relative to that for a sharp-nosed cone. However, if the bluntness is increased further, this trend is reversed, and the transition Reynolds number decreases drastically.

Cone blunting decreases locally the boundary-layer edge Reynolds number, which partially explains the downstream movement of the transition location. The transition reversal is not yet fully understood, but entropy-layer instabilities appear to be a possible cause for it, [32], [33].

Recent experimental and theoretical studies by Dietz and Hein, [34], further support this view. Their visualization of an entropy-layer instability on a flat plate with blunt leading edge is shown in Fig. 8.8. The oblique dark areas in the upper part indicate regions with large density gradients. These are most likely caused by instability waves in the entropy layer. If the entropy layer is swallowed, these disturbances are finally transported by convection into the boundary layer, where they interact with the boundary-layer disturbances [34].

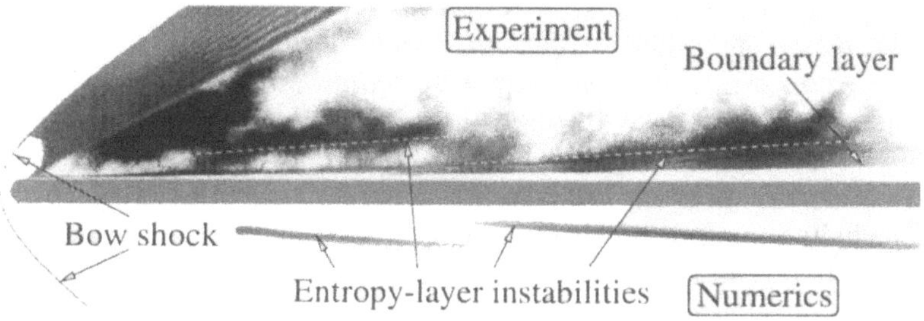

Fig. 8.8. Entropy-layer instability in the flow past a flat plate with blunt leading edge: density gradients in direction normal to the wall in a Schlieren picture (upper part), compared to those of the perturbed flow, which were numerically obtained (lower part) [34]. $M_\infty = 2.5$, $Re^u_\infty = 9.9 \cdot 10^6\ m^{-1}$, adiabatic surface, angle of attack $\alpha = 0.0°$. The broken white lines in the Schlieren picture are the computed locations with maximum $\partial\rho'/\partial y$, ρ' being the density eigenfunction.

Correlations of a large experimental data base regarding bluntness effects on flat plate transition are given in [35], where also effects of hypersonic viscous interaction, Section 9.3, are taken into account. A recent overview regarding work on entropy-layer instability can be found in [36].

Attachment-Line Instability Primary attachment lines exist at the windward side of a flight vehicle with sufficiently flat lower side. At large angles of attack secondary and tertiary attachment lines can be present on the leeward side of the vehicle, Sub-Section 3.3.2. The canonical attachment line situation in aerodynamics corresponds to an attachment line along the leading edge of a swept wing with in the span-wise direction constant symmetric profile at zero angle of attack, or at the windward symmetry line of a circular cylinder at angle of attack or yaw.

At such attachment lines both inviscid and boundary-layer flow diverge symmetrically with respect to the upper and the lower side of the wing, respectively to the left and the right hand side of the cylinder at angle of attack. The infinitely extended attachment line is a useful approximation of reality, which can be helpful for basic considerations and for estimations of flow properties. We have discussed flow properties of such cases in Section 7.2. We have noted that finite flow and hence a boundary layer exists in the direction of the attachment line[20]. This boundary layer can be laminar, transitional or turbulent. On the infinitely extended attachment line only one of these three flow states can exist.

The simplest presentation of an infinite swept attachment-line flow is the swept Hiemenz boundary-layer flow, which is an exact solution of the incompressible continuity equation and the Navier-Stokes equations, [37]. The (linear) stability model for this flow is the Görtler-Hämmerlin model, which in its extended form gives insight into the stability behaviour of attachment-line flow, see, e. g., [38]. Attachment-line flow is the "initial condition" for the, however only initially, highly three-dimensional boundary-layer flow away from the attachment line to the upper and the lower side of the wing or cylinder (see above). The there observed cross-flow instability, see below, has been connected recently by Bertolotti to the instability of the swept Hiemenz flow [39].

An extension of these concepts to general supersonic and hypersonic attachment-line boundary layers would give the basis needed to understand instability and transition phenomena of these flows, including the attachment-line contamination phenomenon which we comment on in the following paragraph. We mention in this regard a recent investigation of swept ($\varphi = 30°$) leading-edge flow at $M_\infty = 8$ by means of a direct numerical simulation [40].

Attachment-Line Contamination Consider transition in the boundary layer on a flat plate or on a wing of finite span. Instability will set in at a certain distance from the leading edge and downstream of it the flow will become fully turbulent by regular transition. If locally at the leading edge a disturbance[21] is present, the boundary layer can become turbulent just behind this disturbance. In that case a "turbulent wedge" appears in the otherwise laminar flow regime, with the typical half angle of approximately

[20] At a non-swept, infinitely long cylinder an attachment line exists, where the flow comes fully to rest. In the two-dimensional picture this attachment line is just the forward (primary) stagnation point.

[21] This can be a dent on the plate's surface, or an insect cadaver at the leading edge of the wing.

7°, which downstream merges with the turbulent flow, Fig. 8.9 a). Only a small part of the laminar flow regime is affected[22].

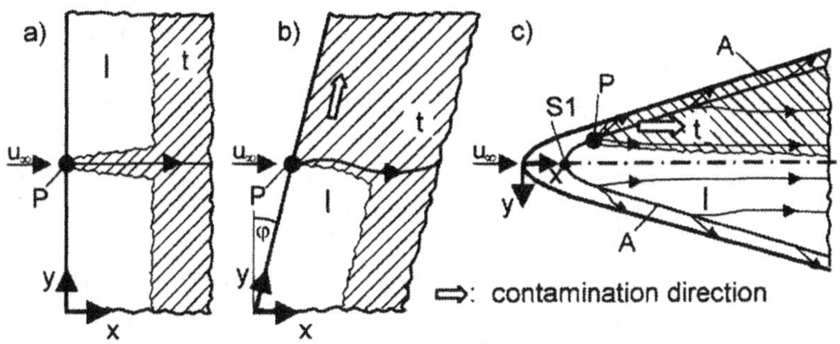

Fig. 8.9. Attachment-line contamination (schematic): transition forced by a surface disturbance P on **a)** flat plate, **b)** leading edge of swept wing, and **c)** primary attachment line on the lower side of a flat blunt-nosed delta wing or fuselage configuration (symbols: l: laminar flow, t: turbulent flow, S_1: forward stagnation point, A: primary attachment line). Note that in this figure the coordinates x, y, z are the usual aerodynamic body coordinates.

At the leading edge of a swept wing the situation can be very different, Fig. 8.9 b). A turbulent wedge can spread out in span-wise direction, "contaminating" the originally laminar flow regime between the disturbance location and the wing tip. On a real aircraft with swept wings it is the turbulent boundary layer of the fuselage which contaminates the otherwise laminar flow at the leading edge.

The low-speed flow criterion, see, e. g., [41]:

$$Re_\theta = \frac{0.4 \sin\varphi\, u_\infty}{\sqrt{\nu(\partial u_e/\partial x^*)|_{l.e.}}}, \tag{8.24}$$

illustrates well the physical background. Here $\sin\varphi\, u_\infty$ is the component of the external inviscid flow along the leading edge in the span-wise (wing-tip) direction, $\partial u_e/\partial x^*)|_{l.e.}$ the gradient of the external inviscid flow in direction normal to the leading edge at the leading edge, and ν the kinematic viscosity. Experimental data show that $Re_\theta \gtrapprox 100 \pm 20$ is the critical value, and that for $Re_\theta \gtrapprox 240$ "leading-edge contamination", as it was termed originally, fully happens.

We see from that criterion the following: the larger the external inviscid flow component in the span-wise (wing-tip) direction, and the smaller the acceleration of the flow normal to the leading edge, the larger the tendency of leading-edge, or more in general, attachment-line contamination. Otherwise

[22] Remember in this context the global characteristic properties of attached viscous flow, Sub-Chapter 7.1.4.

only a turbulent wedge would show up from the location of the disturbance in the chord-wise direction, similar to that shown in Fig. 8.9 a), however skewed.

"Contamination" can happen on general attachment lines, for instance, on those at the lower side of a flat blunt-nosed delta wing or fuselage configuration, Fig. 8.9 c). If, for instance, the TPS of a RV-type flight vehicle has a misaligned tile lying on the attachment line, turbulence can be spread prematurely over a large portion of the lower side of the flight vehicle. This argument was brought forward by Poll [42] in order to explain transition phenomena observed on the Space Shuttle during re-entry, see also the discussion in [13].

The effect of attachment-line contamination in this case would be - temporally, until further down on the trajectory the ordinary transition occurs - large and asymmetric thermal loads, a drag increase (which is not a principle problem for a RV-type flight vehicle), but also a yaw moment, whose magnitude depends on size and location of the contaminated surface part.

Attachment-line contamination in high-speed flows was studied since the sixties, see the overviews in [13] and [10]. Poll [43] made an extensive study of attachment-line contamination at swept leading edges for both incompressible and compressible flows in the seventies. Today, still all prediction capabilities concerning attachment-line contamination rely on empirical data.

Effect of Three-Dimensional Flow - Cross-Flow Instability Three-dimensional boundary-layer flow is characterized by skewed boundary-layer profiles which can be decomposed into a main-flow profile and into a cross-flow profile, Fig. 7.1. With increasing cross flow, i. e. increasing three-dimensionality, the so-called cross-flow instability becomes a major instability and transition mechanism. This observation dates back to the early fifties, when transition phenomena on swept wings became research and application topics. It was found that the transition location with increasing sweep angle of the wing moves forward to the leading edge. The transition location then lies upstream of the location, which is found at zero sweep angle, and which is governed by Tollmien-Schlichting instability. Steady vortex patterns - initially visualized as striations on the surface - were observed in the boundary layer, with the vortex axes lying approximately parallel to the streamlines of the external inviscid flow, see, e. g., [43].

Owen and Randall, [44] proposed a criterion based on the properties of the cross-flow profile. The cross-flow Reynolds number χ reads with the notation used in Fig. 7.1:

$$\chi = \frac{\rho v^{*2}_{max} \delta_q}{\mu}. \tag{8.25}$$

Here v^{*2}_{max} is the maximum cross-flow speed in the local cross-flow profile and δ_q a somewhat vaguely defined boundary-layer thickness found from that profile:

$$\delta_q = \int_0^{\delta} \frac{v^{*2}}{v^{*2}_{max}} dx^3. \tag{8.26}$$

If $\chi \gtrapprox 175$, transition due to cross-flow instability happens. If the main-flow profile becomes unstable first, transition will happen, in the frame of this ansatz, due to the Tollmien-Schlichting instability as in two-dimensional flow.

The phenomenon of cross-flow instability was studied so far mainly for low speed flow, see the discussion in , e. g., [10], [13]. Experimental and theoretical/numerical studies have elucidated many details of the phenomenon. Local and especially non-local stability theory has shown that the disturbance wave vector lies indeed approximately normal to the external inviscid streamlines. The disturbance flow exhibits counter-rotating vortex pairs. Their superposition with the mean flow results in the experimentally observed co-rotating vortices with twofold distance. The critical cross-flow disturbances have wave lengths approximately 2 to 4 times the boundary-layer thickness, compared to the critical Tollmien-Schlichting waves which have wave lengths approximately 5 to 10 times the boundary-layer thickness. This is the reason why non-parallel effects and surface curvature must be regarded, for which non-local stability methods are better suited than local methods.

The subsequent transition to turbulent flow can be due to a mixture of cross-flow and Tollmien-Schlichting instability. Typically cross-flow instability plays a role, if the local flow angle is larger than 30°. Tollmien-Schlichting instability comes into play earliest in the region of an adverse pressure gradient. However also fully cross-flow dominated transition can occur, [45], [46]. For high-speed flow critical χ data, however of questionable generality, have been established from dedicated experiments, see, e. g., [10], [13], [47].

Effect of Shock/Boundary-Layer Interaction Shock/boundary-layer interaction, Section 9.2, happens where the bow shock of the flight vehicle, or embedded shocks interfere with the boundary layer on the vehicle surface. This may happen in the external flow path of an airframe including the outer part of inlets, see, e. g., Figs. 6.4 to 6.6, at control surfaces, or in the internal flow path of propulsion systems, see, e. g., Figs. 6.5 and 6.7.

If a shock wave impinges on a (laminar) boundary layer, the flow properties of the latter change, even if the flow does not become separated locally or globally. We have also seen in Sub-Section 6.6 that the unit Reynolds number across a ramp shock changes, which, of course, changes the flow properties too of the boundary layer across the shock. Depending on Reynolds number, shock strength et cetera the stability properties of the boundary layer will be affected.

Although this phenomenon is potentially of importance for hypersonic flight vehicles, it has found only limited attention so far. For ramp flow, for instance, with control surface heating, including Görtler instability (see below), in the background, experimental data and data from dedicated numerical studies, e. g., [48], have been assembled and summarized in [49].

We note further a recent stability investigation of shock/boundary-layer interaction in a $M_e = 4.8$ flow with linear stability theory and direct numerical simulation (DNS), [50]. It was shown that and how second-mode instability is promoted by the interaction. Linear stability theory taking into account non-parallel effects yielded results in good agreement with DNS for wall-distant disturbance-amplitude maxima with small obliqueness angles. For large obliqueness angles and wall-near amplitude maxima accuracy of linear stability theory deteriorated considerably in comparison to DNS. The results show that both the effect and the related simulation problems warrant further investigations.

High-Temperature Real-Gas Effects Regarding the influence of high-temperature real-gas effects on instability and transition, two basic scenarios appear to play a role. The first is characterized by flow in thermo-chemical equilibrium, and by frozen flow, the second by flow in thermo-chemical non-equilibrium.

In view of the first scenario we note that high-temperature real-gas effects affect properties of attached viscous flow, for instance the temperature and the density distribution in the direction normal to the surface. Hence they will have an indirect influence on instability and transition phenomena in the same way as pressure gradients, the thermal state of the wall, and so on have. For the investigation and prediction of instability and transition of a boundary layer they must therefore be taken into account in order to determine the mean flow properties with the needed high accuracy. We remember that the point-of-inflexion properties of a boundary layer are governed by the first and the second derivative of $u(y)$, and by the first derivative of $T(y)$.

Thermal and chemical non-equilibrium effects on the other hand affect the stability properties of boundary layers directly. Relaxation of chemical non-equilibrium has been shown experimentally and theoretically to stabilize boundary-layer flow, see, e. g. [51], [52]. Relaxation of rotational energy stabilizes, while relaxation of vibrational energy, contrary to what was believed until recently, can destabilize the flow strongly, [53]. This holds especially for boundary layers downstream of a blunt nose or leading edge, which is the standard situation on the hypersonic flight vehicles considered in this book. Important again is that the wind-tunnel situation must be distinguished from the free-flight situation. In the wind-tunnel frozen vibrational non-equilibrium might exist in the free-stream flow of the test section, Fig. 5.9 in Sub-Section 5.5.2, whereas in flight the atmosphere ahead of the flight vehicle is in equilibrium. In [53] it is shown that only for the thin flat plate in the free-flight situation with weak non-equilibrium the influence of vibrational relaxation is slightly stabilizing for second-mode instabilities.

Surface Irregularities We consider surface irregularities as a sub-set of surface properties. In Section 1.1 we have noted the definition of surface properties as one of the tasks of aerothermodynamics. In the context of laminar-turbulent transition "permissible" surface irregularities include surface roughness, waviness, steps, gaps et cetera, which are also important in view of fully turbulent flow, Section 8.3. These surface properties should be "sub-critical" in order to avoid either premature transition or amplification of turbulent transport, which both can lead to unwanted increments of viscous drag, and can affect significantly the thermal state of the surface of a flight vehicle.

On CAV-type flight vehicles all permissible, i. e. sub-critical, values of surface irregularities should be well known, because surface tolerances should be as large as possible in order to minimize manufacturing cost. On RV-type flight vehicles the situation is different in so far as a thermal protection system consisting of tiles or shingles is inherently rough [1], which is not a principal problem with respect to laminar-turbulent transition at altitudes above approximately 40.0 to 60.0 km. Below these altitudes a proper behaviour and prediction of transition is desirable especially in order to avoid especially adverse increments of the thermal state of the surface.

Surface irregularities usually are not of much concern in fluid mechanics and aerodynamics, because flow past hydraulically smooth surfaces usually is at the centre of attention. Surface irregularities are kind of a nuisance which comes with practical applications. Nevertheless, empirical knowledge is available, especially concerning single and distributed, see, e. g., [1], surface roughness.

Surface roughness can be characterized by the ratio k/δ_1, where k is the height of the roughness and δ_1 the displacement thickness of the boundary layer at the location of the roughness. The height of the roughness at which it becomes effective - with given δ_1 - is the critical roughness height k_{cr}, with the Reynolds number at the location of the roughness, Re_k, playing a major role. For $k < k_{cr}$ the roughness does not influence transition, and the surface can be considered as hydraulically smooth. This does not necessarily rule out that the roughness influences the instability behaviour of the boundary layer, and thus regular transition. For $k > k_{cr}$ the roughness triggers turbulence and we have forced transition. The question then is whether turbulence appears directly at the roughness or at a certain, finite, distance behind it.

Since a boundary layer is thin at the front part of a flight vehicle, and becomes thicker in down-stream direction, a given surface irregularity may be critical at the front of the vehicle, and sub-critical further downstream.

Actually the effectiveness of surface irregularities to influence or to force transition depends on several flow parameters, including the Reynolds number and the thermal state of the surface, and on geometrical parameters, like configuration and spacing of the irregularities. Important is the observation that with increasing boundary-layer edge Mach number, the height of a roughness must increase drastically in order to be effective. For $M_e \gtrapprox 5$ to 8 the limit of effectiveness seems to be reached, [20], [13], [10], in the sense that it becomes extremely difficult, or even impossible, to force transition by means of surface roughness.

This has two practical aspects. The first is that on a hypersonic flight vehicle at large flight speed not only the Reynolds number but also the Mach number at the boundary-layer edge plays a role. We remember in this context the different boundary-layer edge flow parameters at RV-type and CAV-type flight vehicles, which operate at vastly different angles of attack, Fig. 1.3 in Section 1.2. At a RV-type vehicle surface roughness thus can be effective to trigger turbulence once the Reynolds number locally is large enough, because the boundary-layer edge Mach numbers are small, i. e., $M_e \lessapprox 2.5$. In fact the laminar-turbulent transition on the windward side of the Space Shuttle with its "rough" tile surface is roughness dominated [1].

Of importance is the case of a single surface roughness. A misaligned tile, for instance, can cause attachment-line contamination (see above). In any case a turbulent wedge will be present downstream of it, which may be dissipated soon, if locally the Reynolds number is not large enough to sustain this - premature - turbulence. However, in high-enthalpy flow such a turbulent wedge can lead to a severe hot-spot situation.

The other aspect is that of turbulence tripping in ground-simulation facilities, if the attainable Reynolds number is too small. Boundary-layer tripping in the lower speed regimes is already a problem[23]. In high Mach-number flows boundary-layer tripping might require roughness heights of the order of the boundary-layer thickness in order to trigger turbulence. In such a situation the character of the whole flow field will be changed (over-tripping). If moreover the Reynolds number is not large enough to sustain turbulent flow, the boundary layer will re-laminarize.

For further details and also surface roughness/tripping effectiveness criteria, also in view of attachment-line contamination, see, e. g., [13], [10], [47], [55].

Görtler Instability The Görtler instability is a centrifugal instability which appears in flows over concave surfaces, but also in other concave flow situations, for instance in the stagnation region of a cylinder. It can lead to high thermal loads in striation form, for instance at deflected control surfaces. However, striation heating can also be observed at other parts of a flight vehicle configuration, [56].

Consider the boundary-layer flow past curved surfaces in Fig. 8.10. Although we have boundary layers with no-slip condition at the surface, we

[23] Major issues are the location of the boundary-layer tripping device (roughness elements, et cetera) on the wind tunnel model, the effectiveness of the device, and the avoidance of over-tripping (e. g. too large roughness height), which would falsify the properties (displacement thickness, wall shear stress, heat flux) of the ensuing turbulent boundary layer, see, e. g., [54].

assume that we can describe the two flow cases with the lowest-order approximation:

$$\rho \frac{U^2}{|R|} = |\frac{dp}{dy}|. \tag{8.27}$$

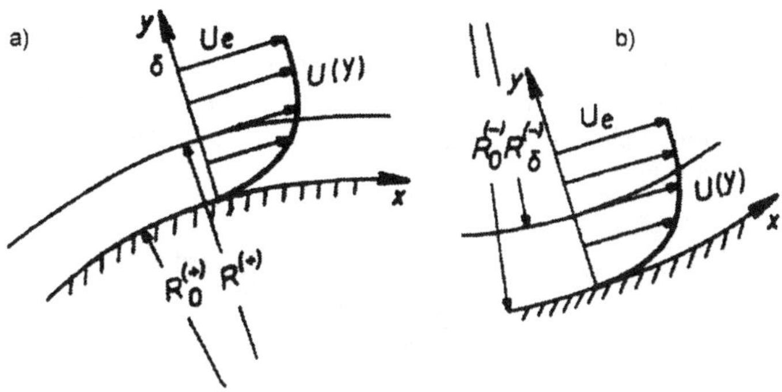

Fig. 8.10. Boundary-layer flow past curved surfaces: **a)** convex surface, and **b)** concave surface.

With assumed constant pressure gradient $| dp/dy |$ and constant density, the term U^2 at a location inside the boundary layer $0 \leq y \leq \delta$ must become larger, if we move from R to $R + \Delta R$. This is the case on the convex surface, Fig. 8.10 a). It is not the case on the concave surface, Fig. 8.10 b). As a consequence in this concave case flow particles at R with velocity U attempt to exchange their location with the flow particles at the location $R + \Delta R$ where the velocity $U - \Delta U$ is present.

In this way a vortical movement inside the boundary layer can be triggered which leads to stationary, counter-rotating pairs of vortices, the Görtler vortices, with axes parallel to the mean flow direction. They were first described by Görtler [57] in the frame of the laminar-turbulent transition problem (influence of surface curvature on flow instability). Results of early experimental investigations are found in [58] and [59].

Görtler vortices can appear in almost all concave flow situations. In our context these are control surfaces, but also inlet ramp flows. They were observed in ground-facility experiments for instance at jet spoilers [60], but also behind reflections of oblique planar shock waves on flat-plate surfaces. They can appear in laminar, transitional and turbulent flow and can lead in each case to appreciable striation-wise heat loads. They are thus not only of interest with regard to laminar-turbulent transition, but especially also with regard to thermal loads on flight-vehicle structures.

To understand the occurrence of striation-wise heat loads consider Fig. 8.11 with three vortex pairs each at the two stations 1 and 2. Shown is the perturbation flow in two cross-sections of the time-averaged re-circulation region of the ramp flow indicated at the top of the figure. The in this case turbulent flow computation was made with the DLR-CEVCATS-N RANS method [61]. The vortices were triggered, like in a related experiment, with artificial surface distortions (turbulators). They appear already in the recirculation regime and extend far downstream behind it.

We see in Fig. 8.11 the cross-flow velocity components, the stream-wise velocity components lie normal to them. Between the vortex pairs the flow is directed towards the ramp surface and inside the vortex pairs away from it. Consequently on the ramp surface attachment lines with diverging flow patterns are present between the vortex pairs, like shown in Fig. 3.5, and detachment lines with converging flow patterns inside the vortex pairs, like shown in Fig. 3.6.

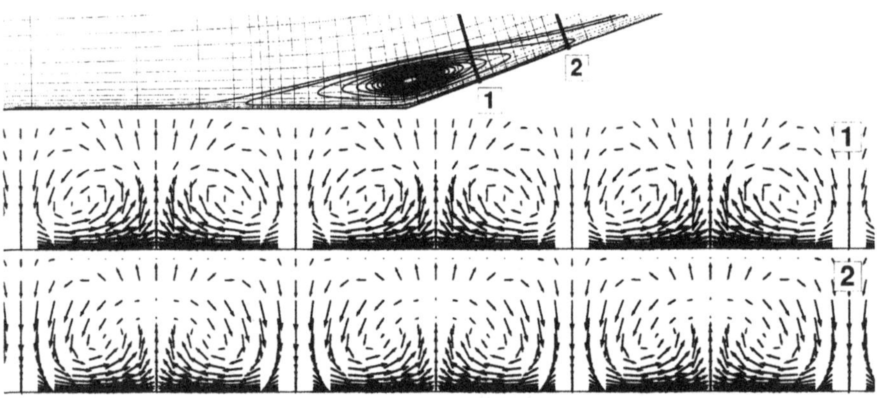

Fig. 8.11. Computed Görtler vortices in nominally 2-D flow past a 20° ramp configuration, $M_\infty = 3$, $Re = 12{\cdot}10^6$ [61].

Following the discussion in Sub-Section 3.2.3, we conclude that on the surface between the vortex pairs heat fluxes - or temperatures in the case of radiation cooling - will be found which can be substantially larger (hot-spot situation) than those found inside the vortex pairs (cold-spot situation), but also in the absence of such vortices. In [62] computed span-wise heat-flux variations of ± 20 per cent for the X-38 body flap are reported.

Such hot/cold-spot situations in streak form are indeed found also in laminar/transitional flow on flat plate/ramp configurations, see, e. g., [63], [64]. There a strong influence of the leading edge of the ramp configuration (entropy-layer effect?) on the ensuing Görtler flow is reported. Results from a numerical/experimental study of the flow past the X-38 configuration with Görtler flow on the extended body flaps are given in [65], see also [62].

The Görtler instability can be treated in the frame of linear stability theory, see, e. g., [13], [14]. The Görtler parameter G_ℓ reads:

$$G_\ell = \frac{\rho_e u_e \ell}{\mu_e} \left(\frac{\ell}{R} \right)^{0.5}, \tag{8.28}$$

where ℓ is a characteristic length, for instance the displacement thickness δ_1 of the mean flow. A modification of this parameter for supersonic and hypersonic flows is proposed in [66]. It seems to be an open question whether Görtler instability acts as an operation modifier on the linear amplification process in the sense of Morkovin, [11], or whether it can also lead directly to transition (streak breakdown?). For an overview see, e. g., [67]. In [10] it is mentioned that transition was found (in experiments) to occur for $G_\ell = 6$ to 10.

Relaminarization A turbulent flow can effectively relaminarize. Narasimha [68] distinguishes three principle types of relaminarization or reverse transition:

- Reynolds number relaminarization, due to a drop of the local (boundary-layer edge based) Reynolds number.
- Richardson relaminarization, if the flow has to work against buoyancy or curvature forces.
- Acceleration relaminarization, if the boundary-flow is strongly accelerated.

For acceleration relaminarization in two-dimensional flow a criterion is, see [69]:

$$K_{crit} = \frac{\nu}{u_e^2} \frac{du_e}{dx} \gtrapprox 2 \cdot 10^6. \tag{8.29}$$

The phenomenon of relaminarization can play a role also in the flow past hypersonic flight vehicles. Consider, for instance, the flow around the leading edge towards the windward side of the Blunt Delta Wing, Section 3.3. The flow accelerates away from the two primary attachment lines towards the leading edges, Fig. 3.16, and is expanding around the latter towards the lee side of the configuration, Fig. 3.17. This is accompanied by a drop of the unit Reynolds number, see the discussion at the end of Sub-Section 3.3.3. Whether the two effects, single or combined, would be strong enough in this case to actually relaminarize a turbulent flow coming from the windward side of a re-entry vehicle is not known. Also it is not known whether other flow situations exist in high-speed flight, where relaminarization can play a role, including the phenomenon of relaminarization with subsequent re-transition, see for instance [70].

8.1.6 Environment Aspects

Under environment we understand either the atmospheric flight environment of a hypersonic flight vehicle or the environment which the sub-scale model of the flight vehicle has in a ground-simulation facility. The question is how the respective environment influences instability and transition phenomena on the flight vehicle or on its sub-scale model [71], [14]. Ideally there should be no differences between the flight environment and the ground-facility environment, but the fact that we have to distinguish between these two environments points already to the fact that these environments have different characteristics and different influences on transition. Both in view of scientific and practical, i. e. vehicle design issues these different influences pose large problems[24].

The atmosphere, through which a hypersonic vehicle flies, poses a disturbance environment. Information about the environment appears to be available for the troposphere, but not so much for the stratosphere. Morkovin suggests, [11], see also [12], as a work hypothesis, that distribution, intensities and scales of disturbances can be assumed to be similar in the troposphere and the stratosphere. Flight measurements in the upper troposphere (11 km altitude) have shown strong anisotropic air motions with very low dissipation and weak vertical velocity fluctuations [72]. How much the flight speed of the vehicle plays a role is not known. This will be partly a matter of the receptivity properties of the boundary layer.

Much is known of the disturbance environment in ground-simulation facilities, see, e. g., [11], [12], [13], [10]. We have mentioned already as major problem noise, i. e. the sound field radiated from the turbulent boundary layers of the tunnel wall[25]. The quest to create in ground-simulation facilities a disturbance environment similar to that of free flight (whatever that is) has led to the concept of the "quiet" tunnel, see, e. g., [12].

A Mach 3.5 pilot quiet tunnel has been built in the Seventies in the US at NASA Langley, [73]. It is characterized by measures to remove the turbulent boundary layer coming from the settling chamber, a new boundary layer developing on the nozzle wall, and finally a sound shield enclosing the test section. The Ludwieg-Tube facility, see, e. g., [74], at the DLR in Göttingen, Germany and the Weise-Tube facility of similar principle at the University of Stuttgart [75], can be considered as quiet tunnels, too. At least the unit Reynolds number effect[26], see, e. g. [11], [12], [71], has been shown by Krogmann not to exist in the Ludwieg tube [76]. The new facility at Purdue University is explicitly called a "Mach-6 quiet-flow Ludwieg tube" [77].

---

[24] In literature often surface irregularities, like surface roughness, and environment aspects are combined under one heading. We have treated surface irregularities in the preceding sub-section as real-vehicle effects.

[25] The influence of thermo-chemical non-equilibrium due to freezing phenomena in the nozzles of high-enthalpy simulation facilities on instability (and transition) is also an environment issue, although not a disturbance environment issue.

[26] The apparent dependence of transition on the unit Reynolds number $Re^u = \rho u/\mu$ is a facility-induced effect.

The disturbance environment of a flight vehicle or of its sub-scale wind-tunnel model is very important, because it provides for regular laminar-turbulent transition:

1. The "initial" conditions in flight and in the ground-simulation facility.
2. The "boundary" conditions in flight (surface conditions, engine noise) and in the ground-simulation facility (tunnel-wall noise, model surface conditions).

In the aerodynamic practice velocity fluctuations u', v', w', which are also called free-stream turbulence, are the entities of interest. The classical measure is the "level of free-stream turbulence":

$$Tu = \sqrt{\frac{\overline{u'^2} + \overline{v'^2} + \overline{w'^2}}{3u_\infty^2}}. \tag{8.30}$$

If $\overline{u'^2} = \overline{v'^2} = \overline{w'^2}$, we call this isotropic free-stream turbulence. At low speed, the level of free-stream disturbances governs strongly the transition process. The free-stream turbulence of wind tunnels even for industrial measurements should be smaller than $Tu = 0.001$.

A rational and rigorous approach to identify types of disturbances is the consideration of the characteristic values of the system of equations of compressible stability theory, see, e. g., [21]. There the following types of disturbances are distinguished:

- Temperature fluctuations, T', also called entropy fluctuations.
- Vorticity fluctuations, ω'_x, ω'_y, ω'_z.
- Pressure fluctuations, p', or acoustic disturbances (noise). These are of large importance in hypersonic wind tunnels for $M \gtrapprox 3$, but also in transonic wind tunnels with slotted or perforated walls.

It is interesting to note that for instance at hypersonic flight a free-stream temperature fluctuation can trigger vorticity and acoustic modes while passing the bow-shock surface ahead of the swept leading edge of the wing of the flight vehicle [40].

The environment (free-stream) disturbance properties are of large importance especially for non-local non-linear instability methods, which are the basis of non-empirical transition prediction methods, see the following Section 8.2. These methods need a receptivity model. Actually all types of disturbance-transport equations (non-linear/non-local theories) need initial values in the form of free-stream disturbances. These are also needed for the direct numerical simulation (DNS) of stability and transition problems.

The state of the art regarding boundary-layer receptivity to free-stream disturbances is discussed in [78]. A comprehensive discussion of the problems of receptivity models, also in view of the influence of flight speed and flow-field deformation in the vicinity of the airframe is still missing.

We note in this context that for the computational simulation of turbulent flows by means of transport-equation turbulence models, for instance of $k-\varepsilon$ or $k-\omega$ type, initial values of the turbulent energy k, the dissipation ε or the dissipation per unit turbulent energy ω as free-stream values are needed, too, see, e. g., [9]. A typical value used in many computational methods for the turbulent energy is $k_\infty \approx (0.005\ u_\infty)^2$, while ω or ε should be "sufficiently small" [79], [80]. Large eddy simulation (LES) of turbulent flow also needs free-stream initial values. The question is whether in non-empirical transition prediction methods for the free-flight situation, apart from surface vibrations and engine noise (relevance of both?), this kind of "white noise" approach is a viable approach. For the ground-facility situation of course the environment, which the facility and the model pose, must be determined and incorporated in a prediction method [12].

8.2 Prediction of Stability/Instability and Transition in High-Speed Flows

In this section we wish to acquaint the reader with the possibilities to actually predict stability/instability and transition in high-speed flows. No review is intended, but a general overview is given with a few references to prediction models and criteria. Because recent developments in non-local and non-linear instability prediction are mainly made for transonic flows, these developments will be treated too in order to show their potential also for the high-speed flows of interest here. Transition-prediction theory and methods based on experimental data (ground-simulation facility or free-flight data) are treated under the headings "semi-empirical" and "empirical" transition prediction in Sub-Section 8.2.2.

8.2.1 Stability/Instability Theory and Methods

Theory and methods presented here are in any case methods for compressible flow but not necessarily for flow with high-temperature real-gas effects. The thermal boundary conditions are usually only the constant surface temperature or the adiabatic-wall condition. The radiation-adiabatic wall condition is implemented only in few methods, although it is in general straight forward to include it into a method. Likewise, it is not a problem to include adequate high-temperature real-gas models into stability/instability methods.

Linear and Local Theory and Methods The classical stability theory is a linear and local theory. It describes only the linear growth of disturbances (stage 2 - see the footnote on page 268 - in branch IIa, Fig. 8.1). Neither the receptivity stage is covered[27], nor the saturation stage and the last two stages of transition. Extensions to include non-parallel effects are possible and have been made. The same is true for curvature effects. However, the suitability of such measures appears to be questionable, see, e. g., [81].

Linear and local theory is, despite the fact that it covers only stage 2, the basis for the semi-empirical e^n transition prediction methods, which are discussed in Sub-Section 8.2.2.

Linear and local stability methods for compressible flows are for instance COSAL (Malik, 1982 [82]), COSTA (Ehrenstein and Dallmann, 1989 [83]), COSMET (Simen, 1991 [84], see also Kufner [27]), CASTET (Laburthe, 1992 [85]), SHOOT (Hanifi, 1993 [86]), LST3D (Malik, 1997 [87]), COAST (Schrauf, 1992 [88], 1998 [89]).

Non-Local Linear and Non-Linear Theory and Methods Non-local theory takes into account the wall-normal and the downstream changes of the mean flow as well as the changes of the amplitudes of the disturbance flow and the wave numbers. Non-local and linear theory also describes only stage 2 in branch IIa, Fig. 8.1. However, non-parallelism and curvature are consistently taken into account which makes it a better basis for e^n methods than local linear theory.

Non-linear non-local theory on the other hand describes all five stages, especially also stage 1, the disturbance reception stage, the latter however not in all respects. Hence, in contrast to linear theory, form and magnitude of the initial disturbances must be specified, i. e., a receptivity model must be employed, Sub-Section 8.1.6.

Non-local methods are (downstream) space-marching methods that solve a system of disturbance equations, which must have space-wise parabolic character. Hence such methods are also called "parabolized stability equations (PSE)" methods. We don't discuss here the parabolization and solution strategies and refer the reader instead to the review article of Herbert [90] and to the individual references given in the following.

Non-local linear stability methods for compressible flow are for instance xPSE (Bertolotti, linear and non-linear (the latter incompressible only), 1991 [91]), PSE method (linear and non-linear) of Chang et al., 1991 [92], NOLOS (Simen, 1993 [93]).

Non-local non-linear stability methods for compressible flow are for instance, COPS (Herbert et al., 1993 [94]), NOLOT/PSE (Simen et al., 1994 [95], see also Hein [36]), CoPSE (Mughal and Hall, 1996 [96]), PSE3D (Malik, 1997 [87], with chemical reactions also see [97]), xPSE with rotational and vibrational non-equilibrium (Bertolotti, 1998 [53]), NELLY (Salinas, 1998 [98]).

---

[27] Note that the result of linear stability theory is the relative growth of (unstable) disturbances of unspecified small magnitude, eq. (8.11) or (8.12), only.

8.2.2 Transition Models and Criteria

We have today a rather good knowledge of instability and transition phenomena, and many new and good stability/instability methods are available. However, the accurate and reliable prediction of the shape, the extent and the location of the transition zone[28], i. e., the transition sub-branch IIb, Fig. 8.1, for real-life flight vehicles of the vehicle classes considered here is not possible. This holds partly also for flows in the other speed regimes.

We distinguish three classes of means for transition prediction, namely

- non-empirical,
- semi-empirical,
- empirical

methods and criteria. Of these the latter two rely partly or fully on experimental data.

The major problem is that a suitable high-speed experimental data base is not available. This, in principle, could come either from ground-simulation facilities or from free-flight measurements. For the first it is noted that a hypersonic ground-simulation facility is not able to duplicate the atmospheric flight environment and also not the relevant boundary-layer properties, i. e., the profiles of the tangential velocity and the density normal to the surface in presence of radiation cooling, as they are present on the real vehicle [6]. This holds especially for regular but also for forced transition. With regard to free-flight measurements we must observe that the few available transition data sets obtained at actual high-speed flight are not the necessary multi-parametric data bases, which are required to identify the major instability and transition phenomena involved, including the relevant flow and surface properties.

We give now a short overview over the three classes of methods and criteria for transition prediction.

Non-empirical Transition Prediction It appears that non-local non-linear instability methods have the real potential to become the needed non-empirical transition prediction methods for practical purposes. Direct numerical simulation (DNS) as well as large eddy simulation (LES), due to the needed very large computer power, will have, for a long time to come, their domain of application in numerical experiments of research only.

The present state of development of non-local non-linear methods appears to permit the prediction of the location of stage 5, i. e., the begin of sub-branch IIb, Fig. 8.12. However, two different combinations of disturbance modes (receptivity problem) lead to small, but significant differences between the solutions (location and initial shape of sub-branch IIb).

[28] The transition zone in reality is an arbitrarily shaped surface with rather small downstream extent wrapped around the configuration.

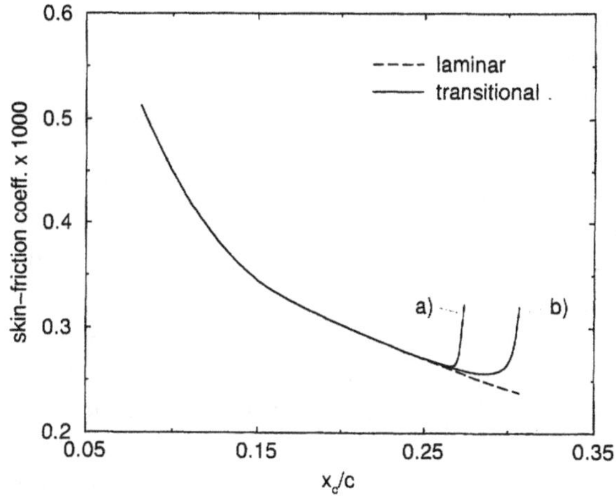

Fig. 8.12. Result of a non-local non-linear method: rise of the skin-friction coefficient in stage 5, i. e., at the begin of sub-branch IIb. Swept wing, $\varphi_{LE} = 21.75°$, $M_\infty = 0.5$, $Re_\infty = 27{\cdot}10^6$, two different disturbance mode combinations [36].

The result of Fig. 8.12 is for a low-speed case, similar results for the high-speed flows of interest are available to a certain extent. It can be expected, in view of the references given in Sub-Section 8.2.1, that at least results similar to those shown can be obtained, after additional research has been conducted, especially also with regard to the receptivity problem.

Very encouraging is in this context that the problem of surface irregularities (transition triggering, permissible properties) seems to become amenable for non-empirical prediction methods, see [99], at least for transonic flow past swept wings[29]. That would allow to take into account the influence of weak surface irregularities on regular transition, but also to model transition forced, for example, by the rough tile surface of a TPS.

Semi-empirical Transition Prediction Semi-empirical transition prediction methods go back to J. L. van Ingen [101], and A. M. O. Smith and N. Gamberoni [102]. They observed independently in the frame of local linear stability theory that, for a given boundary-layer mean flow, the envelope of the most amplified disturbances, see Fig. 8.7 for (two) examples, correlates observed transition locations.

For airfoils it turned out, that on average the value, see Sub-Section 8.1.2,

$$\ln \frac{A}{A_0} = \int_{t_0}^{t} \omega_i dt = 9, \tag{8.31}$$

[29] An alternative approach with a large potential also for optimization purposes, e. g., to influence the instability and transition behaviour of the flow by passive or active means, is to use adjoint equation systems [100].

best correlates the measured data, hence the name e^9 criterion. Unfortunately later the "universal constant" 9 turned out to be a "universal variable". Already in the data of Smith and Gamberoni the scatter was up to 20 per cent. Now we speak of the e^n criterion, which for well defined two-dimensional low-speed flow classes with good experimental data bases can be a reliable and accurate transition prediction tool, see, e. g. [103].

For three-dimensional flows the situation becomes complicated, see, e. g., [81]. Non-local linear stability theory with n-factors for Tollmien-Schlichting modes (n_{TS}) and for cross-flow modes (n_{CF}), based on in-flight measured data, are used with mixed success in transonic swept-wing flows.

Although the e^n method does not take into account the environment[30] and does not describe at least the initial stage of transition, i. e., the beginning of sub-branch IIb, it has been extensively applied to mainly two-dimensional high-speed flows [10], [103], [47]. It is a valuable tool to study instability phenomena and their influence on transition, but also to perform parametric studies. To reliably predict transition on hypersonic flight vehicles is primarily a matter of a reliable, multi-parametric experimental data base, which is not available. To question is also whether an e^n method permits to treat the effect of weak surface irregularities on regular transition, but also to describe forced transition.

Empirical Transition Prediction Empirical transition prediction is based on criteria derived from experimental data which are obtained in ground-simulation facilities but also in free flight. Such criteria have been applied and are applied in the design of the Space Shuttle, of BURAN, of re-entry bodies, launchers, and missiles.

The empirical criteria usually are local criteria, i. e., they employ local integral boundary-layer and boundary-layer edge-flow properties. This means that all the phenomena discussed in Sub-Sections 8.1.3 to 8.1.5 are not taken into account explicitly. At best some of them are implicitly regarded via an employed boundary-layer integral quantity. Data from ground-simulation facilities are possibly falsified by tunnel noise effects. The lower and the upper transition location often are not explicitly given, i. e., length and shape of the transition sub-branch IIb are not specified. This holds also for the overshoot at the end of this branch.

Due to the nature of the criteria, the predictions based on them are of questionable quality. If error bars are given, uncertainties of describing data and hence design margins can be established. In any case parametric guesses can be made.

Empirical criteria are discussed in depth in, for instance, [10] and [47]. In [47] available high-speed flow criteria, in this case basically two-dimensional criteria, are grouped into four classes:

---

[30] In [104] it is proposed to take into account the free-stream turbulence Tu of a wind tunnel by introducing $n_{tr} = n_{tr}(Tu)$.

- Smooth body criteria, i. e. criteria to predict regular transition. A typical criterion from this class is the Thyson criterion [105]:

$$Re_{\delta_2,tr} = 200\ e^{0.187\ M_e}. \tag{8.32}$$

$Re_{\delta_{2,tr}}$ is the Reynolds number based on the boundary-layer edge data and the momentum thickness δ_2, M_e is the edge Mach number. The correlation data are mainly from ground-simulation facilities and hence the noise problem exists.
To this class of criteria also the criterion by Kipp and Masek [106] is counted. Written in the general form

$$\frac{Re_{\delta_2,tr}}{M_e} = const., \tag{8.33}$$

it can be applied to different configurations, although it was originally developed for the Space Shuttle design. Stetson [10] discusses this criterion in detail and states that it cannot be applied in general.

- Rough body criteria, i. e. criteria to predict transition forced by single or distributed (TPS surface) roughnesses. As example a criterion for distributed roughness, developed in the frame of the HERMES project with data from Space Shuttle flights, is given [107], [47]:

$$Re_{k,tr} = \frac{\rho_e u_e k_{corr}}{\mu_e} \frac{1}{1+\frac{\gamma-1}{2}M_e^2} = 200. \tag{8.34}$$

The corrected roughness height k_{corr} is defined by

$$k_{corr} = \frac{1}{1+350\ k/R}. \tag{8.35}$$

The roughness is efficient if located less than 6.0 m from the vehicle nose and $k/\delta_1 = 0.1$ to 0.5.
The criterion takes into account the roughness height k, the boundary-layer edge data ρ_e, u_e, μ_e, the displacement thickness δ_1, the edge Mach number M_e, and the longitudinal curvature R. It is defined for the lower symmetry line of the vehicle and duplicates well the Space Shuttle data there. The need for a truly three-dimensional approach is acknowledged by the authors.

- Combined rough/smooth body criteria, i. e. criteria with parameters to distinguish sub-critical and super-critical roughness regimes. An example is the criterion given by Goodrich, Derry and Bertin [1] in order to improve the transition prediction for the Space Shuttle.
- Global criteria take into account implicitly all phenomena of interest for a given vehicle class. In [108] for instance data are correlated for ballistic re-entry vehicles with ablative nose tips, however with mixed success.

Criteria, which take into account effects of three-dimensionality, for instance, cross-flow instability or attachment-line contamination are discussed too in [47]. We have treated them in Sub-Section 8.1.5.

8.2.3 Determination of Permissible Surface Properties

In Section 1.3 we have noted the necessity to define permissible surface properties. In the present context these are surface irregularities like roughness, waviness, steps, gaps et cetera. They must be "sub-critical" in order to avoid unwanted increments of viscous drag, and of the thermal state of the surface. This holds in view of both laminar-turbulent transition and the fully developed turbulent flow.

Permissible surface properties in the sense that clear-cut criteria for sub-critical behaviour in the real-flight situation are given are scarce. Usually they are included when treating distributed roughness effects on transition, see, e. g., [1], [35] and also the overviews and introductions [11], [12], [13], [10]. Some systematic work on the influence of forward and backward facing steps and surface waviness on transition was performed in a FESTIP study, see [109].

Permissible surface properties for low-speed turbulent boundary layers are given for instance in [20]. Data for supersonic and hypersonic turbulent boundary layers are not known. As a rule the height of a surface irregularity must be smaller than the viscous sub-layer thickness in order to have no effect on the wall shear stress and the heat flux in the gas at the wall.

8.2.4 Concluding Remarks

The topic of laminar-turbulent transition was given much room in this chapter. This was deemed necessary because of its large importance for hypersonic vehicle design.

A definitive improvement of the capabilities to predict laminar-turbulent transition, but also to optimize surface shapes in order to, for instance, delay transition, is mandatory for CAV-type flight vehicles as well as for ARV-type vehicles. Vehicles of these classes are viscous-effect dominated, which, as was discussed and shown in several of the preceding chapters, regards the thermal state of the surface, the drag of the vehicle, the thermal loads in view of the structure and materials concept of the vehicle, and issues of aerothermodynamic propulsion integration.

For RV-type flight vehicles the improvement of transition prediction is highly desirable, because the effectiveness of these vehicles must be improved by minimizing the mass of the thermal protection system. In this regard laminar-turbulent transition on the lower branch of the re-entry trajectory, as well as on alternative lower altitude trajectories than preferred today, including contingency trajectories, is of great importance and demands a more accurate and reliable prediction than is possible today.

The general knowledge about transition phenomena in high-speed flows is already rather good, and the development of new prediction methods is encouraging. Here non-local and non-linear theory appears to have the necessary potential. The use of such methods in aerothermodynamic numerical simulations and optimizations also in industrial design work will be no problem in view of the still strongly growing computer capabilities, Section 10.2.

Necessary to achieve the improvements of transition prediction is continuous concerted research, extension and use of "quiet" ground-simulation facilities, and in-flight measurements on ad-hoc experimental vehicles or in passenger experiments on other vehicles. Ground and flight measurements need also the careful recording of the thermal state of the surface by means of a suitable hot experimental technique, Section 10.3. A combination of analytical work, computational simulation, ground-facility simulation, and in-flight simulation (unified approach [110]), in a transfer-model ansatz, which takes also into account the coupling of the flow to the vehicle surface, is considered to be necessary, [2], in order to advance this scientifically and technically fascinating and challenging field.

8.3 Turbulence Modeling for High-Speed Flows

In Chapter 7 we have discussed attached high-speed turbulent flow by means of a very simple description, the $\frac{1}{7}$-power law with the reference-temperature extension. Our understanding was and is that such a description can be used only for the establishment of general insights, for trend considerations on typical configuration elements, and for the approximate quantification of attached viscous flow effects, namely boundary layer integral properties, the thermal state of the surface, and the skin friction.

For the "exact" quantification of viscous flow effects the methods of numerical aerothermodynamics are required. These methods employ statistical turbulence models, which permit a description of high-speed turbulent flows with acceptable to good accuracy, as long as the flow is attached. Turbulent flow in the presence of strong interaction phenomena still poses large problems for statistical turbulence models, which hopefully can be diminished in the future.

In this section we give only a short overview of issues of turbulence modeling for attached viscous flows on high-speed vehicles. On RV-type vehicles the boundary layer due to the large angle of attack on a large part of the lower trajectory, Sub-Section 1.2, is a subsonic, transonic and low supersonic boundary layer with large boundary-layer edge temperatures and steep temperature gradients normal to the surface due to the surface radiation cooling, and hence steep density gradients with opposite sign. On CAV-type vehicles we have true hypersonic boundary layers at relatively small boundary-layer edge temperatures[31] but also steep gradients of temperature and density, respectively, also due to the surface radiation cooling. In the literature transonic, supersonic, and hypersonic boundary-layer flows are all together called compressible boundary layers flows. For detailed introductions to the topic see, e. g., [9], [8], [111], [112], [113].

A compressible boundary-layer flow is a flow in which non-negligible density changes occur. These changes can occur even if stream-wise pressure changes are small, and also in low-speed flows, if large temperature gradients normal to the surface are present.

In turbulence modeling density fluctuations in a turbulent boundary layer can be neglected, if they are small compared to the mean-flow density: $\rho' \ll \rho_{mean}$. Morkovin's hypothesis [114] states that this holds for boundary-layer edge-flow Mach numbers $M_e \lessapprox 5$ in attached viscous flow.

The hypothesis does not hold, see, e. g., [115], [116], for flows with large heat transfer, free shear flows, and turbulent combustion. It also does not hold if the turbulent flow crosses a shock wave, for instance at the boundary-layer edge in case of an incident shock wave, see, e. g., [117], and in strong interaction situations (shock/boundary-layer interaction, Section 9.2), see, e. g., [118]. In flows with edge Mach numbers $M_e \gtrapprox 5$, with shock/boundary-layer interactions, et cetera, hence explicit compressibility corrections must be applied.

For compressible flows besides the continuity and the momentum equations also the energy equation must be regarded. Hence in turbulence modeling not only velocity and pressure fluctuations must be taken into account but besides the density fluctuations also temperature fluctuations. This brings us from the Reynolds-averaging to the Favre-averaging, the latter being a mass-averaging process, which also necessitates further closure assumptions, [9], [113].

Special issues appear due to turbulent heat conduction and turbulent mass diffusion, the latter in the case of non-equilibrium flow. Analogously to the Prandtl and Schmidt numbers in laminar flows, turbulent Prandtl and Schmidt numbers are introduced. These are usually taken as constant[32]. For the Prandtl number it is known since long that it is not constant in attached high speed turbulent flows [119]. With measured turbulent Prandtl numbers in attached flow $0.8 \lessapprox Pr_{turb} \lessapprox 1$ usually a mean constant Prandtl number $Pr_{turb} = 0.9$ is employed in turbulence models. It is advisable to check with parametric variations whether the solution for a given flow class reacts sensitively to the choice of the (constant) Prandtl number. The same holds for the Schmidt number.

[31] At the (small) blunt nose of such a vehicle of course the edge temperatures are large, but there the boundary layer is not yet turbulent.

[32] We remember that the laminar Prandtl and Schmidt number are in fact functions of the temperature, Section 4.1.

The turbulence models used are in general two-equation models, see, e. g., [113]. They allow to compute with good accuracy attached turbulent two-dimensional and three-dimensional high-speed boundary layers with pressure gradients, surface heat transfer, surface radiation cooling, surface roughness, and high-temperature real-gas effects as long as the transition location is known or the flow is insensitive to the transition location.

Once the flow separates, the usual prediction problems, present already in the low-speed regime, see, e. g., [80], arise. Especially in the case of shock-induced separation (shock/boundary-layer interaction) it is observed, see also [9], that the upstream influence is wrongly computed, the primary separation location is not met, the computed pressure in the interaction zone is different from the measured one, while downstream of the interaction zone the relaxation of pressure, skin friction and heat transfer (usually overestimated already in the interaction zone) is different from the measured one. This all despite the use of compressibility corrections.

Instructive computational results are given in [120]. There, several two-equation models with compressibility corrections were applied to flat-plate boundary layers with cooled and adiabatic wall in the Mach number regime $1.2 \leqq M_\infty \leqq 10$, to a hypersonic compression corner with cooled surface (two-dimensional control surface) with an onset flow Mach number $M = 9.22$, and to the flow past the X-38 configuration with extended body flap at $M_\infty = 6$ and $\alpha = 40°$ in the wind-tunnel situation. We will come back to the topics of separation and strong interaction in Chapter 9.

References

1. W. D. GOODRICH, S. M. DERRY, J. J. BERTIN. "Shuttle Orbiter Boundary-Layer Transition: A Comparison of Flight and Wind-Tunnel Data". AIAA-Paper 83-0485, 1983.
2. E. H. HIRSCHEL. "The Technology Development and Verification Concept of the German Hypersonics Technology Programme". AGARD R-813, 1986, pp. 12-1 to 12-15.
3. J. F. SHEA. "Report of the Defense Science Board Task Force on the National Aerospace Plane (NASP)". Office of the Under Secretary of Defense for Acquisition, Washington, D. C., 1988.
4. R. M. WILLIAMS. "National Aerospace Plane: Technology for America's Future". Aerospace America, Vol. 24, No. 11, 1986, pp. 18 - 22.
5. W. STAUDACHER, J. WIMBAUER. "Design Sensitivities of Airbreathing Hypersonic Vehicles". AIAA-Paper 93-5099, 1993.
6. E. H. HIRSCHEL. "Thermal Surface Effects in Aerothermodynamics". *Proc. Third European Symposium on Aerothermodynamics for Space Vehicles, Noordwijk, The Netherlands, November 24 - 26, 1998.* ESA SP-426, 1999, pp. 17 - 31.

7. E. H. HIRSCHEL. "Aerothermodynamic Phenomena and the Design of Atmospheric Hypersonic Airplanes". *J. J. Bertin, J. Periaux, J. Ballmann (eds.), Advances in Hypersonics, Vol. 1, Defining the Hypersonic Environment.* Birkhäuser, Boston, 1992, pp. 1 - 39.
8. J. G. MARVIN, T. J. COAKLEY. "Turbulence Modeling for Hypersonic Flows". *J. J. Bertin, J. Periaux, J. Ballmann (eds.), Advances in Hypersonics, Vol. 2, Modeling Hypersonic Flows.* Birkhäuser, Boston, 1992, pp. 1 - 43.
9. D. C. WILCOX. "Turbulence Modelling for CFD". DCW Industries, La Cañada, CAL., USA, 1998.
10. K. F. STETSON. "Hypersonic Boundary-Layer Transition". *J. J. Bertin, J. Periaux, J. Ballmann (eds.), Advances in Hypersonics, Vol. 1, Defining the Hypersonic Environment.* Birkhäuser, Boston, 1992, pp. 324 - 417.
11. M. V. MORKOVIN. "Critical Evaluation of Transition from Laminar to Turbulent Shear Layers with Emphasis on Hypersonically Travelling Bodies". Air Force Flight Dynamics Laboratory, Wright-Patterson Air Force Base, AFFDL-TR-68-149, 1969.
12. E. RESHOTKO. "Boundary-Layer Stability and Transition". Annual Review of Fluid Mechnics, Vol. 8, 1976, pp. 311 - 349.
13. D. ARNAL. "Laminar-Turbulent Transition Problems in Supersonic and Hypersonic Flows". AGARD R-761, 1988, pp. 8-1 to 8-45.
14. P. J. SCHMID, D. S. HENNINGSON. "Stability and Transition in Shear Flows". Springer-Verlag, New York/Berlin/Heidelberg, 2001.
15. G. B. SCHUBAUER, H. K. SKRAMSTADT. "Laminar Boundary Layer Oscillations and Transition on a Flat Plate". NACA Rep. 909.
16. J. M. KENDALL. "Supersonic Boundary-Layer Stability Experiments ". *W. D. McCauley (ed.), Proceedings of Boundary Layer Transition Study Group Meeting.* Air Force Report No. BSD-TR-67-213, Vol. II, 1967.
17. L. M. MACK. "Boundary-Layer Stability Theory". JPL 900-277, 1969.
18. L. M. MACK. "Stability of the Compressible Laminar Boundary Layer According to a Direct Numerical Solution". AGARDograph 97, Part I, 1965, pp. 329 - 362.
19. H. MCDONALD, R. W. FISH. "Practical Calculations of Transitional Boundary Layers". Int. Journal of Heat and Mass Transfer, Vol. 16, No. 9, 1973, pp. 25 - 53.
20. H. SCHLICHTING. "Boundary Layer Theory". 7th edition, McGraw-Hill, New York, 1979.
21. L. M. MACK. "Boundary-Layer Linear Stability Theory". AGARD R-709, 1984, pp. 3-1 to 3-81.
22. D. ARNAL. "Laminar Turbulent Transition". *T. K. S. Murthy (ed.), Computational Methods in Hypersonic Aerodynamics.* Computational Mechanics Publications and Kluwer Academic Publishers, 1991, pp. 233 - 264.
23. M. LANG, O. MARXEN, U. RIST, S. WAGNER. "A Combined Numerical and Experimental Investigation of Transition in a Laminar Separation Bubble". *S. Wagner, M. Kloker, U. Rist (eds.), Recent Results in Laminar-Turbulent Transition.* Notes on Numerical Fluid Mechanics and Multidisciplinary Design, NNFM 86, Springer, Berlin/Heidelberg/New York, 2004, pp. 149 - 164.

24. L. LEES, C. C. LIN. "Investigation of the Stability of the Boundary Layer in a Compressible Fluid". NACA TN 1115, 1946.

25. M. R. MALIK, R. E. SPALL. "On the Stability of Compressible Flow Past Axisymmetric Bodies". High Technology Corporation, Hampton, VA, Report No. HTC-8905, 1989.

26. L. LEES. "The Stability of the Laminar Boundary Layer in a Compressible Fluid". NACA TN 876, 1947.

27. E. KUFNER. "Numerische Untersuchungen der Strömungsinstabilitäten an spitzen und stumpfen Kegeln bei hypersonischen Machzahlen (Numerical Investigations of the Flow Instabilities at Sharp and Blunt Cones at Hypersonic Mach Numbers)". Doctoral Thesis, Universität Stuttgart, Germany, 1995.

28. P. PERRIER. "Concepts of Hypersonic Aircraft". *J. J. Bertin, J. Periaux, J. Ballmann (eds.), Advances in Hypersonics, Vol. 1, Defining the Hypersonic Environment.* Birkhäuser, Boston, 1992, pp. 40 - 71.

29. L. M. MACK. "Stability of Axisymmetric Boundary Layers on Sharp Cones at Hypersonic Mach Numbers". AIAA-Paper 87-1413, 1987.

30. M. R. MALIK. "Prediction and Control of Transition in Hypersonic Boundary Layers". AIAA-Paper 87-1414, 1987.

31. A. FEZER, M. KLOKER. "DNS of Transition Mechanisms at Mach 6.8 - Flat Plate versus Sharp Cone". *D. E. Zeitoun, J. Periaux, J.-A. Désidéri, M. Marini (eds.), West East High Speed Flow Fields 2002.* CIMNE Handbooks on Theory and Engineering Applications of Computational Methods, Barcelona, Spain, 2002.

32. E. RESHOTKO, M. M. S. KHAN. "Stability of the Laminar Boundary Layer on a Blunted Plate in Supersonic Flow". *Proc. IUTAM Symposium Laminar-Turbulent Transition, Stuttgart, 1979.* Springer, Berlin/Heidelberg/New York, 1980, pp. 181 - 200.

33. A. V. FEDOROV. "Instability of the Entropy Layer on a Blunt Plate in Supersonic Gas Flow". Journal Appl. Mech. Tech. Phys., Vol. 31, No. 5, 1990, pp. 722 - 728.

34. G. DIETZ, S. HEIN. "Entropy-Layer Instabilities over a Blunted Flat Plate in Supersonic Flow". Physics of Fluids, Vol. 11, No. 1, 1999, pp. 7 - 9.

35. G. SIMEONIDES. "Correlation of Laminar-Turbulent Transition Data over Flat Plates in Supersonic/Hypersonic Flow Including Leading Edge Bluntness Effects". Shock Waves, Vol. 12, No. 6, 2003, pp. 497 - 508.

36. S. HEIN. "Nonlinear Nonlocal Transition Analysis". Doctoral Thesis, Universität Stuttgart, Germany, 2004.

37. G. B. WHITHAM. "The Navier-Stokes Equations of Motion". *L. Rosenhead, (ed.), Laminar Boundary Layers.* Oxford Univ. Press, 1963, pp. 114 - 162.

38. V. THEOFILIS, A. V. FEDOROV, D. OBRIST, U. CH. DALLMANN. "The Extended Görtler-Hämmerlin Model for Linear Instability of Three-Dimensional Incompressible Swept Attachment-Line Boundary-Layer Flow". J. Fluid Mechanics, Vol. 487, 2003, pp. 271 - 313.

39. F. P. Bertolotti. "On the Connection between Cross-Flow Vortices and Attachment-Line Instability". *H. F. Fasel, W. S. Saric (eds.), Laminar-Turbulent Transition.* Springer, Berlin/Heidelberg/New York, 2000, pp. 625 - 630.

40. J. Sesterhenn, R. Friedrich. "Numerical Receptivity Study of an Attachment Boundary Layer in Hypersonic Flow". *J.-P. Dussauge, A. A. Chikhaoui (eds.), Aerodynamics and Thermochemistry of High Speed Flow.* Euromech 440, Marseille, France, 2002.

41. N. A. Cumpsty, M. R. Head. "The Calculation of Three-Dimensional Turbulent Boundary Layers. Part II: Attachment Line Flow on an Infinite Swept Wing". The Aeronautical Quarterly, Vol. XVIII, Part 2, 1967, pp. 99 - 113.

42. D. I. A. Poll. "Boundary Layer Transition on the Windward Face of Space Shuttle During Re-Entry". AIAA-Paper 85-0899, 1985.

43. D. I. A. Poll. "Some Aspects of the Flow near a Swept Attachment Line with Particular Reference to Boundary Layer Transition". Doctoral thesis, Cranfield, U. K., C of A Report 7805./L 1978.

44. P. R. Owen, D. G. Randall. "Boundary Layer Transition on a Swept Back Wing". R. A. E. TM 277, 1952, R. A. E. TM 330, 1953.

45. H. Bippes. "Basic Experiments on Transition in Three-Dimensional Boundary Layers Dominated by Crossflow Instability". Progress in Aerospace Sciences, Elsevier Science Ltd, Oxford, Vol. 35, No. 3-4, 1999, pp. 363 - 412.

46. W. S. Saric, H. L. Reed , E. B. White. "Stability and Transition of Three-Dimensional Boundary Layers". Annual Review of Fluid Mechnics, Vol. 35, 2003, pp. 413 - 440.

47. D. I. A. Poll, Ph. Tran, D. Arnal. "Capabilities and Limitations of Available Transition Prediction Tools". Aerospatiale TX/AP no. 114 779, 1994.

48. G. Simeonides, W. Haase. "Experimental and Computational Investigations of Hypersonic Flow About Compression Corners". J. Fluid Mechanics, Vol. 283, 1995, pp. 17 - 42.

49. G. Simeonides. "Laminar-Turbulent Transition Promotion in Regions of Shock Wave Boundary Layer Interaction". *ESA/ESTEC MSTP Code Validation Workshop.* March 25 - 27, 1996, ESA/ESTEC EWP-1880, 1996.

50. A. Pagella, U. Rist, S. Wagner. "Numerical Investigations of Small-Amplitude Disturbances in a Boundary Layer with Impinging Shock Wave at M = 4.8". Physics of Fluids, Vol. 14, No. 7, 2002, pp. 2088 - 2101.

51. P. D. Germain, H. G. Hornung. "Transition on a Slender Cone in Hypervelocity Flow". Exps. Fluids, Vol. 22, 1997, pp. 183 - 190.

52. M. L. Hudson, N. Chokani, G. V. Candler. "Linear Stability of Hypersonic Flow in Thermochemical Non-Equilibrium". AIAA J., Vol. 35, 1997, pp. 958 - 964.

53. F. P. Bertolotti. "The Influence of Rotational and Vibrational Energy Relaxation on Boundary-Layer Stability". J. Fluid Mechnics, Vol. 372, 1998, pp. 93 - 118.

54. N. N. "Boundary Layer Simulation and Control in Wind Tunnels". AGARD-AR-224, 1988.

55. D. I. A. POLL. "Laminar-Turbulent Transition". AGARD-AR-319, Vol. I, 1996, pp. 3-1 to 3-20.

56. H. W. KIPP, V. T. HELMS. "Some Observations on the Occurance of Striation Heating". AIAA-Paper 85-0324, 1985.

57. H. GÖRTLER. "Über den Einfluß der Wandkrümmung auf die Entstehung der Turbulenz". ZAMM, Vol. 20, 1940, pp. 138 - 147.

58. H. W. LIEPMANN. "Investigation of Boundary-Layer Transition on Concave Walls". NACA ACR 4J28, 1945.

59. I. TANI, Y. AIHARA. "Görtler Vortices and Boundary-Layer Transition". ZAMP, Vol. 20, 1969, pp. 609 - 618.

60. F. MAURER. "Three-Dimensional Effects in Shock-Separated Flow Regions Ahead of Lateral Control Jets Issuing from Slot Nozzles of Finite Length". AGARD-CP-4, Part 2, 1966, pp. 605 - 634.

61. H. LÜDECKE, E. SCHÜLEIN. "Simulation of Streamwise Vortices on Turbulent Hypersonic Ramps". Proc. Second International Conference on CFD, Sydney, Australia, 2002.

62. H. LÜDECKE. "Untersuchung von Längswirbeln in abgelösten hypersonischen Strömungen (Investigation of Longitudinal Vortices in Separated Hypersonic Flows)". Doctoral Thesis, Technische Universität Braunschweig, Germany, 2002.

63. G. SIMEONIDES, J. P. VERMEULEN, S. ZEMSCH. "Amplification of Disturbances and the Promotion of Laminar-Turbulent Transition Through Regions of Hypersonic Shock Wave Boundary Layer Interaction". *R. Brun, A. A. Chikhaoui (eds.), Aerothermochemistry of Spacecraft and Associated Hypersonic Flows.* Proc. IUTAM Symposium, Marseille, France, 1992, pp. 344 - 351.

64. L. DE LUCA, G. CARDONE, D. AYMER DE LA CHEVALERIE, A. FONTENEAU. "Viscous Interaction Phenomena in Hypersonic Wedge Flow". AIAA J., Vol. 33, No. 12, 1995, pp. 2293 - 2298.

65. H. LÜDECKE, P. KROGMANN. "Numerical and Experimental Investigations of Laminar/Turbulent Boundary Layer Transition". Proc. ECCOMAS 2000, Barcelona, Spain, 2000.

66. R. E. SPALL, M. R. MALIK. "Görtler Vortices in Supersonic and Hypersonic Boundary Layers". Physics of Fluids, Vol. 1, 1989, pp. 1822 - 1835.

67. W. S. SARIC. "Görtler Vortices". Annual Review of Fluid Mechnics, Vol. 26, 1994, pp. 379 - 409.

68. R. NARASIMHA. "The Three Archetypes of Relaminarisation". Proc. 6th Canadian Conf. of Applied Mechanics, Vol. 2, 1977, pp. 503 - 518.

69. F. M. WHITE. "Viscous Fluid Flow". Second edition, McGraw-Hill, New York, 1991.

70. R. MUKUND, P. R. VISWANATH, J. D. CROUCH. "Relaminarization and Retransition of Accelerated Turbulent Boundary Layers on a Convex Surface". *H. F. Fasel , W. S. Saric (eds.), Laminar-Turbulent Transition.* Springer-Verlag, Berlin/Heidelberg/New York, 2000, pp. 243 - 248.

71. E. RESHOTKO. "Environment and Receptivity". AGARD R-709, 1984, pp. 4-1 to 4-11.

72. U. SCHUMANN, P. KONOPKA, R. BAUMANN, R. BUSEN, T. GERZ, H. SCHLAGER, P. SCHULTE, H. VOLKERT . "Estimate of Diffusion Parameters of Aircraft Exhaust Plumes near the Tropopause from Nitric Oxide and Turbulence Measurements". J. of Geophysical Research, Vol. 100, No. D7, 1995, pp. 14,147 - 14,162.

73. I. E. BECKWITH. "Development of a High-Reynolds Number Quiet Tunnel for Transition Research". AIAA J., Vol. 13, 1997, pp. 300 - 316.

74. W. KORDULLA, R. RADESPIEL, P. KROGMANN, F. MAURER. "Aerothermodynamic Activities in Hypersonics at DLR". AIAA-Paper 92-5032, 1992.

75. A. WEISE, G. SCHWARZ. "Der Stosswindkanal des Instituts für Aerodynamik und Gasdynamik der Universität Stuttgart". Zeitschrift für Flugwissenschaften (ZFW), Vol. 21, No. 4, 1973, pp. 121 - 131.

76. P. KROGMANN. "An Experimental Study of Boundary Layer Transition on a Slender Cone at Mach 5". AGARD-CP-224, 1977, pp. 26-1 to 26-12.

77. S. P. SCHNEIDER. "Development of a Mach-6 Quiet-Flow Ludwieg Tube for Transition Research". *H. F. Fasel , W. S. Saric (eds.), Laminar-Turbulent Transition.* Springer-Verlag, Berlin/Heidelberg/New York, 2000, pp. 427 - 432.

78. W. S. SARIC, H. L. REED , E. J. KERSCHEN. "Boundary-Layer Receptivity to Freestream Disturbances". Annual Review of Fluid Mechnics, Vol. 34, 2002, pp. 291 - 319.

79. F. R. MENTER. "Influence of Freestream Values on $k - \omega$ Turbulence Model Predictions". AIAA J., Vol. 33, No. 12, 1995, pp. 1657 - 1659.

80. A. CELIC. "Performance of Modern Eddy-Viscosity Turbulence Models". Doctoral Thesis, Universität Stuttgart, Germany, 2004.

81. G. SCHRAUF. "Industrial View on Transition Prediction". *S. Wagner, M. Kloker, U. Rist (eds.), Recent Results in Laminar-Turbulent Transition.* Notes on Numerical Fluid Mechanics and Multidisciplinary Design, NNFM 86, Springer, Berlin/Heidelberg/New York, 2004, pp. 111 - 122.

82. M. R. MALIK. "COSAL - A Black Box Compressible Stability Analysis Code for Transition Prediction in Three-Dimensional Boundary Layers". NASA CR 165925, 1982.

83. U. EHRENSTEIN, U. DALLMANN. "Ein Verfahren zur linearen Stabilitätsanalyse von dreidimensionalen, kompressiblen Grenzschichten". DFVLR IB 221-88 A 20, 1988.

84. M. SIMEN. "COSMET, a DLR-Dornier Computer Program for Compressible Stability Analysis with Local Metric". DFVLR IB 221-91 A 09, 1991.

85. F. LABURTHE. "Problème de stabilité linéaire et prévision de la transition dans des configurations tridimensionelles, incompressibles et compressibles". Doctoral Thesis, ENSAE, Toulouse, France, 1992.

86. A. HANIFI. "Stability Characteristics of the Supersonic Boundary Layer on a Yawed Cone". Licentiate Thesis, TRITA-MEK, TR 1993:6, Royal Institute of Technology, Stockholm, Sweden, 1993.

87. M. R. MALIK. "Boundary-Layer Transition Prediction Toolkit". AIAA-Paper 97-1904, 1997.

88. G. Schrauf. "Curvature Effects for Three-Dimensional, Compressible Boundary Layer Stability". Zeitschrift für Flugwissenschaften und Weltraumforschung (ZFW), Vol. 16, No. 2, 1992, pp. 119 - 127.

89. G. Schrauf. "COAST3 - A Compressible Stability Code. User's Guide and Tutorial". Deutsche Airbus, TR EF-040/98, 1998.

90. Th. Herbert. "Parabolized Stability Equations". Annual Review of Fluid Mechanics, Vol. 29, 1997, pp. 245 - 283.

91. F. P. Bertolotti. "Linear and Nonlinear Stability of Boundary Layers with Streamwise Varying Properties". Doctoral Thesis, Ohio State University, USA, 1991.

92. C. L. Chang, M. R. Malik, G. Erlebacher, M. Y. Hussaini. "Compressible Stability of Growing Boundary Layers Using Parabolized Stability Equations". AIAA-Paper 91-1636, 1991.

93. M. Simen. "Lokale und nichtlokale Instabilität hypersonischer Grenzschichtströmungen (Local and Non-local Stability of Hypersonic Boundary-Layer Flows)". Doctoral Thesis, Universität Stuttgart, Germany, 1993.

94. Th. Herbert, G. K. Stuckert, N. Lin. "Nonparallel Effects in Hypersonic Boundary Layer Stability". WL-TR-93-3097, 1993.

95. M. Simen, F. P. Bertolotti, S. Hein, A. Hanifi, D. S. Henningson, U. Dallmann. "Nonlocal and Nonlinear Stability Theory". *S. Wagner, J. Periaux, E. H. Hirschel (eds.), Computational Fluid Dynamics '94*. John Wiley and Sons, Chichester, 1994, pp. 169 - 179.

96. M. S. Mughal, P. Hall. "Parabolized Stability Equations and Transition Prediction for Compressible Swept-Wing Flows". Imperial College for Science, Technology and Medicine, final report on DTI contract ASF/2583U, 1996.

97. C. L. Chang, H. Vinh, M. R. Malik. "Hypersonic Boundary-Layer Stability with Chemical Reactions using PSE". AIAA-Paper 97-2012, 1997.

98. H. Salinas. "Stabilité linéaire et faiblement non linéaire d'une couche limite laminaire compressible tridimensionelle par l'approache PSE". Doctoral Thesis, ENSAE, Toulouse, France, 1998.

99. F. P. Bertolotti. "The Equivalent Forcing Model for Receptivity Analysis with Application to the Construction of a High-Performance Skin-Perforation Pattern for Laminar Flow Control". *S. Wagner, M. Kloker, U. Rist (eds.), Recent Results in Laminar-Turbulent Transition.* Notes on Numerical Fluid Mechanics and Multidisciplinary Design, NNFM 86, Springer, Berlin/Heidelberg/New York, 2004, pp. 25 - 36.

100. D. C. Hill. "Adjoint Systems and their Role in the Receptivity Problem for Boundary Layers". J. Fluid Mechnics, Vol. 292, 1995, pp. 183 - 204.

101. J. L. van Ingen. "A Suggested Semi-Empirical Method for the Calculation of the Boundary-Layer Transition Region". Reports UTH71 and UTH74, Delft, The Netherlands, 1956.

102. A. M. O. Smith, N. Gamberoni. "Transition, Pressure Gradient and Stability Theory". Douglas Report No. ES 26388, 1956.

103. D. Arnal. "Boundary-Layer Transition: Predictions Based on Linear Theory". AGARD-R-793, 1994, pp. 2-1 to 2-63.

104. L. M. MACK. "Transition Prediction and Linear Stability Theory". AGARD CP-224, 1977, pp. 1-1 to 1-22.

105. N. A. THYSON, K. K. CHEN. "Extension of the Emmon's Spot Theory to Flows on Blunt Bodies". AIAA J., Vol. 9, No. 5, 1971, pp. 821 - 825.

106. H. W. KIPP, R. V. MASEK. "Aerodynamic Heating Constraints on Space Shuttle Vehicle Design". ASME pub. 70 HT/SpT 45, 1970.

107. P. HERNANDEZ, PH. TRAN. "Modèle global de déclenchement de la transition sur la ligne centrale intrados d'un planteur hypersonique". Aerospatiale H-NT-1-650-AS, 1991.

108. A. MARTELLUCCI, A. M. BERKOWITZ, C. L. KYRISS. "Boundary-Layer Transition - Flight Test Observations". AIAA-Paper 77-0125, 1977.

109. A. VELÁZQUEZ. "FESTIP Technology Developments in Aerothermodynamics for Reusable Launch Vehicles". SENER Doc. No. P215809-FR, 2000.

110. E. H. HIRSCHEL. "Present and Future Aerodynamic Process Technologies at DASA Military Aircraft". ERCOFTAC Industrial Technology Topic Meeting, Florence, Italy, 1999.

111. E. F. SPINA, A. J. SMITS, S. K. ROBINSON. "The Physics of Supersonic Turbulent Boundary Layers". Annual Review of Fluid Mechanics, Vol. 26, 1994, pp. 287 - 319.

112. J.-P. DUSSAUGE, H. H. FERNHOLZ, R. W. SMITH, P. J. FINLEY, A. J. SMITS, E. F. SPINA. "Turbulent Boundary Layers in Subsonic and Supersonic Flow". AGARDograph 335, 1996.

113. B. AUPOIX. "Introduction to Turbulence Modelling for Turbulent Flows". *C. Benocci , J. P. A. J. van Beek, (eds.), Introduction to Turbulence Modeling.* VKI Lecture Series 2002-02, VKI, Rhode Saint Genèse, Belgium, 2002.

114. M. V. MORKOVIN. "Effects of Compressibility on Turbulent Flows". Colloque International CNRS No. 108, Mécanique de la Turbulence, Editions CNRS, 1961.

115. P. G. HUANG, G. N. COLEMAN, P. BRADSHAW. "Compessible Turbulent Channel Flows - DNS Results and Modeling". J. Fluid Mechnics, Vol. 305, 1995, pp. 185 - 218.

116. TH. MAEDER. "Numerical Investigation of Supersonic Turbulent Boundary Layers". Doctoral Thesis, ETH Zürich, Switzerland, 2000, Fortschritts-Berichte VDI, Reihe 7, Strömungstechnik, Nr. 394, 2000.

117. R. FRIEDRICH, F. P. BERTOLOTTI. "Compressibility Effects Due to Turbulence Fluctuations". Applied Scientific Research, Vol. 57, 1997, pp. 165 - 194.

118. D. S. DOLLING. "Fifty Years of Shock-Wave/Boundary-Layer Interaction Research: What Next?". AIAA J., Vol. 39, No. 8, 2001, pp. 1517 - 1531.

119. H. U. MEIER, J. C. ROTTA. "Temperature Distributions in Supersonic Turbulent Boundary Layers". AIAA J., Vol. 9, No. 11, 1971, pp. 2149 - 2156.

120. C. WEBER, R. BEHR, C. WEILAND. "Investigation of Hypersonic Turbulent Flow over the X-38 Crew Return Vehicle". AIAA-Paper 2000-2601, 2000.

9 Strong Interaction Phenomena

In Sub-Section 7.1.4 we have noted that if the displacement properties of a boundary layer are of $O(1/\sqrt{Re_{ref}})$, it influences the pressure field, i. e. the inviscid flow field, only weakly. We call this "weak interaction" between the attached viscous flow and the inviscid flow[1]. If, however, the boundary layer separates, the inviscid flow is changed and we observe a "strong interaction". This phenomenon is present in all Mach number regimes, especially also in the subsonic regime.

"Shock/boundary-layer interaction", present in the transonic, supersonic, and hypersonic regimes, also leads through thickening or even separation of the boundary layer to strong interaction. "Shock/shock interaction" with the associated interaction with the boundary layer is a strong interaction phenomenon, too.

In the high supersonic and the hypersonic regime strong interaction happens also if the attached boundary layer becomes very thick, which is the case with large Mach numbers and small Reynolds numbers at the boundary-layer edge. This is the "hypersonic viscous interaction". Associated with hypersonic viscous interaction are "rarefaction effects" which appear in the continuum flow regime with slip effects, Section 2.3. They are, as we will see, directly related to large Mach numbers and small Reynolds numbers at the boundary-layer edge, too.

We consider the strong-interaction phenomena in general in their two-dimensional appearance. In should be noted that for instance shock/boundary-layer interaction usually is less severe in three-dimensional cases compared to strictly two-dimensional cases. The computation of turbulent three-dimensional interactions in general is also less problematic concerning turbulence models. This holds also for ordinary (turbulent) flow separation. On a (two-dimensional) airfoil separation at angles of attack of, say, above approximately 15°, is characterized by vortex shedding. The flow is highly unsteady. This is in contrast to three-dimensional separation on delta wings or fuselage-like bodies. Here the lee-side separation, beginning at angles of attack of, say, approximately 5°, is macroscopically steady up to, say, approximately 50°.

[1] Depending on the given problem it can be advisable in a computation scheme to take iteratively into account the displacement properties of the boundary layer with, for instance, perturbation coupling, see, Sub-Section 7.2.1.

This shows that an extended classification of separated flows, and strong-interaction flows in general, is desirable in view of turbulence modeling [1].

Unsteady pressure loads (dynamic pressure loads), due to separation phenomena with vortex shedding, further the intersection of vortex wakes with configuration components (leading to, for instance, fin vibration), and especially also due to unsteadiness of shock/boundary-layer interaction, are of large concern in flight vehicle design. Like noise they can lead to material fatigue and thus endanger structural integrity. In hypersonic flows they are usually combined with large thermal loads, which make them the more critical.

All the strong interaction phenomena, which we will treat in the following sections, can have large, even dramatic influence on the surface pressure field and hence on the aerodynamic forces and moments acting on the flight vehicle, as well as on the thermal state of the surface, and hence on thermal surface effects and on thermal loads.

We aim for a basic understanding of these phenomena and also of the related computational and ground-facility simulation problems.

9.1 Flow Separation

Flow separation is defined by the local violation of the boundary-layer criteria (constant pressure in wall-normal direction, approximately surface-parallel flow), by the convective transport of vorticity away from the body surface and the subsequent formation of vortex sheets and vortices behind or above/behind a flight vehicle. This definition goes beyond the classical view on separation. It implies that separation can be present at any flight vehicle just because any flight vehicle is of finite length. However, it is justified, because behind or above/behind every flight vehicle with aerodynamic lift always vortex phenomena (vortex sheets and vortices) are present, see Fig. 9.1.

We distinguish two principle types of separation [2]:

- type a: flow-off separation,

- type b: squeeze-off separation (the classical separation).

Flow-off separation happens at sharp edges, which can be leading edges or trailing edges, Fig. 9.1 a). Squeeze-off separation typically occurs at regular surface portions, due to the pressure field of the external inviscid flow, Fig. 9.1 a) and b)).

In both cases always two boundary layers are involved. In the case of flow-off separation the boundary layers from the two sides of the sharp edge (at a wing trailing or leading edge those from above and below the wing) flow off the edge and merge into the near-wake. In the case of squeeze-off separation the boundary layers squeeze each other off the same surface [2].

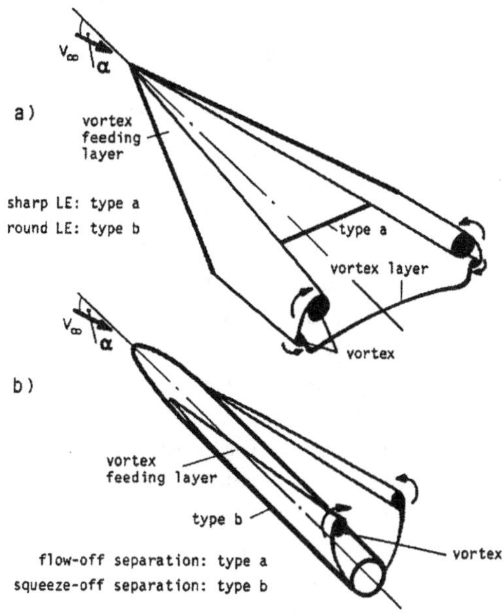

Fig. 9.1. Schematic of separation types, and resulting vortex sheets and vortices at basic configurations at large - typical for re-entry flight - angles of attack [7]: **a)** wing with large leading-edge (LE) sweep, **b)** fuselage. In both cases secondary and higher-order separation phenomena on the lee side are not indicated.

While in two-dimensional flow squeeze-off separation is defined by vanishing skin friction - actually by the change of sign of the wall shear stress τ_w - such a clear criterion is not available for three-dimensional flow. In three-dimensional attached viscous flow τ_w vanishes only in a few singular points on the body surface[2], for instance at the Blunt Delta Wing (BDW) in the forward stagnation point, see, e. g., Fig. 3.16, but also at singular points in the separation region [4]. Along three-dimensional separation lines, as well as attachment lines, Figs. 3.16 and 3.17, skin friction does not vanish.

In [5] and [2] criteria for three-dimensional separation are proposed. A practical indication for a three-dimensional squeeze-off separation line is the converging skin-friction line pattern, Fig. 3.6, and for a three-dimensional attachment line the diverging pattern, Fig. 3.5, see also Figs. 3.16 and 3.17.

Fig. 9.1 shows where on configuration elements, typical for hypersonic flight vehicles, the two types of separation occur[3]. On a delta wing with

[2] Only in such points streamlines actually terminate at (attach) or leave from (detach or separate) the body surface. This holds also in two dimensions, where the separation point is such a singular point, where the separation streamline leaves the surface under a finite angle [3].

[3] Note that secondary and higher-order separation phenomena on the lee side of the generic configurations are not indicated in the figure.

large leading-edge sweep and sufficiently large angle of attack [6], we observe squeeze-off separation at round leading edges, which are typical for hypersonic vehicles, Fig. 9.1 a). The primary squeeze-off separation line can lie well on the upper side of the wing, as we have seen in the case of the BDW, which has a rather large leading-edge radius, Figs. 3.17 and 3.19. If the leading-edge radius is very small - the sharp swept leading edge is the limiting case - we have flow-off separation at the leading edge, Fig. 9.1 a), type a. At the sharp trailing edge of such a generic configuration we find flow-off separation, which we find also at sharp trailing edges of stabilization and control surfaces. At round fuselages, Fig. 9.1 b), and also at round trailing edges, always squeeze-off separation occurs. If local separation (see below) occurs, we have squeeze-off separation with subsequent reattachment.

The near-wake resulting from flow-off separation at the trailing edge of a lifting wing in steady sub-critical[4] flight has the general properties indicated in Fig. 9.2. The profile of the velocity component $v^{*1}(x^3)$ resembles the classical wake behind an airfoil at sub-sonic speed. It is characterized by kinematically inactive vorticity, and represents locally the viscous drag (due to the skin friction) and the pressure drag (form drag), i. e. the total drag of the airfoil, [7].

This kinematically inactive wake type would appear also if we have a blunt trailing edge with a von Kármán type of vortex shedding. Such trailing edges are sometimes employed at hypersonic wings or control surfaces in order to cope locally with the otherwise high thermal loads. This can be done without much loss of aerodynamic efficiency, if the boundary layers, which flow off the trailing edge, are sufficiently thick.

In Fig. 9.2 the profile of the velocity component $v^{*2}(x^3)$, in contrast to $v^{*1}(x^3)$, is kinematically active and represents locally the induced drag. The angles Ψ_{e_u} and Ψ_{e_l} vary in the span-wise direction of the wing's trailing edge. In sub-critical flow we have $\Psi_{e_u} = -\Psi_{e_l}$.

These near-wake properties are found in principle regardless of the type of separation (flow-off or squeeze-off separation), and whether vortex layers or vortex feeding layers are present, Fig. 9.1.

Flow separation can also be categorized according to its extent:

- local separation,
- global separation.

Local separation is given, for instance, if the flow separates and reattaches locally (separation bubbles or separation occurring at a ramp or a two-dimensional control surface, e. g., Fig. 9.6 c)), or if it is confined otherwise to a small region. Shock/boundary-layer interaction, if the shock is

[4] Sub-critical flight means without total-pressure loss due to an embedded shock wave. The schematic holds, however, also for flow with total-pressure loss due to shock waves.

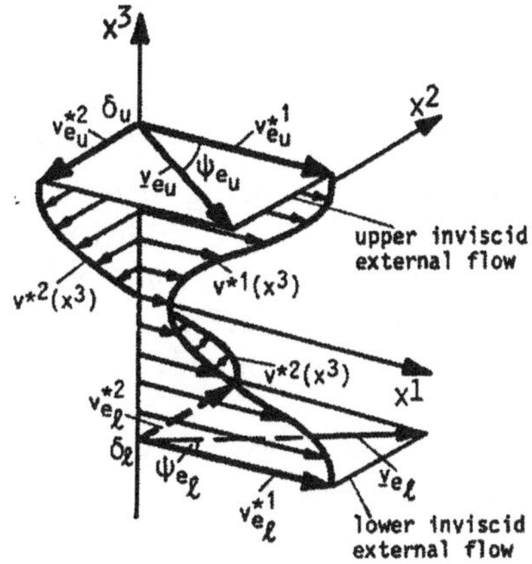

Fig. 9.2. Schematic of a three-dimensional near-wake of a lifting wing in steady sub-critical flight [7]. The coordinates x^1 and x^2 are tangential to the skeleton surface of the wake, x^3 is straight and normal to it. v^{*1} and v^{*2} are the tangential velocity components. δ_u and δ_l denote the upper and the lower edge of the wake, where the inviscid flow vectors $\underline{v}_{e_u}$ and $\underline{v}_{e_l}$ have the inclinations Ψ_{e_u} and Ψ_{e_l} against the x_1-axis.

strong enough, leads to local separation. In any case local separation is characterized by rather local influence on the flow past a flight vehicle[5].

Flow-off separation always is global separation with global influence on the vehicle's flow field. The canonical phenomena are the induced angle of attack and the induced drag, respectively, which appear on a lifting wing.

The basis of this categorization is the locality principle [6], [7], [8]. It says basically that a minor change of the flow field, for instance due to local separation, but also of the body geometry, affects the flow field only locally and downstream of that location (see in this context also Sub-Section 7.1.4). Of course, due to the strong interaction the inviscid flow field is changed, but then it is a question of its spatial characteristic properties (elliptic, hyperbolic), whether or not upstream changes are induced. Such changes can be significant, if the wake of the body is kinematically active. Then we have global separation, which leads, for instance, to the induced drag of classical aerodynamics.

At a lifting delta wing, the lee-side vortices, Fig. 9.1 a), induce an additional lift increment, the non-linear lift. This does not violate the locality principle, because this is not an upstream effect. At hypersonic flight with large angle of attack (re-entry flight) the lee-side vortices are present, too, see, e. g., Section 3.3, but the hypersonic shadow effect (see the introduc-

[5] Local separation, however, can have a global influence on the aerodynamic forces and moments of a flight vehicle, if, for instance, the effectiveness of the control surface or the trim flap is affected.

tion to Chapter 6), annuls the non-linear lift. Also phenomena like vortex breakdown, see, e. g., [9], [10], don't play a role at such flight.

The location of squeeze-off separation (type b) primarily is governed by the Reynolds number of the flow. This is important especially for the primary separation at the well rounded wing leading edges of RV-type flight vehicles. The Mach number may have an effect as well as possibly also noise and surface vibrations, see, e. g., the numerical study of local separation cases at the border of the continuum regime in [11]. High-temperature real-gas effects probably have an influence on separation only via their influence on the flow field as such.

The thermal state of the surface, especially the surface temperature also influences the location of separation. However, only few investigations have been made so far. In Fig. 9.3 results of a combined experimental and numerical study, [12], [13], of the influence of the surface temperature on (local) separation and attachment of a $M = 7.7$ flow past a 15° ramp configuration (flat plate and wedge) illustrate the effect for two-dimensional flow. The flow parameters are given in Table 9.1.

Table 9.1. Flow parameters of the ramp computation case [12], [13].

M_∞	T_t $[K]$	Re^u_∞ $[m^{-1}]$
7.7	1,500.0	$4.2{\cdot}10^6$

The wall temperature ratio is $0.2 \leqq T_w/T_t \leqq 0.666$ with $T_t = 1{,}500.0\ K$. With increasing surface temperature the length of the separation region L_{sep} becomes larger. The separation point moves upstream and the reattachment point downstream. The upstream movement of the separation point is due to the fact that the hotter the surface, the smaller the density in the flow near the surface. This reduces the momentum of the boundary layer and hence its ability to negotiate the adverse pressure gradient induced by the ramp.

We will discuss local separation further in connection with shock/boundary-layer interaction in the next section.

Global separation regions, for instance the lee-side flow field of the basic configurations shown in Fig. 9.1, exhibit also embedded secondary and even tertiary separation (and attachment) phenomena, as well as cross-flow shocks, which can interact with the attached surface-flow portions, causing local separation phenomena[6]. A classification of such flows is given, e. g., in [14], [15].

In closing some words about the simulation of separated flows. In ground-facility simulation it is a question for both global and local separation, whether the Mach and Reynolds number capabilities of the facility permit to

[6] See also the BDW case study, Section 3.3.

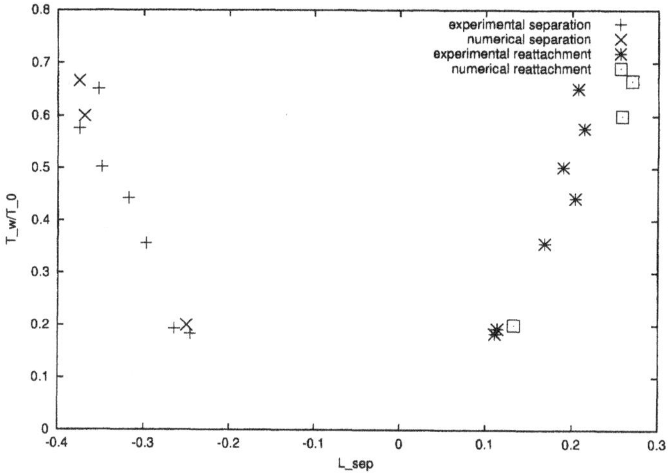

Fig. 9.3. Experimentally and numerically determined separation and reattachment points on a ramp configuration as function of the surface temperature T_w/T_0 ($T_0 \equiv T_t$) [12], [13]. The ramp angle is $\delta = 15°$, $L_{sep} = 0$ defines the corner point of the ramp, which lies 10.5 cm downstream of the sharp leading edge.

get proper results, and whether the flow to be simulated in reality is laminar or turbulent. The surface temperature has an influence, but the question is, how large it is for which flow classes[7].

Computational simulation today is able to give results of high accuracy as long as the flow is laminar throughout the whole flow field. It also permits to quantify the influence of the surface temperature, if it is governed by radiation cooling. However, if laminar-turbulent transition and/or turbulent separation are present, computational capabilities become limited. The influence of the location and form of the transition zone may possibly be parameterized, at least for simple configurations, but turbulent separation is still nearly untreatable.

An almost not known tool to investigate the structures of ordinary separation flows, but also of shock/boundary-layer interaction flows, and to establish credibility of results of experimental and numerical simulations are topological considerations, see, e. g. [4], [16] and also [8]. Consider the skin-friction patterns on the windward and on the lee-side of the HERMES configuration in Figs. 9.4 and 9.5. The singular points on the vehicle surface can be connected via topological rules. A surface normal to the longitudinal axis of the vehicle yields a Poincaré surface in which attachment and separation lines (lines with diverging and converging skin-friction patterns in Figs. 9.4

[7] Remember that the lift to drag ratio of the Space Shuttle was underestimated in all Mach number regimes, Section 3.1. How far the insufficient simulation of separation played a role has not been determined yet.

and 9.5) show up as half-saddles, which also, together with off-surface singular points, can be connected with the help of topological rules, see, e. g., Sub-Section 3.3.2. In this way the plausibility of results, missing information, and indications of, for instance, hot-spot situations at attachment lines on radiation cooled surfaces can be gained.

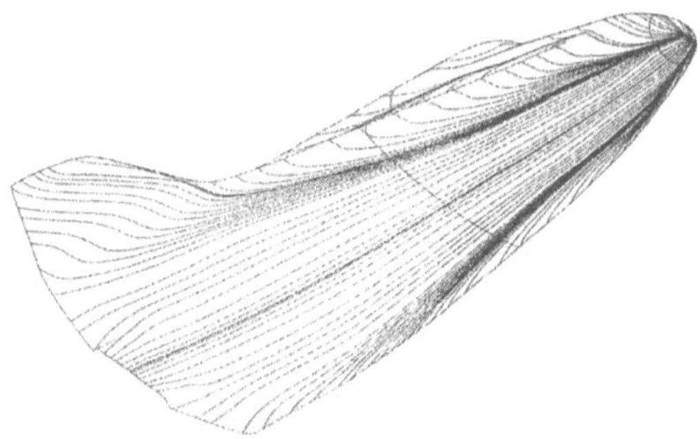

Fig. 9.4. Computed skin-friction lines on the windward side of the HERMES configuration in a ground-simulation facility situation at $M_\infty = 10$, $p_t = 10.0\ bar$, $T_t = 1{,}100.0\ K$, $T_w = 300.0\ K$, $\alpha = 30°$ [17].

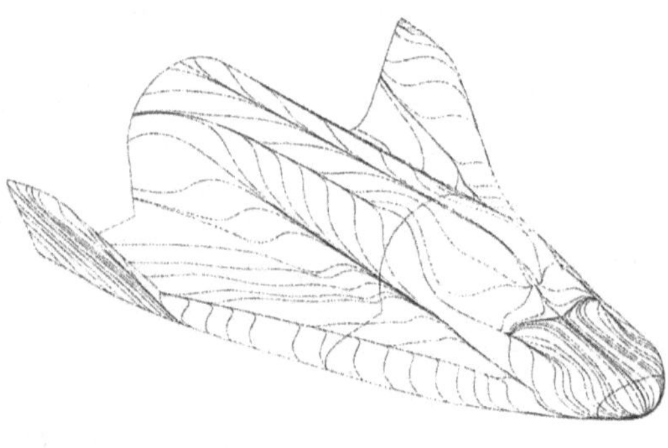

Fig. 9.5. Computed skin-friction lines on the lee-side of the HERMES configuration in a ground-simulation facility situation at $M_\infty = 10$, $p_t = 10.0\ bar$, $T_t = 1{,}100.0\ K$, $T_w = 300.0\ K$, $\alpha = 30°$ [17].

9.2 Shock/Boundary-Layer Interaction Phenomena

We summarize under this title several phenomena which are of interest in hypersonic flight vehicle design. They can be found for instance:

- at wings, Fig. 6.4, or stabilizers,
- at a the cowl lip of an inlet, Fig. 6.6,
- at the struts of a scramjet, Fig. 6.7.
- at a control surface, in two dimensions this is a ramp, Fig. 6.1 d) or ahead of a canopy (canopy shock, Fig. 6.3),
- on ramps of the external part of the inlet, Fig. 6.5,
- in the internal part of an inlet (oblique shock reflections), e. g., the shock-train in Fig. 6.5, or in a scramjet, Fig. 6.7,
- at the side walls of the internal part of an inlet, or a scramjet (glancing interaction),
- in the longitudinal corners of the internal part of an inlet, or a scramjet (corner flow),
- at the (flush) nozzle of a reaction-control system.

Shock/boundary-layer interaction phenomena can occur combined with local separation, but can also be connected with global separation. An example for the latter are cross-flow shocks in the lee-side flow field of a body at large angle of attack. Interaction phenomena reduce the effectiveness of control surfaces and inlets by thickening of the boundary layer or causing flow separation. Glancing shocks induce longitudinal vortex separation, oblique reflecting shocks can result in a Mach reflection, see, e. g., [18], and also [19]. In the attachment region usually an increase of the heat flux in the gas at the wall occurs. Moreover also very large and very concentrated heat flux and pressure peaks can be found locally. The interaction can support laminar-turbulent transition and can induce flow unsteadiness.

We will not discuss here all the mentioned phenomena, and refer instead the reader to the overviews [21], [22], [20], and especially [23]. We will concentrate on ramp-type (Fig. 6.1 d)) and on nose/leading-edge-type (Figs. 6.4 and 6.6) interactions. These are Edney-type V and VI, and Edney type III and IV interactions, respectively.

Shock/shock interaction with the associated boundary-layer interaction probably was first observed on the pylon of a ramjet model that was carried by a X-15. B. Edney [24] at FFA in Sweden was prompted by this event to make his by now classical investigations of the phenomenon. He identified and studied experimentally six interaction types, the type IV interaction, see Fig. 9.15 in Sub-Section 9.2.2, being the most severe one.

9.2.1 Ramp-Type (Edney Type V and VI) Interaction

We study first the control-surface problem and discuss a few critical modeling items with the help of results of mainly computational simulations for two-dimensional flat-plate/wedge flows, which we call for convenience simply ramp flows.

Basic experiments on laminar and turbulent supersonic and hypersonic flows of this kind were performed by Holden [25], who established the influence of the main parameters ramp angle, Mach number, Reynolds number.

The basic flow phenomena of laminar separated ramp flow are shown schematically in Fig. 9.6 c). The flow field is very complex compared to that of the inviscid flow case, Fig. 9.6 a), where we have simply the flow deflection through an oblique shock, addressed in Sub-Section 6.3.2, Fig. 6.9, and that of the non-separated viscous flow case, Fig. 9.6 b). Note here that the shock wave reaches down into the supersonic portion of the boundary layer.

The interaction shown in Fig. 9.6 c) is a Edney type VI interaction[8] The flow field harbors three shocks. The separation shock, induced by the flow deflection due to the separation regime, and the inner reattachment shock, induced by the final reflection of the flow at the ramp[9] do not cross each other, because they belong to the same family, see Sub-Section 6.3.2. They meet in the triple point T and form a single stronger shock, the outer reattachment shock. A slip line and an expansion fan originate at the triple point T. In the turbulent flow case a basically similar flow field results.

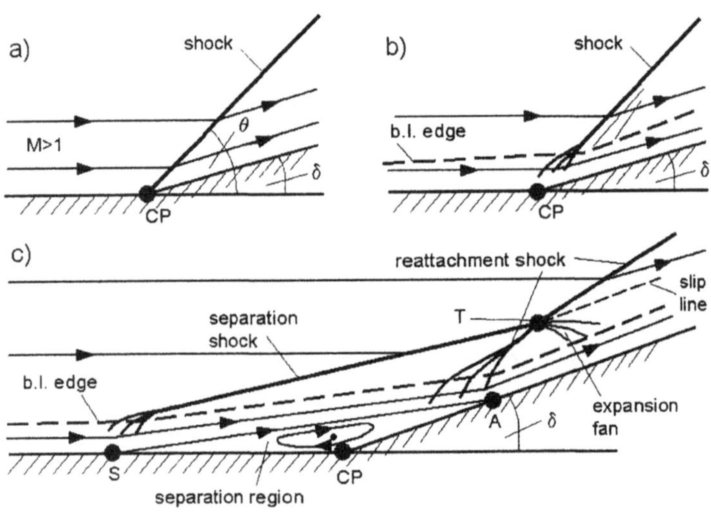

Fig. 9.6. Schematics of two-dimensional flows over a ramp (flat plate/wedge) configuration: **a)** inviscid flow, **b)** viscous flow with non-separating boundary layer, **c)** laminar viscous flow with (local) separation. S denotes the separation, A the reattachment, T the triple point, and CP the corner point.

[8] The type V interaction looks similar, but the reattachment shock below the triple point T would have a λ like structure, and the expansion fan would originate below T.

[9] This can be idealized to the flow past a two-wedge configuration.

The upstream influence of the ramp increases with increasing ramp angle, decreases with increasing Mach number, and is affected by the Reynolds number, however only weakly in turbulent flow. The length of the separation region, Fig. 9.6 c) (distance between the locations A and S), which is a measure of the strength of the interaction, increases with increasing ramp angle and decreases with increasing Mach number. In general laminar and turbulent flows are affected in the same way, the interaction being weaker for turbulent flows. Bluntness of the leading edge of the flat-plate/wedge configuration reduces the pressure and the heat flux in the gas at the wall.

We study now the influence of the surface temperature on the ramp flow field, which we discussed already shortly in the preceding section. The laminar flow over the 15° ramp configuration, for which the results in Fig. 9.3 were given, is one of the cases also investigated in [26], where the radiation-adiabatic surface was included, too, however without non-convex effects, Sub-Section 3.2.4. The flow parameters are given in Table 9.2.

Table 9.2. Flow parameters of the ramp computation cases [26].

$Case$	M_∞	T^0 $(\equiv T_t)$ $[K]$	Re^u_∞ $[m^{-1}]$	γ
I	7.7	1,500.0	$4.2 \cdot 10^6$	1.4
II	7.4	2,500.0	$4.0 \cdot 10^6$	1.4

Fig. 9.7 gives the computed $T_w(X)$ for two radiation-adiabatic cases ($T_{rad.eq} \equiv T_{ra}$) with the total temperatures ($T^0 \equiv T_t$ = 1,500.0 K and 2,500.0 K), as well as two adiabatic wall cases for the same total temperatures. Two prescribed isothermal wall temperatures, T_w = 300.0 K and 700.0 K are indicated, too.

The computations in [26] were made assuming perfect gas, although the total temperatures are large. The two adiabatic results are varying with x, each having a first weak maximum shortly downstream of the separation point, and a second weaker maximum shortly downstream of the reattachment point[10].

The radiation-adiabatic temperatures ahead of the separation points behave as expected for flat-plate laminar flow ($T_{ra} \sim x^{-0.125}$), Sub-Section 3.2.1. Downstream of the reattachment points, however, they follow that trend only approximately. Their ratio does not scale well, Sub-Section 3.2.5.

Both radiation-adiabatic temperatures ($T_{rad,eq} \equiv T_{ra}$) lie well below their respective adiabatic temperatures and are strongly dependent on x. Neither the adiabatic temperatures ($T_{aw} \equiv T_a$) nor constant wall temperatures would be representative for them. The well pronounced dips in the separation regions indicate an enlargement of the characteristic thicknesses Δ, eq. (3.15).

[10] We remember from Section 3.1 that the recovery temperature is constant only in the frame of the flat-plate relation given there.

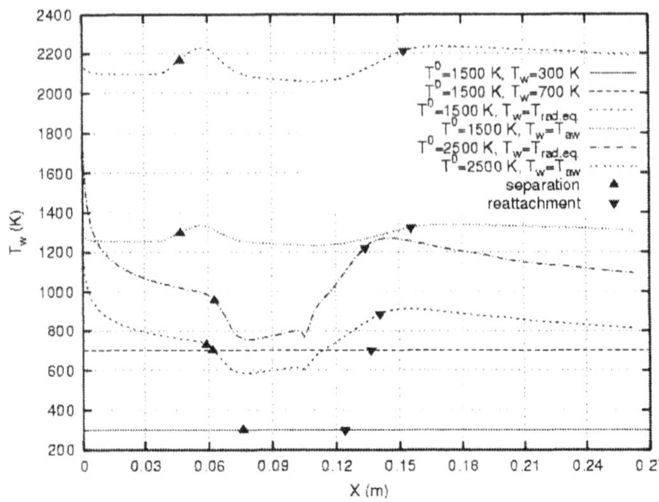

Fig. 9.7. Wall temperature distributions $T_w(x)$ on a ramp configuration and separation and attachment points [26]. Ramp angle $\delta = 15°$, the corner point of the ramp lies 10.5 cm downstream of the sharp leading edge, $T^0 \equiv T_t$, $T_{rad.eq} \equiv T_{ra}$.

The subsequent rise to larger temperatures in both cases is due to the rise of the unit Reynolds numbers downstream of the interaction zones, Section 6.6, which leads to a thinning of the boundary layer.

The distance of the separation point from the leading edge is largest for $T_w = 300.0\ K$. For $T_w = 700.0\ K$ it is shorter, but nearly the same as that for the two radiation-adiabatic cases. For the two adiabatic cases it is even shorter. Here also the separation length is largest, while it is similar for the two radiation-adiabatic cases and the $T_w = 700.0\ K$ case. It is shortest for $T_w = 300.0\ K$.

This all is reflected in the pressure-coefficient distributions in Fig. 9.8, each having a pronounced increase at the separation location (due to the separation shock, Fig. 9.6 c)) and in the reattachment region (due to the reattachment shock, Fig. 9.6 c)). The pressure has in each case also a relative maximum at the leading edge which is a result of hypersonic viscous interaction, Section 9.3. It drops then for all wall-temperature cases, the two adiabatic wall cases showing the smallest drop. These have also the most upstream located separation points, the largest and highest pressure plateau ahead of the corner point and then the smallest slope, but again the largest pressure plateau behind the reattachment point. The maximum pressures there are nearly the same for all cases. The separation point for the $T_w = 300.0\ K$ case lies closest to the corner point, the pressure rises then with the steepest slope to the smallest and lowest plateau. Then again it rises most steeply to the smallest plateau behind the reattachment point. The two radiation-adiabatic cases and the $T_w = 700.0\ K$ case lie close together with slightly different plateau extensions behind the reattachment points.

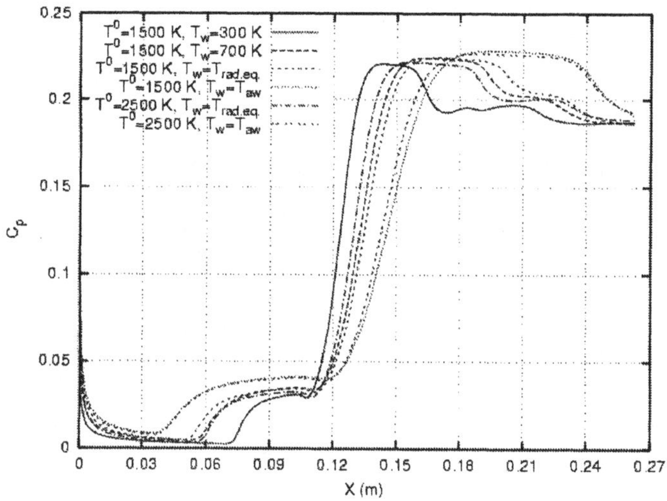

Fig. 9.8. Wall pressure distributions $c_p(x)$ on a ramp configuration as function of the surface temperature [26]. Ramp angle $\delta = 15°$, the corner point of the ramp lies 10.5 cm downstream of the sharp leading edge, $T^0 \equiv T_t$, $T_{rad.eq} \equiv T_{ra}$.

We see from these results that the surface temperature indeed is an important parameter. Whether it can be correlated with the help of the Reynolds number with reference-temperature extension, Sub-Section 7.1.6, has not been established yet. Of course, high-temperature real-gas effects play a role, too, and so does the state of the boundary layer, if the flight occurs at altitudes where laminar-turbulent transition occurs.

Finally it is a question how the different pressure distributions due to different surface temperatures affect the flap effectiveness. In [27], for instance, this was investigated for a FESTIP RV-type configuration. The flow past the body contour in the symmetry plane, i. e. a two-dimensional flow, with blunt nose, angle of attack ($\alpha = 19.2°$) and deflected body flap ($\delta = 5$ to $30°$) was simulated with a thermal and chemical non-equilibrium Navier-Stokes solver for the $M_\infty = 9.9$ flight at 51.9 km altitude. Laminar and turbulent flow, assuming transition at $Re_{\delta_2,tr}/M_e = 100$ to 200, eq. (8.33), was computed. For turbulent flow a $k - \epsilon$ model was employed.

The results are in short, with due reservations regarding the turbulent cases because of the problems associated with turbulence models in such flow situations:

- For laminar flow increasing flap deflection increases the pressure rise in the interaction region and the thermal loads.

- For laminar flow the flap effectiveness is only weakly affected by the surface temperature, it rises, also the thermal loads, up to $\delta \approx 15°$ and stays then approximately constant.
- Turbulent flow rises the flap effectiveness beyond $\delta \approx 15°$, increasing also the thermal loads.
- For turbulent flow the flap effectiveness is significantly affected by the surface temperature.

These are results for a single two-dimensional re-entry case at already low speed and altitude. In the three-dimensional reality the flow field past a deflected control surface can be quite different, see, e. g., [28], [29], and the effect of the surface temperature may also be different at larger speeds and altitudes.

In view of CAV-type flight vehicles we close the section by looking at the problem of turbulence modeling for the computational simulation of control-surface flow in, e. g., [30], [31]. The ramp configuration chosen there has a rather large ramp angle ($\delta = 38°$) which leads to a very strong compression. The experimental data are from a gun tunnel operated with nitrogen [32]. The flow was computed with several two-equation turbulence models with a variety of length scale and compression corrections. Perfect gas was assumed. The flow parameters are given in Table 9.3.

Table 9.3. Flow parameters of the turbulent ramp computation case [30].

M_∞	T_∞ $[K]$	Re^u_∞ $[m^{-1}]$	T_w $[K]$
9.22	59.44	$4.7 \cdot 10^7$	295.0

The computed flow field in Fig. 9.9 exhibits well the separation shock ahead of the recirculation regime, and the final reattachment shock.

The computed surface pressure in general rises too late (separation shock) and thus indicates the primary separation too much downstream, Fig. 9.10. None of the turbulence models reaches the measured peak pressure, however the pressure relaxation behind it to the inviscid pressure level $p_w/p_\infty \approx 58.8$ is sufficiently well met by all turbulence models.

The computed heat transfer in the gas at the wall rises for all turbulence models also too late and too high compared to the experimental data, Fig. 9.11. Only one model comes close to the measurements. Although this is a very demanding case, the results are very unsatisfactory. In practice, however, flows past control surfaces with such large deflection angles will be highly three-dimensional, which is likely to change the picture. The situation is similarly difficult if the flow is transitional [33].

Ramp-type interaction can occur also on wings. In [34] the Edney type VI interaction was studied numerically at a vehicle with the double-delta

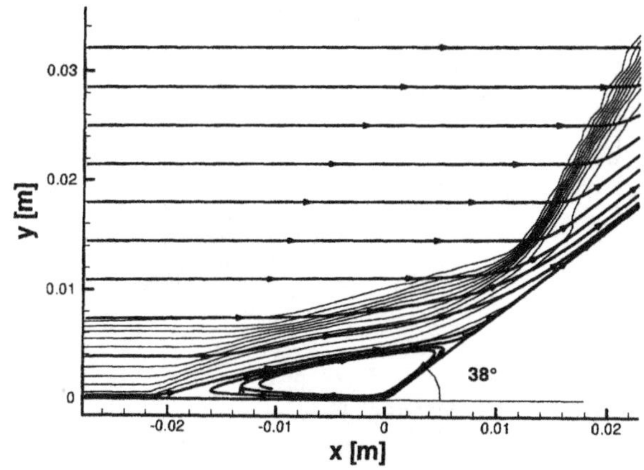

Fig. 9.9. Computed Mach number isolines and streamlines in the interaction region of a 38° ramp configuration [30]. Turbulent flow, low Reynolds number $k-\omega$ model.

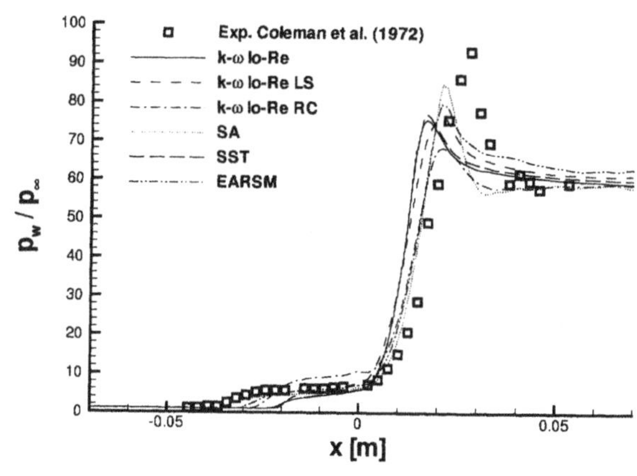

Fig. 9.10. Surface pressure $p_w(x)$ computed with several turbulence models, and experimental data [30].

planform type shown in Fig. 6.4, in this case the HALIS configuration at an angle of attack $\alpha = 40°$. The flow parameters can be found in Tab. 5.7 in Section 5.6. We see in Fig. 9.12 the plan view and the computed density isolines. The sketch to the left shows the type VI interaction of the bow shock with the embedded wing shock.

Computed radiation-adiabatic temperatures along the leading edge of the wing's second delta are given in Fig. 9.13 for different catalytic wall models. The fully catalytic wall gives, as expected, the highest radiation-adiabatic temperatures and the non-catalytic wall the smallest with the results for the

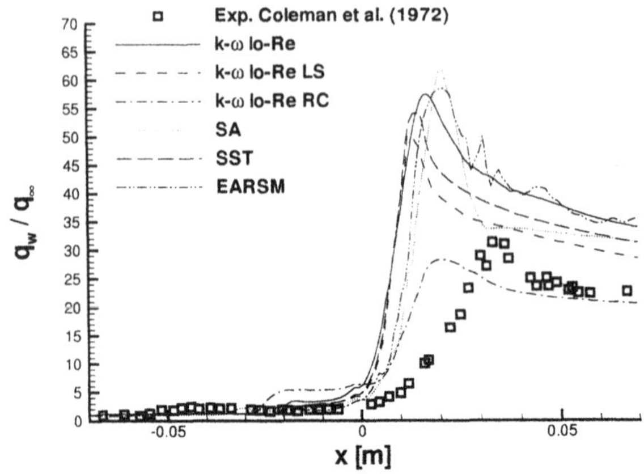

Fig. 9.11. With q_∞ non-dimensionalised heat transfer in the gas at the wall $q_w(x)$ ($\equiv q_{gw}(x)$) computed with several turbulence models, and experimental data [30].

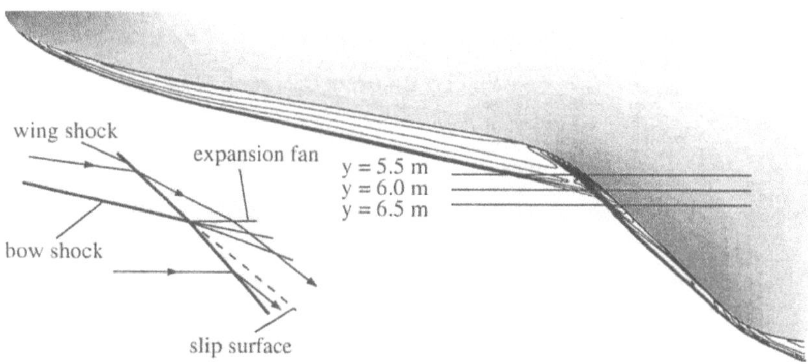

Fig. 9.12. Plan view of the HALIS configuration with sketch of the type VI shock/shock interaction [34].

finite catalytic wall lying just above the latter. The fully catalytic wall has kind of a temperature plateau lying mostly below the stagnation point value, while for the other catalytic wall models the temperature plateaus lie above the temperature at the stagnation point. At the end of the leading edge of the first wing delta ($y = 4.0\ m$) the temperatures are approximately 400.0 K lower than those plateau temperatures. At $y \approx 6.0\ m$ for each catalytic model a small temperature maximum is discernible which is due to the shock/shock interaction[11]. At $y \approx 10.0\ m$ for each catalytic wall model a second maximum is present, which is attributed to the impingement of the slip surface on the leading edge.

[11] With increasing angle of attack the excess thermal loads along the leading edge are becoming smaller, [22].

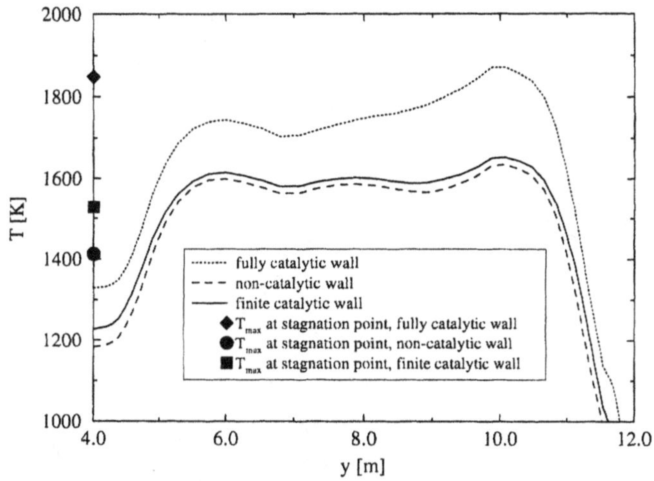

Fig. 9.13. Computed radiation-adiabatic temperatures $T(y)$ ($\equiv T_{ra}(y)$) along the wing's leading edge of the HALIS configuration [34].

A proper modeling of high-temperature real-gas effects is necessary in any case, as we see in Fig. 9.14 [35]. The axisymmetric hyperboloid-flare configuration has been derived for validation purposes from the HERMES 1.0 configuration at approximately 30° angle of attack [36]. The computations in [35] were made with different gas models for the flight situation with the flow parameters given in Table 9.4.

Table 9.4. Flow parameters of the shock/shock interaction study on the hyperboloid-flare configuration [35].

M_∞	H $[km]$	Re^u_∞ $[m^{-1}]$	T_∞ $[K]$	T_w $[K]$	L_{ref} $[m]$
25	77.0	$1.46 \cdot 10^4$	192.0	700.0	10.0

The use of different high-temperature real-gas models results in different interaction types over the flare, which has an angle $\delta = 49.6°$ against the body axis. This is due to the different shock stand-off distances at the blunt nose of the hyperboloid-flare configuration. We have noted in Sub-Section 6.4.1 that the stand-off distance Δ_0 is largest for perfect gas and becomes smaller with increasing high-temperature real-gas effects.

At the left side of Fig. 9.14 the perfect-gas result is given, case a). In this case the bow shock lies most forward and its intersection with the ramp shock results in a type V interaction. Fully equilibrium flow gives the most aft position of the bow shock (smallest bow-shock stand-off distance, Sub-Section 6.4.1) which results in a type VI interaction, case d). In between lie the other two models which also lead to type VI interactions, cases b) and c). Of course the wall pressure and the heat flux in the gas at the wall vary strongly from case to case, the fully equilibrium case giving the largest surface loads, see also [37].

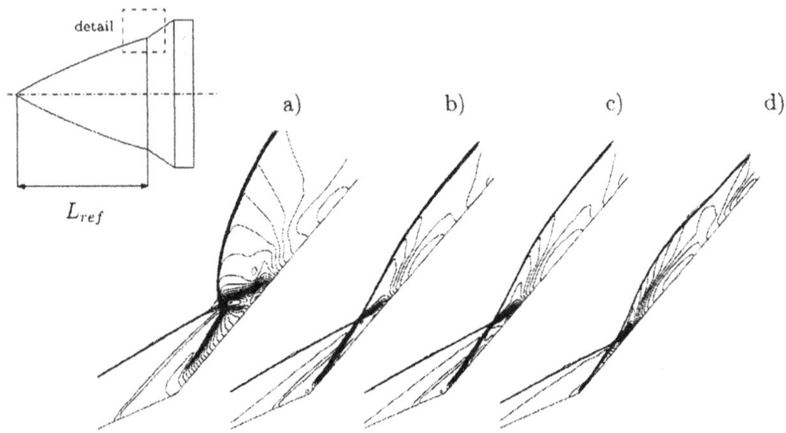

Fig. 9.14. Influence of the modeling of high-temperature real-gas effects on the shape of the interaction region and the pressure distribution (isobars) on the hyperboloid-flare configuration (upper left) [35]: **a**) perfect gas, **b**) thermal and chemical non-equilibrium, **c**) thermal equilibrium and chemical non-equilibrium, **d**) thermal and chemical equilibrium.

The influence of high-temperature real-gas effects on interaction phenomena is one issue of concern. While the interaction locations in some instances, e. g., at inlet cowl lips and engine struts, are more or less fixed, they shift strongly also with the flight Mach number, and the flight vehicle's angles of attack and yaw in the case of wings and stabilizers[12]. The understanding of all aspects of interaction phenomena, the prediction of their locations and effects on an airframe therefore are very important.

9.2.2 Nose/Leading-Edge-type (Edney Type III and IV) Interaction

Shock/shock interaction (with associated boundary-layer interaction) of Edney type IV, Fig. 9.15, is the most severe strong-interaction phenomenon.

[12] For the HERMES configuration it was demanded that the winglets were lying, for all re-entry flight attitudes, completely, i. e., without interference, within the bow-shock surface [38]. For the Space Shuttle configuration, but also for a CAV-type vehicle configuration as shown in Fig. 6.4, the interaction cannot be avoided and suitable measures must be taken in order to ensure the structural integrity of the airframe.

The type III interaction is less severe. Both can pose particular problems for instance at cowl lips, Fig. 6.6, but also at unswept pylons, struts et cetera, because they can lead to both very large and very localized thermal and pressure loads.

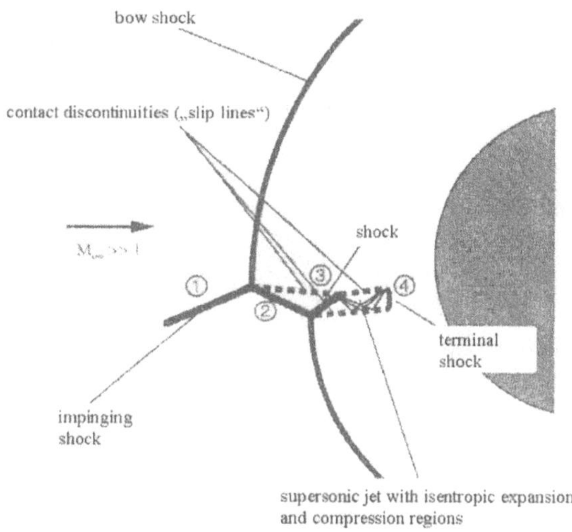

Fig. 9.15. Schematic of the Edney type IV shock/shock boundary-layer interaction [39].

We will give in the following some results for type IV interaction found on a cylinder [39]. For the interpretation of the flow structures shown schematically in Fig. 9.15, we use the flow parameters given in Table 9.5. They are from the experiment of Wieting and Holden [40], supplemented in [39] by data from [41]. For a recent review and data see Holden et al [42], and for a recent numerical study, e. g., [43].

Table 9.5. Flow parameters of the shock/shock boundary-layer interaction on a cylinder ($D = 76.2\ mm$) computation case [39].

M_∞	$T_\infty\ [K]$	$Re_{D,\infty}$	$T_w\ [K]$	$\theta_{imp}\ [^\circ]$	γ
8.03	122.11	387,500.0	294.44	18.11	1.4

Consider now the flow structures shown in Fig. 9.15. The basic flow field would be characterized by a smooth bow shock, whose distance from the surface of the cylinder can be estimated with the help of eq. (6.120) in Sub-Section 6.4.1 to be $\Delta_0 \approx 0.23\ R_b$. The impinging shock wave with the shock angle $\theta_{imp} = 18.11°$ divides the free-stream flow ahead of the cylinder into two regimes. The upper regime, above the impinging shock, contains the original undisturbed flow. The lower regime, below the impinging shock, contains an upward deflected uniform "free-stream" flow with a Mach number $M_{defl} = 5.26$, which is smaller than the original M_∞. The flow-deflection angle, which is identical with the ramp angle that causes the impinging oblique shock (the boundary-layer displacement thickness on the ramp neglected), is $\delta = 12.5°$.

In the lower regime the "free-stream" flow has a density $\rho_{defl} = 3.328\rho_\infty$. The density increase across a normal shock then would be $\rho_{2,defl}/\rho_\infty = 16.91$, compared to $\rho_{2,orig}/\rho_\infty = 5.568$. This would give with eq. (6.120) for the isolated deflected flow a shock stand-off distance $\Delta_{0,defl} \approx 0.1\ R_b$. Of course, eq. (6.120) cannot be applied in our problem to estimate the shock stand-off distances in the upper and the lower regime. However, the ratio of these, found with the help of eq. (6.120) ($\Delta_{0,defl}/\Delta_{0,orig} \approx 0.44$), and that of the smallest computed stand-off distances measured in Fig. 9.16 ($\Delta_{0,defl}/\Delta_{0,orig} \approx 0.5$), are not so far away from each other.

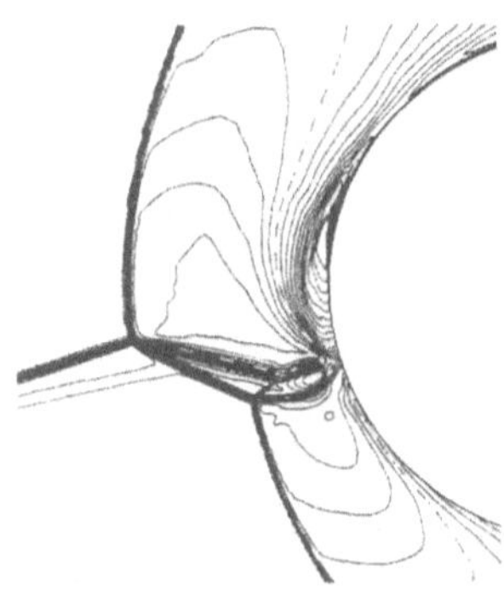

Fig. 9.16. Computed Mach number isolines of the Edney type IV shock/shock boundary-layer interaction [39].

Edney's different interaction types are characterized by the location in which the impinging shock wave meets the bow shock of the body. If this point lies rather low then two shocks of different families meet and cross each other, see Fig. 6.13 b). This is the type I interaction. Our type IV interaction case obviously is characterized by an intersection of shocks of the same family near the location where the original bow shock lies normal to the free-stream, Figs. 9.15 and 9.16, although this is not fully evident from the figures. Actually the finally resulting pattern shows two of these intersections of shocks of the same family, each well marked by the emerging third shock and the slip line.

Between these two slip lines, Fig. 9.15, a supersonic jet penetrates deep into the subsonic domain between the deformed bow shock and the body surface. The final, slightly curved, normal shock, Figs. 9.15 and 9.16, equivalent to a Mach disk in a round supersonic jet, leads to a large density increase close to the body surface.

The supersonic jet has a particular characteristic, Fig. 9.17. It captures, due to the upward deflection of the flow behind the impinging shock wave by $\delta = 12.5°$ in the lower regime, a considerable total enthalpy flux $(\rho u h_t)H_{in}$, which then is discharged by the final normal shock towards the body surface $(H_{out} \approx 0.2 H_{in})$.

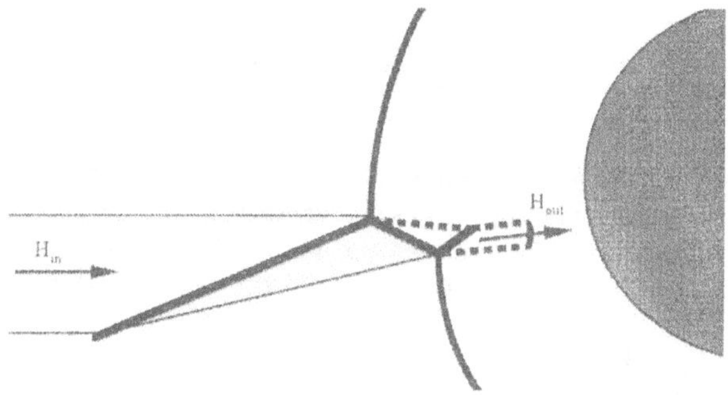

Fig. 9.17. Capture area of the supersonic jet in a Edney type IV shock/shock interaction [39].

Because of this concentrated enthalpy flux, and the close vicinity of the final, slightly curved normal shock to the surface, a very localized pressure peak results as well as a very large heat flux in the gas at the (cold) wall, Figs. 9.18 and 9.19. The maximum computed pressure in the stagnation point at the cylinder surface is $p_w \approx 775.0\, p_\infty$, and the maximum density $\rho_w \approx 310.0\, \rho_\infty$. The large heat flux in the gas at the wall is due to the locally very small thickness of the thermal boundary layer, δ_T, which in turn is due to the large density there [34].

The maximum computed pressure coefficient $c_p/c_{p_0} \approx 9.4$ is somewhat higher than the measured one. The computed heat flux meets well the measured one. The reference data c_{p_0} and q_0 are the stagnation-point data for the flow without the impinging shock. The agreement of computed and measured data is not fully satisfactory. In any case a very strong local grid refinement was necessary parallel and normal to the cylinder surface in order to capture the detailed flow structures.

Another problem is the slightly unsteady behaviour found both in the computation (the computation in [39] was not performed with a time-accurate code) and in the experiment. Similar observations and a more detailed analysis can be found in [43]. In [34], [44] results are given regarding the wall pressure, the heat flux in the gas at the wall, and the radiation-adiabatic temperature for the Edney interaction types III, IV and IVa, the latter be-

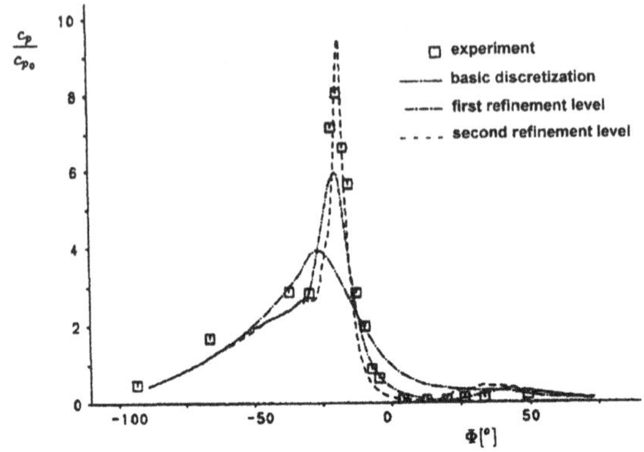

Fig. 9.18. Comparison of computed and measured wall pressure coefficients $c_p(\Phi)$ ($\equiv c_{p_w}(\Phi)$) for a Edney type IV shock/shock interaction [39] (c_{p_0} = 1.8275). Φ is the angular distance on the cylinder surface, with $\Phi = 0$ being the horizontal position, Fig. 9.15, $\Phi > 0$ the upper, and $\Phi < 0$ the lower contour.

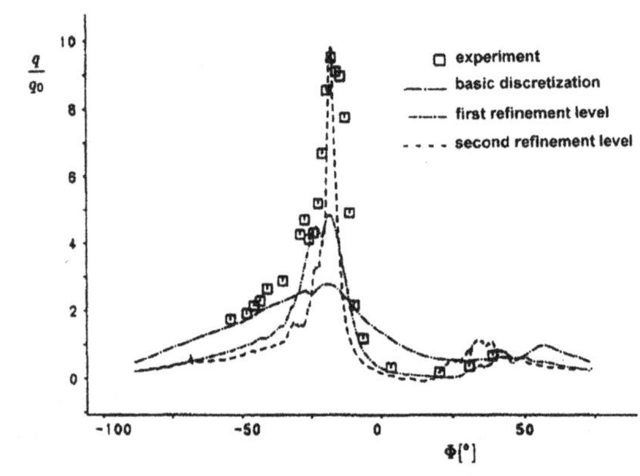

Fig. 9.19. Comparison of computed and measured heat fluxes in the gas at the wall $q(\Phi)$ ($\equiv q_{gw}(\Phi)$) for a Edney type IV shock/shock interaction [39] (q_0 = -6.8123$\cdot 10^5$ W/m^2).

ing a there newly defined interaction type due to high-temperature real-gas effects. The results show that in all cases the type IV interaction gives the largest pressure and heat transfer peaks, but that no sharp boundaries between the various interaction types exist.

9.3 Hypersonic Viscous Interaction

If attached viscous flow is of boundary-layer type, we can treat the inviscid flow field and the boundary layer separately, because we have only a weak interaction between them[13], Chapter 7. In Sub-Section 7.1.7 we have seen, however, that for large Mach numbers and small Reynolds numbers the attached viscous flow is no more of boundary-layer type. We observe in such flows, for instance that past a flat plate, just downstream of the leading edge a very large surface pressure, which is in contrast to high-Reynolds number boundary-layer flows, where the surface pressure indeed is the free-stream pressure. This phenomenon is called "hypersonic viscous interaction". We use the flow past the infinitely thin flat plate, the canonical case, to gain a basic understanding of the phenomenon and its detail issues.

Hypersonic viscous interaction can occur on sharp-nosed slender bodies and, of course, also on slender bodies with small nose bluntness, i. e., on CAV-type flight vehicles, and can lead to large local pressure loads and, in asymmetric cases, also to increments of global forces and moments. In [45] this interaction is called "pressure interaction" in contrast to the "vorticity interaction", which is observed on blunt-nosed bodies only. We have met this phenomenon in the context of entropy-layer swallowing, Sub-Chapter 6.4.2.

The cause of hypersonic viscous interaction is the large displacement thickness of the initial "boundary layer", which makes the free-stream flow "see" a virtual body of finite thickness instead of the infinitely thin flat plate. Consequently the incoming flow is deflected by the virtual body, a pressure gradient normal to the surface is present, eq. (7.76) in Sub-Section 7.1.7, and a slightly curved oblique shock wave is induced. Actually this shock appears at the end of the shock-formation region (in the classical terminology this is the "merged layer"). We depict this schematically in Fig. 9.20 [46]. The end of the shock-formation region is characterized by the separation of the emerging shock wave from the viscous flow, which then becomes boundary-layer like. Since the oblique shock wave is slightly curved, the inviscid flow between it and the boundary layer is rotational. In Fig. 9.20 also slip flow and temperature jump are indicated in the shock-formation region. As we will see in Section 9.4, attached viscous flow at large Mach numbers and small Reynolds numbers can exhibit such low density effects.

Before we discuss some interaction criteria, we illustrate the flow phenomena in the shock-formation region and downstream of it [47]. The data shown in the following figures were found by means of a numerical solution of eqs. (7.74) to (7.77), where the pressure-gradient term in eq. (7.75) has been omitted[14], because the criterion eq. (7.87) holds:

$$\frac{\partial \rho u}{\partial x} + \frac{\partial \rho v}{\partial y} = 0, \tag{9.1}$$

$$\rho u \frac{\partial u}{\partial x} + \rho v \frac{\partial u}{\partial y} = \frac{\partial}{\partial y}\left(\mu \frac{\partial u}{\partial y}\right), \tag{9.2}$$

[13] As we have already noted, it may be necessary to take iteratively into account the displacement properties of the boundary layer, Sub-Section 7.2.1.

[14] This system of equations was first solved by Rudman and Rubin [48].

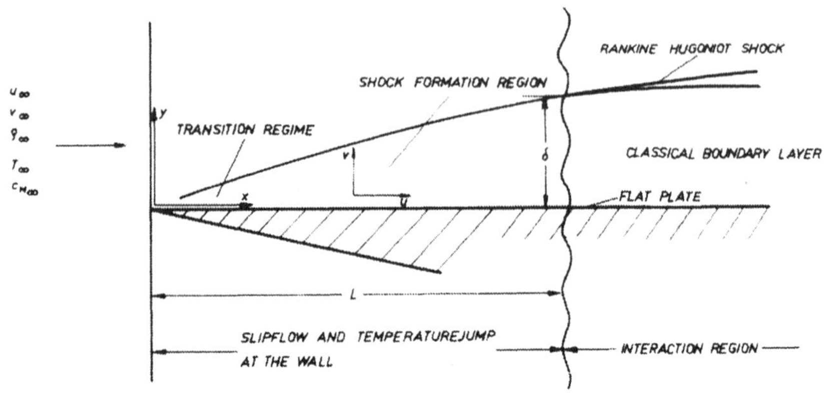

Fig. 9.20. Schematic of hypersonic viscous interaction flow at a flat plate [46].

$$\frac{\gamma_{ref} M_{ref}^2}{Re_{ref}} \left(\rho u \frac{\partial v}{\partial x} + \rho v \frac{\partial v}{\partial y} \right) = - \frac{\partial p}{\partial y} + \frac{\gamma_{ref} M_{ref}^2}{Re_{ref}} \left\{ \frac{\partial}{\partial y} \left[(\frac{4}{3}\mu + \right. \right.$$
$$\left. \left. + \kappa) \frac{\partial v}{\partial y} \right] - \frac{\partial}{\partial y} \left[\frac{2}{3}(\mu - \kappa) \frac{\partial u}{\partial x} \right] + \frac{\partial}{\partial x} \left(\mu \frac{\partial u}{\partial y} \right) \right\}, \tag{9.3}$$

$$c_p \left(\rho u \frac{\partial T}{\partial x} + \rho v \frac{\partial T}{\partial y} + \right) = \frac{1}{Pr_{ref}} \frac{\partial}{\partial y} \left(k \frac{\partial T}{\partial y} \right) + \frac{\gamma_{ref} - 1}{\gamma_{ref}} \left(u \frac{\partial p}{\partial x} + v \frac{\partial p}{\partial y} \right) +$$
$$+ (\gamma_{ref} - 1) M_{ref}^2 \mu \left(\frac{\partial u}{\partial y} \right)^2 + (\gamma_{ref} - 1) \frac{M_{ref}^2}{Re_{ref}} \left\{ (\frac{4}{3}\mu + \kappa) \left[\left(\frac{\partial u}{\partial x} \right)^2 + \right. \right.$$
$$\left. \left. + \left(\frac{\partial v}{\partial y} \right)^2 \right] - 2(\frac{2}{3}\mu - \kappa) \frac{\partial u}{\partial x} \frac{\partial v}{\partial y} + 2\mu \frac{\partial u}{\partial y} \frac{\partial v}{\partial x} \right\}. \tag{9.4}$$

These equations are solved for the whole flow domain including the oblique shock wave, whose structure is fully resolved. This is permitted, because the shock angle θ is small everywhere, Sub-Section 6.3.3.

The free-stream parameters are those of an experimental investigation of hypersonic flat-plate flow with argon as test gas [49], Table 9.6. The length of the shock-formation region L_{sfr} was found with the criterion eq. (9.47) given by Talbot [50], Section 9.4. The stream-wise coordinate x is made dimensionless with it. The viscous interaction region thus is present at $x \gtrapprox 1$.

The thickness δ of the computed initial viscous layer can approximately be set equal to the location of the pressure maxima in direction normal to the plate's surface, which are shown in Fig. 9.21. The Knudsen number Kn

Table 9.6. Free-stream parameters of the hypersonic flat-plate argon flow with viscous hypersonic interaction [49], [47].

M_∞	T_∞ $[K]$	Re^u_∞ $[cm^{-1}]$	λ_∞ $[cm]$	T_w $[K]$	L_{sfr} $[cm]$
12.66	64.5	985.45	0.021	285.0	6.676

$= \delta/\lambda_\infty$, then is $Kn \approx 0.125$ at $x = 0.1$ and $Kn \approx 0.0125$ at $x = 1$. Thus also the Knudsen number shows that we are in the slip flow regime for $0.1 \lessapprox x \lessapprox 1$, and in the continuum regime for $x \gtrapprox 1$, Section 2.3. In Fig. 9.20 the transition regime, where the flow is becoming a continuum flow, is indicated schematically downstream of the leading edge. In this area the use of the Navier-Stokes equations becomes questionable, but experience shows that they can be strained to a certain degree.

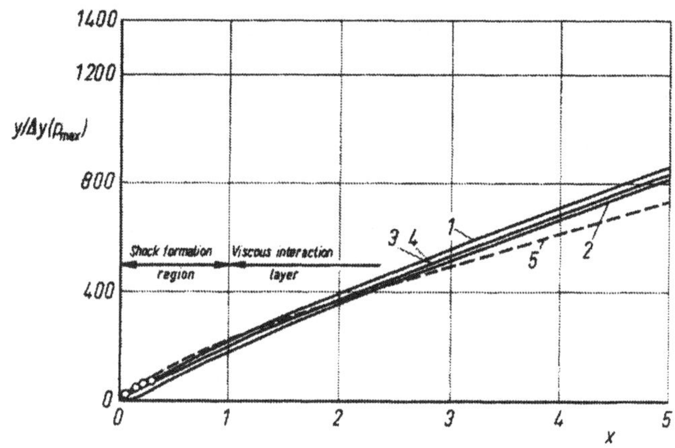

Fig. 9.21. Location of the maximum static pressure in y-direction as function of x [47] ($\Delta y = 0.00817$ cm). For the symbols see Table 9.7.

One may argue that strong interaction effects and low-density effects are mixed unduly in this example. However at an infinitely thin flat plate we will always have this flow situation just behind the leading edge, even at much larger Reynolds numbers as we have in this case.

Because we are in the slip-flow regime, too, the slip flow boundary conditions, eq. (4.45) in Sub-Section 4.3.1 without the second term, and eq. (4.79) in Sub-Section 4.3.2, are employed. In [47] the influence of the choice of the reflection/accommodation coefficients σ and α was investigated, hence the following figures have curves with different symbols, which are specified in Table 9.7. Also the symbols for the strong-interaction limit [51] and the experimental results from [49] are given there.

Table 9.7. Symbols used in the figures [47].

Symbol	$\sigma = \alpha$
1	1
2	1 - $0.6e^{-5x}$
3	0.825
4	0.825 - $0.425e^{-60x}$
5	strong-interaction limit [51]
o	experiment [49]

We discuss now some of the results in view of hypersonic viscous interaction. Fig. 9.22 shows the static pressure p_w at the surface of the flat plate non-dimensionalized by the free-stream pressure p_∞. The measured data reach in the shock-formation region a maximum of $p_w \approx 11$, i. e., the pressure is there eleven times larger than the free-stream static pressure. The pressure decreases with increasing x in the viscous interaction region. At $x = 5$ the pressure is still four times p_∞. The agreement between measured and computed data is reasonably good for the curves 3 and 4, the strong interaction limit lies too high for $x \lessapprox 5$, which probably is due to the choice of the exponent in the power-law relation of the viscosity in [47].

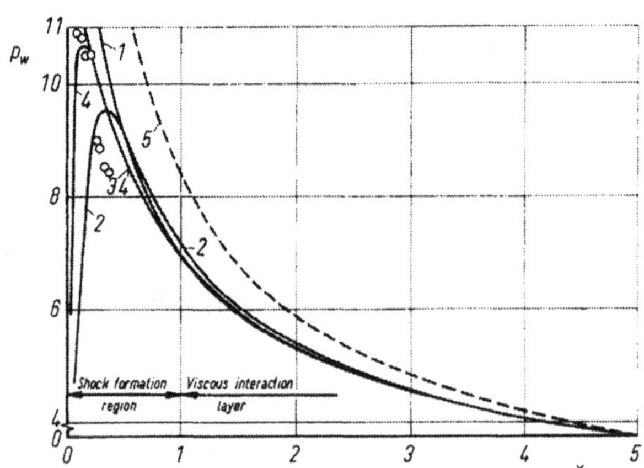

Fig. 9.22. Non-dimensional static pressure $p_w(x)$ at the plate's surface [47]. For the symbols see Table 9.7.

Very interesting are some features of the flow field, in particular the y-profiles of the static pressure p, Fig. 9.23, and the static temperature T, Fig. 9.24, at different x-locations on the plate. The pressure increase from the undisturbed side of the flow (from above in the figure) and the pressure maximum at the top of each curve in Fig. 9.23 indicate for $x \gtrapprox 1$ the high pressure side of the oblique shock wave, which is fully resolved. With the help of the locations of the pressure maxima in the y-direction, Fig. 9.21, the shock angle θ as function of x can be determined, too. It is possible then to compute with eq. (6.84) the Rankine-Hugoniot pressure jump across the shock wave, which is indicated for each curve in Fig. 9.23, too. This pressure jump is well duplicated for $x \gtrapprox 1$. We see also for $x \gtrapprox 1$ the pressure plateau (zero normal pressure gradient) which indicates that the flow there indeed is of boundary-layer type. For $x \lessapprox 1$ this does not hold. We are there in the shock-formation region, where we cannot speak even of a shock wave and a boundary layer which are actually merged.

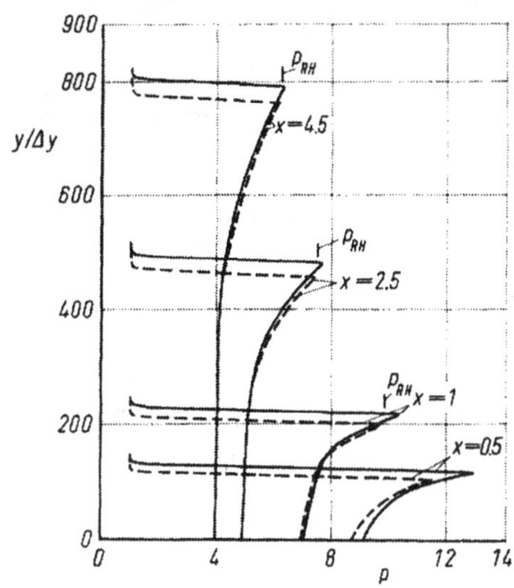

Fig. 9.23. Profiles of the non-dimensional static pressure $p(y)$ at different locations x [47] (full lines: $\sigma = \alpha = 1$, broken lines: $\sigma = \alpha = 0.825$), RH = Rankine-Hugoniot.

All this is reflected in the y-profiles of the static temperature in Fig. 9.24. The temperature there is non-dimensionalised with T_∞. The temperature rise in the shock wave compares well for $x \gtrapprox 1$ with the Rankine-Hugoniot temperature jump, found with eq. (6.83). For $x \gtrapprox 1$ we see also the typical temperature plateau between the (thermal) boundary layer and the shock wave, which coincides with the non-constant portion of the pressure field. In the boundary layer we find the temperature maxima typical for the cold-wall situation, however with the initially strong wall temperature jump which can appear in the slip-flow regime. For $x \lessapprox 1$, in the shock-formation region, again neither a boundary layer nor a shock wave are identifiable.

The classical hypersonic viscous interaction theory, see, e. g., [45], assumes that a boundary layer exists, whose displacement properties (displacement thickness $\delta_1(x)$) induce an oblique shock wave. This means that the boundary

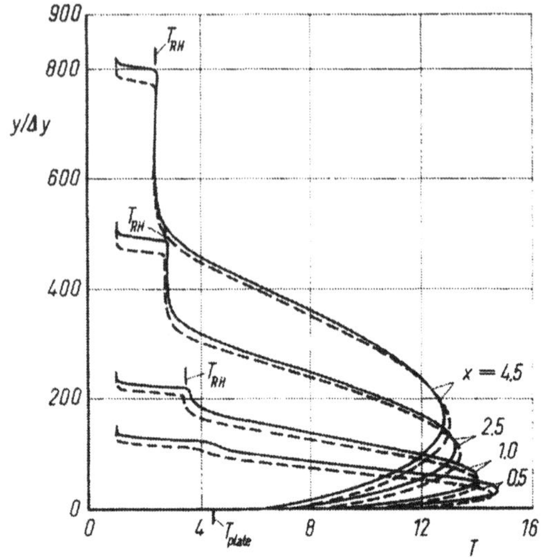

Fig. 9.24. Profiles of the non-dimensional static temperature $T(y)$ at different locations x [47] (full lines: $\sigma = \alpha = 1$, broken lines: $\sigma = \alpha = 0.825$), RH = Rankine-Hugoniot.

layer, downstream of the shock-formation region, is considered as a (convex) ramp, Fig. 6.9, however with a flow deflection angle δ_{defl} which is initially large and decreases then in downstream direction:

$$\tan \delta_{defl} = \frac{d\delta_1}{dx}(x). \tag{9.5}$$

The flow deflection angle δ_{defl} can be expressed with the help of eq. (6.111) as function of the Mach number M_1, the shock angle θ, and the ratio of the specific heats γ. We assume small deflection angles δ_{defl} and small shock angles θ, because $M \gg 1$, and reduce eq. (6.111) to:

$$\delta_{defl} = \frac{M_1^2\theta^2 - 1}{\frac{\gamma+1}{2} M_1^2 \theta}. \tag{9.6}$$

Eq. (6.84), which describes the pressure jump across the oblique shock, reads, also for small angles θ:

$$\frac{p_2}{p_1} = 1 + \frac{2\gamma(M_1^2\theta^2 - 1)}{\gamma + 1}. \tag{9.7}$$

Eq. (9.6) is now rewritten to yield:

$$\left(\frac{\theta}{\delta_{defl}}\right)^2 - \frac{\gamma + 1}{2}\left(\frac{\theta}{\delta_{defl}}\right) - \frac{1}{M_1^2 \delta_{defl}^2} = 0. \tag{9.8}$$

We met in Section 6.8 the hypersonic similarity parameter K, which we identify now in the denominator of the last term:

$$K = M_1 \delta_{defl}. \tag{9.9}$$

Solving eq. (9.8) for θ/δ_{defl}, substituting $M_1\delta_{defl}$ by K, and putting back the solution with positive sign of the radical into eq. (9.8) gives:

$$M_1^2\theta^2 = \frac{\gamma+1}{2}\left[\frac{\gamma+1}{4} + \sqrt{\left(\frac{\gamma+1}{4}\right)^2 + \frac{1}{K^2}}\right] K^2 + 1. \tag{9.10}$$

In the classical hypersonic viscous interaction theory the pressure p_2 behind the oblique shock wave is assumed to be the pressure p_e at the edge of the boundary layer. In view of Fig. 9.23 this is not correct. The pressure drops considerably in negative y-direction, however downstream with decreasing strength, before the plateau of the boundary-layer pressure is reached. Despite this result, we stick here with the classical theory, substitute on the left-hand side of eq. (9.7) p_2/p_1 by p_e/p_∞ and introduce eq. (9.10) to finally obtain the pressure increase due to the displacement properties in terms of the hypersonic similarity parameter eq. (9.9):

$$\frac{p_e}{p_\infty} = 1 + \frac{\gamma(\gamma+1)}{4}K^2 + \gamma\left[\sqrt{\left(\frac{\gamma+1}{4}\right)^2 + \frac{1}{K^2}}\right] K^2. \tag{9.11}$$

We have now to express the boundary-layer displacement thickness δ_1, eq. (7.114), in terms of the interaction pressure p_e. Assuming flow past an infinitely thin flat plate at zero angle of attack, we write with the reference-temperature extension, Sub-Section 7.1.6:

$$\delta_1 = c\frac{x}{\sqrt{Re_{\infty,x}}}\sqrt{\frac{\rho_\infty \mu^*}{\rho^* \mu_\infty}}, \tag{9.12}$$

where $Re_{\infty,x} = \rho_\infty u_\infty x/\mu_\infty$. Without viscous interaction we would have in this relation:

$$\frac{\rho_\infty}{\rho^*}|_{p^*=p_\infty} = \frac{T^*}{T_\infty}. \tag{9.13}$$

However, with viscous interaction we get:

$$\frac{\rho_\infty}{\rho^*}|_{p^*=p_e} = \frac{T^*}{T_\infty}\frac{p_\infty}{p_e}. \tag{9.14}$$

Following [52] we assume a linear relationship between the viscosity ratio and the temperature ratio:

$$\frac{\mu^*}{\mu_\infty} = C_\infty \frac{T^*}{T_\infty}, \tag{9.15}$$

and choose C_∞ according to the boundary-layer situation:

$$\frac{\mu_w}{\mu_\infty} = C_\infty \frac{T_w}{T_\infty}. \tag{9.16}$$

With p = const. in the boundary layer, see Fig. 9.23, we obtain finally the Chapman-Rubesin linear viscosity-law constant:

$$C_\infty = \frac{\mu_w T_\infty}{\mu_\infty T_w}. \tag{9.17}$$

This gives then for the displacement thickness a reference temperature extension of the form:

$$\delta_1 = c \frac{x}{\sqrt{Re_{\infty,x}}} \sqrt{C_\infty \left(\frac{T^*}{T_\infty}\right)^2 \frac{p_\infty}{p_e}}. \tag{9.18}$$

The classical hypersonic viscous interaction theory assumes $T^*/T_\infty \sim M_\infty^2$ [45]. With that we get a Mach number dependence in analogy to the Mach number dependence of the boundary-layer thickness at an adiabatic wall, eq. (7.103):

$$\delta_1 \sim x \frac{M_\infty^2 \sqrt{C_\infty}}{\sqrt{Re_{\infty,x}}} \sqrt{\frac{p_\infty}{p_e}}, \tag{9.19}$$

and its x-derivative:

$$\frac{d\delta_1}{dx} \sim \frac{M_\infty^2 \sqrt{C_\infty}}{\sqrt{Re_{\infty,x}}} \sqrt{\frac{p_\infty}{p_e}}. \tag{9.20}$$

In [53] and [54] it is shown, see also [45], that the pressure interaction can be described for two (asymptotic) limits on, for instance, a flat plate. These are the strong and the weak interaction limit. In the strong interaction limit the streamline deflection in the viscous layer is large and hence also the induced pressure gradient across the viscous layer. In the weak interaction limit the streamline deflection is small and hence also the induced pressure gradient.

We treat first the weak interaction limit, where $d\delta_1/dx$ is small, hence $K < 1$, and $K^2 \ll 1$. From eq. (9.11) we get for the pressure, which we interpret now as the induced wall pressure p_w:

$$\frac{p_w}{p_\infty} = 1 + \gamma K. \tag{9.21}$$

With $p_e \approx p_\infty$ and $tan\delta_{defl} = \delta_{defl} = d\delta_1/dx$, the hypersonic similarity parameter becomes:

$$K = \frac{M_\infty^3 \sqrt{C_\infty}}{\sqrt{Re_{\infty,x}}}. \tag{9.22}$$

It is now called viscous interaction parameter and written either as:

$$\chi = \frac{M_\infty^3}{\sqrt{Re_{\infty,x}}} \tag{9.23}$$

or as:

$$\overline{\chi} = \frac{M_\infty^3 \sqrt{C_\infty}}{\sqrt{Re_{\infty,x}}}. \tag{9.24}$$

For the infinitely thin flat plate we get from [45], [55] the pressure in the weak interaction limit with $Pr = 0.725$:

$$\frac{p_w}{p_\infty} = 1 + 0.578 \frac{\gamma(\gamma-1)}{4} \left[1 + 3.35 \frac{T_w}{T_t}\right] \overline{\chi}. \tag{9.25}$$

This result shows us that the pressure in the weak interaction limit is p_∞ disturbed by a term of approximately $O(\overline{\chi})$. For the cold wall, $T_w/T_t \ll 1$, and we get with $\gamma = 1.4$ for air:

$$\frac{p_w}{p_\infty} = 1 + 0.081 \overline{\chi}. \tag{9.26}$$

For the hot wall, $T_w \approx T_t$, Hayes and Probstein, [45] retain a term $O(\overline{\chi}^2)$ and arrive at the second-order weak interaction result:

$$\frac{p_w}{p_\infty} = 1 + 0.31 \overline{\chi} + 0.05 \overline{\chi}^2. \tag{9.27}$$

In hypersonic flows viscous interaction is a low Reynolds number phenomenon, hence we have so far tacitly assumed that the flow is laminar. However in turbulent flow past slender configurations viscous interaction can happen, if the transition location is close enough to the nose region. In this case the interaction parameter for the weak interaction limit reads [56]:

$$\overline{\chi}_{turb} = \left[\frac{M_\infty^9 C_\infty}{Re_{\infty,x}}\right]^{0.2}. \tag{9.28}$$

At the infinitely thin flat plate the pressure in the turbulent weak interaction limit is, with $Pr = 1$ and $\gamma = 1.4$:

$$\frac{p_w}{p_\infty} = 1 + 0.057 \left[\frac{1 + 1.3(T_w/T_t)}{[1 + 2.5(T_w/T_t)]^{0.6}}\right] \overline{\chi}_{turb}. \tag{9.29}$$

In the strong interaction limit, now again for laminar flow, we have $d\delta_1/dx$ large and therefore $K^2 \gg 1$. Eq. 9.11 becomes then:

$$\frac{p_e}{p_\infty} = 1 + \frac{\gamma(\gamma+1)}{2} K^2 \approx \frac{\gamma(\gamma+1)}{2} K^2. \tag{9.30}$$

Because in the strong interaction region $p_e \neq p_\infty$, we substitute first of all p_∞/p_e in eq. (9.19) with the help of eq. (9.30) in order to obtain a relation for δ_1:

$$\delta_1 \frac{d\delta_1}{dx} \sim \frac{M_\infty \sqrt{C_\infty}}{\sqrt{Re^u_\infty}} x^{0.5}, \tag{9.31}$$

where $Re^u_\infty = \rho_\infty u_\infty / \mu_\infty$ is the unit Reynolds number.

This equation is integrated to yield:

$$\delta_1^2 = \frac{M_\infty \sqrt{C_\infty}}{\sqrt{Re^u_\infty}} x^{1.5}, \tag{9.32}$$

respectively:

$$\delta_1 = \left[\frac{M_\infty \sqrt{C_\infty}}{\sqrt{Re_{\infty,x}}} \right]^{0.5} x. \tag{9.33}$$

We note in passing that the term in brackets resembles the square root of the co-factors in the y-momentum equation (9.3).

We differentiate now δ_1, eq. (9.33), with respect to x and put the result into eq. (9.30) to find finally:

$$\frac{p_e}{p_\infty} \sim \frac{M_\infty^3 \sqrt{C_\infty}}{\sqrt{Re_{\infty,x}}} = \overline{\chi}. \tag{9.34}$$

This results shows that in the strong interaction region the induced pressure is directly proportional to $\overline{\chi}$, and hence falls approximately $\sim x^{-0.5}$ in the downstream direction, see Fig. 9.22. The displacement thickness and hence also the thickness of the viscous layer is $\sim x^{0.75}$, eq. (9.33), compared to $\sim x^{0.5}$ in the weak interaction regime.

For the infinitely thin flat plate Kemp [51] obtains for the wall pressure in the strong interaction limit:

$$\frac{p_w}{p_\infty} = \frac{\sqrt{3}}{4} \sqrt{\frac{\gamma + 1}{2\gamma}} \gamma(\gamma - 1) \left[0.664 + 1.73 \frac{T_w}{T_t} \right] \overline{\chi}. \tag{9.35}$$

For the cold wall, $T_w \ll T_t$, and $\gamma = 1.4$ we get from this:

$$\frac{p_w}{p_\infty} = 0.149 \overline{\chi}, \tag{9.36}$$

and for the hot wall, $T_w \approx T_t$:

$$\frac{p_w}{p_\infty} = 0.537 \overline{\chi}. \tag{9.37}$$

The interaction parameter $\overline{V}$, arising if the hypersonic viscous interaction effect is described in terms of the pressure coefficient c_p, eq. (6.32), is[15]:

$$\overline{V} = \frac{M_\infty \sqrt{C_\infty}}{\sqrt{Re_{\infty,x}}} = \frac{\overline{\chi}}{M_\infty^2} \sim c_p. \tag{9.38}$$

This relation results from eq. (6.32) for $p/p_\infty \gg 1$ with:

$$c_p = \frac{2}{\gamma M_\infty^2} \frac{p}{p_\infty}, \tag{9.39}$$

which is then combined with eq. (9.34).

Finally we consider the consequences of our results. The viscous interaction parameters $\overline{\chi}$ or $\overline{V}$ can be employed to check whether strong interaction can appear on a configuration[16], or to correlate locally pressure, skin friction or the thermal state of the surface on both CAV-type and RV-type flight vehicles, see, e. g., [57].

If the check by means of one of the interaction parameters shows that the occurrence of strong hypersonic viscous interaction is likely, numerical or ground-facility simulation must be employed to quantify it to the needed degree of accuracy[17]. In ground-simulation facilities this will be a matter of Mach number, Reynolds number and T_w/T_∞ similarity. Numerical simulation is no problem for laminar flow. If the flow is turbulent, the transition location can be of large influence, except probably for local hypersonic turbulent viscous interaction effects at inlet ramps et cetera, [21].

While applying the interaction criteria one has to keep in mind their x-dependence, which we write explicitly:

$$\overline{\chi} = \frac{M_\infty^3 \sqrt{C_\infty}}{\sqrt{Re_\infty^u}} \frac{1}{\sqrt{x}} = \frac{\overline{\chi^u}}{\sqrt{x}}, \tag{9.40}$$

where $\overline{\chi^u}$ is the unit interaction parameter.

Strong interaction occurs if $\overline{\chi}$ is larger than a critical value $\overline{\chi}_{crit}$. This can be due to a large value of M_∞ combined with a small value of Re_∞^u (C_∞ usually being of O(1)) , or with given M_∞ and Re_∞^u for a small value of x. In this case we find the strong interaction limit for $x < x_{crit}$, with x_{crit} being:

$$x_{crit} = \left(\frac{\overline{\chi^u}}{\overline{\chi}_{crit}}\right)^2, \tag{9.41}$$

and likewise the weak interaction limit for $x > x_{crit}$.

15 The symbol $\overline{V}$ is also called "rarefaction parameter", see Section 9.4. Again see the co-factors in eq. (9.3).

16 To include the influence of a nose radius or an inclination of the surface, see, e. g., [51].

17 In approximate methods corrections of the inviscid wall pressure can be made with the relations given above.

From eqs. (9.26), (9.27), (9.36), and (9.37) we can deduce critical $\overline{\chi}$ values for the infinitely thin flat plate in air with $\gamma = 1.4$, if we take $p \gtrapprox 2\ p_\infty$ as criterion[18]:

- weak interaction:
  cold wall: $\overline{\chi} < \overline{\chi}_{crit} \approx 11$,
  hot wall: $\overline{\chi} < \overline{\chi}_{crit} \approx 3$,
- strong interaction:
  cold wall: $\overline{\chi} > \overline{\chi}_{crit} \approx 13$,
  hot wall: $\overline{\chi} > \overline{\chi}_{crit} \approx 4$.

Applying the hot wall strong interaction criterion to the flat plate case shown in Figs. 9.22 to 9.24, we find with the flow parameters from Table 9.6 the critical length $x_{crit} = 180.1\ cm$. This, however, appears to be somewhat large in view of the numerical data. We have assumed for argon $\mu \sim T^{0.75}$, which holds for $50.0\ K \lessapprox T \lessapprox 500.0\ K$.

We apply finally the strong interaction criterion to the flow (laminar) past the SÄNGER forebody, Section 7.3. From Table 7.9 we take the flow parameters and choose from Fig. 7.9 the wall temperature in the nose region to be $T_w \approx 1{,}500.0\ K$. We find then, assuming $\mu \sim T^{0.65}$, eq.(4.15), the critical length $x_{crit} = 0.0002\ m$, which shows that no strong hypersonic viscous interaction phenomena are to be expected at the SÄNGER forebody.

9.4 Low-Density Effects

Low-density effects, also called rarefaction effects, occur, if the flow past the hypersonic flight vehicle under consideration - or a component of it - is outside of the continuum regime, Section 2.3. We have seen in Fig. 2.6, that the hypersonic flight vehicles of the major classes (RV-type and CAV-type vehicles) are flying essentially in the continuum regime, at most approaching the slip-flow regime. Therefore we will only touch upon the topic of low-density effects here. We give some references for a deeper study, show some results connected to the results of the flat plate case shown in Figs. 9.22 to 9.24 of the preceding section, and discuss shortly issues of the viscous shock layer around blunt bodies.

Reviews of and theories about the transition from continuum flow to free molecular flow, especially regarding the intermediate regimes, Section 2.3, can be found in, e. g., [45], [58], [59], [60], [61].

We have seen in the preceding section that on an infinitely thin flat plate the strong interaction limit lies just downstream of the shock-formation region. We assume that the Navier-Stokes equations are valid in this region,

[18] Remember that the weak and the strong interaction limits are asymptotic limits with an intermediate area between them, for which however also a description can be provided [45].

although with slip-flow boundary conditions, Section 4.3. This means, that we have to demand $Kn \lessapprox 0.1$, Section 2.3. For viscous layers with the thickness δ in [58] the condition:

$$0.01 \lessapprox Kn_{\infty,x} = \frac{M_\infty}{\sqrt{Re_{\infty,x}}} \lessapprox 0.1 \tag{9.42}$$

is proposed.

For $M_\infty/\sqrt{Re_{\infty,x}} \gtrapprox 0.1$ the continuum regime is left. Therefore the part of the flow just downstream of the leading edge, which is marked as "transition regime" in Fig. 9.20, in principle cannot be prescribed by the Navier-Stokes equations[19].

The fraction in eq. (9.42) is also called rarefaction parameter V:

$$V = \frac{M_\infty}{\sqrt{Re_{\infty,x}}}, \tag{9.43}$$

and together with the Chapman-Rubesin constant C_∞:

$$\overline{V} = \frac{M_\infty\sqrt{C_\infty}}{\sqrt{Re_{\infty,x}}}, \tag{9.44}$$

which we met already in the preceding section.

H. K. Cheng [62] proposes a Chapman-Rubesin constant:

$$C_\infty = \frac{\mu^* T_\infty}{\mu_\infty T^*} \tag{9.45}$$

with the reference temperature:

$$T^* = \frac{1}{6}(T_t + 3T_w), \tag{9.46}$$

which he found from experiments and Monte Carlo simulations. He gives for the end of the slip-flow or shock-formation region $\overline{V} \approx 0.1$, and for the (upstream lying) end of the transition region, where the validity of the Navier-Stokes equations begins, $\overline{V} \approx 0.4$.

Now back to the example in the preceding section. The length of the shock-formation region was determined with the criterion [50]:

$$L_{sfr} \approx \frac{(\overline{V^u})^2}{0.1^2 \; to \; 0.3^2}, \tag{9.47}$$

with $\overline{V^u} = M_\infty\sqrt{C_\infty}/\sqrt{Re^u_\infty}$.

At the end of the shock-formation region the wall slip velocity u_w is supposed to be nearly zero. Actually it reaches zero only asymptotically, as we see in Fig. 9.25, where u_w is non-dimensionalised with u_∞.

---

[19] However, we have noted above that experience shows that the Navier-Stokes equations can be strained to a certain degree.

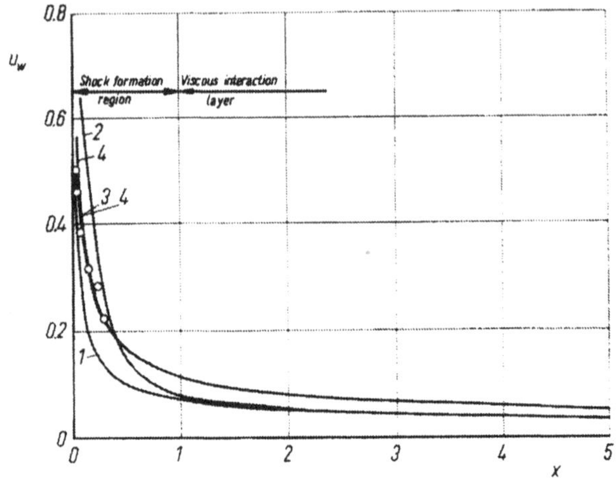

Fig. 9.25. Non-dimensional wall slip velocity $u_w(x)$ [47]. For the symbols see Table 9.7.

This holds also for the wall temperature jump, Fig. 9.26, which reflects the results given in Fig. 9.24 (the temperatures are non-dimensionalised with T_∞). The temperature of the gas at the wall initially is much larger than the wall temperature.

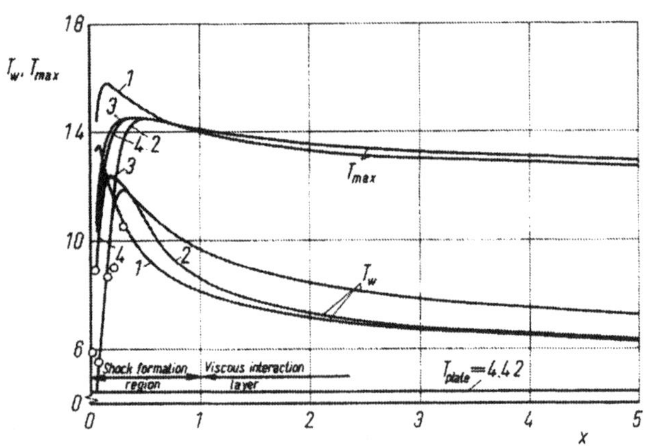

Fig. 9.26. Non-dimensional static wall temperature $T_w(x)$ and maximum static temperature $T_{max}(x)$ in the viscous layer [47]. For the symbols see Table 9.7.

In closing this topic we discuss now some experimental [63] and numerical [64] results for the transition region, in particular the highly rarefied hypersonic flow of helium past an infinitely thin flat plate, Table 9.8, which of course in the experiment is realized as plate of finite thickness with a sharp leading edge[20]. From the graph of V ($\equiv \overline{V}$) in Fig. 9.27 we see that the Navier-Stokes equations are valid only for $x \gtrapprox 1.5\ cm$, if we apply the criterion of H. K. Cheng: $\overline{V} \lessapprox 0.4$. The end of the shock-formation region, $\overline{V} \approx 0.1$, lies at approximately $x = 22.0\ cm$, outside of the scale of Fig. 9.27.

Table 9.8. Free-stream parameters of the hypersonic flat-plate helium flow [63].

M_∞	T_∞ $[K]$	Re^u_∞ $[cm^{-1}]$	λ_∞ $[cm]$	T_w $[K]$
8.92	10.7	111.9	0.1286	296.0

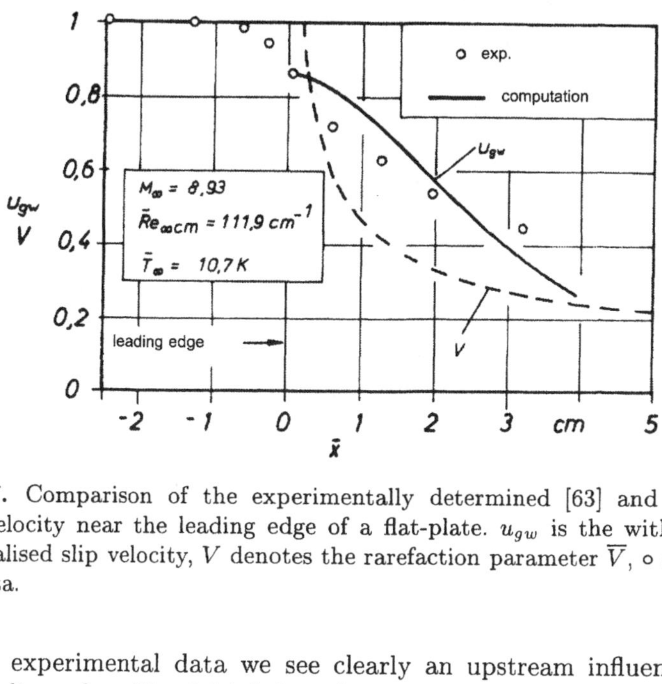

Fig. 9.27. Comparison of the experimentally determined [63] and computed [64] slip velocity near the leading edge of a flat-plate. u_{gw} is the with u_∞ non-dimensionalised slip velocity, V denotes the rarefaction parameter $\overline{V}$, ∘ are experimental data.

In the experimental data we see clearly an upstream influence of the plate's leading edge. The initial data for the computation were taken from measured flow profiles, but obviously they are not consistent, because u_w does not decrease monotonically with x as it should be expected.

The accommodation coefficients employed in the computation [64] for the tangential momentum, σ, and for the energy, α, were derived in [63] with the help of second-order conditions, [65]. They are initially small and increase with increasing x, and show partly wrong tendencies. This is probably due to the fact, that at the large degree of rarefaction in the leading-edge region no

[20] This holds also for the experimental realization of the example in the preceding section.

Maxwellian distribution exists. Hence the mean free path λ_{gw} at the surface of the plate is only approximately representative for the determination of the accommodation coefficients.

The computed temperature jump, however, compares rather well with the experimental data, Fig. 9.28. Near the leading edge the computed data are somewhat smaller than the measured data, which also points to inconsistencies in the initial data. Further downstream the solution also away from the plate's surface shows some deficiencies despite the good agreement at the wall, [64].

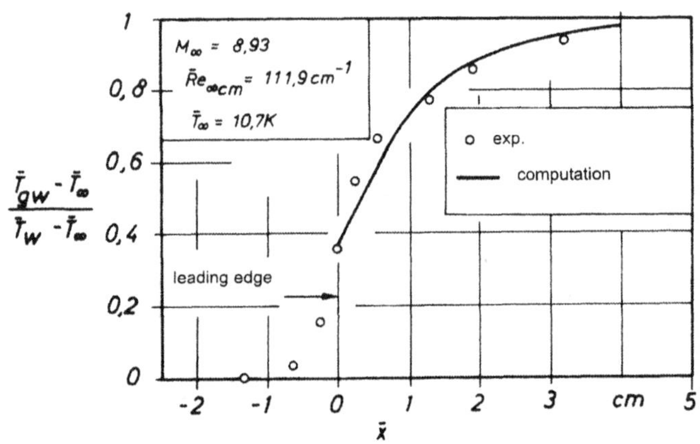

Fig. 9.28. Comparison of the experimentally determined [63] and computed [64] wall temperature jump near the leading edge of a flat-plate. $\overline{T}_{gw}$ is the temperature of the gas at the wall, $\overline{T}_w$ the surface temperature of the flat plate, $\overline{T}_\infty$ the static free-stream temperature, o are experimental data.

We conclude that the transition from the disturbed molecular flow regime to the slip flow regime at the flat plate can approximately be described with the help of the Navier-Stokes equations. However, depending on the problem at hand, the validity of such an approach must be verified.

The counterpart of the flow past an infinitely thin flat plate is the flow past a blunt body. We are interested in the front part of the blunt body. Here two Knudsen numbers are to be considered [66]. The first is formulated in terms of the bow-shock stand-off distance Δ_0, Fig. 6.17, and the second in terms of the boundary-layer thickness δ. As long as for both Knudsen

$$Kn_s = \frac{\lambda_\infty}{\Delta_0} \lessapprox 0.01, \tag{9.48}$$

holds, and

$$Kn_{bl} = \frac{\lambda_\infty}{\delta} \lessapprox 0.01, \tag{9.49}$$

the flow at the front part of the blunt body is a continuum flow. Of course, then also $\delta < \Delta_0$.

We have noted before that a shock wave has a finite thickness of a few mean free paths, say $\delta_s \approx 8\ \lambda_\infty$, see Fig. 6.15 in Sub-Section 6.3.3. We are in the slip flow regime, if either:

$$0.01 \lessapprox Kn_s \lessapprox 0.1, \tag{9.50}$$

or

$$0.01 \lessapprox Kn_{bl} \lessapprox 0.1. \tag{9.51}$$

In this regime the boundary layer will be no more thinner than the shock layer/shock stand-off distance:

$$\delta \lessapprox O(\Delta_0), \tag{9.52}$$

which holds also for the bow-shock wave:

$$\delta_s \lessapprox O(\Delta_0). \tag{9.53}$$

At still larger Knudsen numbers the shock layer ahead of the blunt body becomes a "merged layer". In analogy to the shock-formation layer at the flat plate, the merged layer regime at the blunt body is defined by both a non-distinguishable bow-shock wave and a non-distinguishable boundary layer, the latter with slip-flow and temperature jump at the body surface. The Rankine-Hugoniot conditions are not valid in this case.

The flow field in the slip-flow regime is also called "viscous shock layer" [67]. It encompasses the thick boundary layer with Navier-Stokes slip-flow conditions, see, e. g. [68], and an inviscid flow portion between the bow-shock wave and the boundary layer[21]. The "thick" bow-shock wave is still considered as a field discontinuity, however, shock-slip conditions, [67], [69], must be applied.

Viscous shock layers can be described by a system of governing equations of parabolic/hyperbolic type in space. Numerical methods based on these equations are a cheap alternative to the Navier-Stokes methods, although their applicability is limited, see, e. g. [70].

Slip flow and temperature jump, at least at blunt bodies with cold surfaces, are usually small. However, the wall pressure can appreciably be affected by low-density effects. Approximate theories as well as experimental data about blunt and other bodies in the slip-flow regime can be found, e. g., in [45], [58], [66].

[21] Flow fields of this kind are present at RV-type and AOTV-type flight vehicles at large speeds and altitudes. Hence it is the rule that high-temperature real-gas effects are present in such flow fields.

9.5 Problems

Problem 9.1 In Fig. 9.18 a reference pressure coefficient $c_{p_0} = 1.8275$ is given. How large is the coefficient at the stagnation point, c_{p_s}, of the blunt body (symmetric case, perfect gas) for $M_\infty =$ a) 8.03, b) ∞?

Solution: $c_{p_s} =$ a) 1.827, b) 1.839. Why are these values so close to each other. Formulate the principle in the background.

Problem 9.2 Assume that c_{p_s} of Problem 9.1 can be considered as constant above $M_\infty = 6$ and chosen to be $c_{p_s} = 1.8392$. How large is the actual wall pressure (perfect gas) at the stagnation point, c_{p_s} in terms of p_∞ for $M_\infty =$ a) 6, b) 8.03, c) 10?

Solution: $c_{p_s} =$ a) 47.34 p_∞, b) 84.00 p_∞, c) 129.73 p_∞. Discuss this result and quantify the excess pressure in the Edney type IV interaction case, Fig. 9.18.

Problem 9.3 How large is the Mach number normal to the shock wave, M_{∞_N}, of the hypersonic argon flat-plate flow at $x = 2$, Fig. 9.21?

Solution: $M_{\infty_N} = 2.58$. This is slightly above the permissible value given in Sub-Section 6.3.3. Are the flat-plate results nevertheless acceptable? Discuss.

Problem 9.4 Compute the interaction parameter $\overline{\chi}$, and the rarefaction parameter $\overline{V}$ for the argon flat-plate flow at $x = 1$. Assume $\mu_{argon} \sim T^{0.75}$. How large is x_{crit}?

Solution: $\overline{\chi} = 20.77$, $\overline{V} = 0.1296$, $x_{crit} = 180.09\ cm$. Discuss the result.

Problem 9.5 Compute at $x = 1.0\ m$ the interaction parameter $\overline{\chi}$, the rarefaction parameter $\overline{V}$, and the critical length x_{crit} for the CAV-type flight vehicle data given in Table 7.9. Assume laminar flow, $T_w = 1{,}500.0\ K$ as wall temperature (hot wall), and $\mu_{air} \sim T^{0.65}$.

Solution: $\overline{\chi} = 0.185$, $\overline{V} = 0.004$, $x_{crit} = 0.002\ m$. Discuss the result.

Problem 9.6 Compute at $x = 1.0\ m$ the interaction parameter $\overline{\chi}$, the rarefaction parameter $\overline{V}$, and the critical length x_{crit} for a CAV-type vehicle flying with $M_\infty = 10$ at 50.0 km altitude. Assume laminar flow, $T_w = 1{,}500.0$ K as wall temperature (hot wall), and use eq. (4.15) to determine μ_{air}.

Solution: $\overline{\chi} = 1.694$, $\overline{V} = 0.0169$, $x_{crit} = 0.179\ m$. Discuss the result.

References

1. E. H. Hirschel. "Present and Future Aerodynamic Process Technologies at DASA Military Aircraft". ERCOFTAC Industrial Technology Topic Meeting, Florence, Italy, 1999.
2. E. H. Hirschel. "Evaluation of Results of Boundary-Layer Calculations with Regard to Design Aerodynamics". AGARD R-741, 1986, pp. 5-1 to 5-29.

3. K. OSWATITSCH. "Die Ablösebedingungen von Grenzschichten". *H. Görtler (ed.), Boundary-Layer Research.* Springer, Berlin/Göttingen/Heidelberg, 1958, pp. 357 - 367. Also: "The Separation Conditions of Boundary Layers", NASA TTF-15200.

4. D. J. PEAKE, M. TOBAK. "Three-Dimensional Interaction and Vortical Flows with Emphasis on High Speeds". AGARDograph 252, 1980.

5. E. H. HIRSCHEL, W. KORDULLA. "Shear Flow in Surface-Oriented Coordinates". *Notes on Numerical Fluid Mechanics, Vol. 4.* Vieweg, Braunschweig/Wiesbaden, 1981.

6. A. EBERLE, A. RIZZI, E. H. HIRSCHEL. "Numerical Solutions of the Euler Equations for Steady Flow Problems". *Notes on Numerical Fluid Mechanics, NNFM 34.* Vieweg, Braunschweig/Wiesbaden, 1992.

7. E. H. HIRSCHEL. "Vortex Flows: Some General Properties, and Modelling, Configurational and Manipulation Aspects". AIAA-Paper 96-2514, 1996.

8. U. DALLMANN, T. HERBERG, H. GEBING, WEN-HAN SU, HONG-QUAN ZHANG. "Flow-Field Diagnostics: Topological Flow Changes and Spatio-Temporal Flow Structures". AIAA-Paper 95-0791, 1995.

9. H. J. LUGT. "Introduction to Vortex Theory". Vortex Flow Press, Potomac, Maryland, 1996.

10. S. I. GREEN (ED.). "Fluid Vortices". Kluwer Academic Publisher, 1995.

11. G. A. BIRD. "The Effect of Noise and Vibration on Separated Flow Regions in Hypersonic Flow". *N. G. Barton, J. Periaux (eds.), Coupling of Fluids, Structures and Waves in Aeronautics.* Notes on Numerical Fluid Mechanics and Multidisciplinary Design, NNFM 85, Springer, Berlin/Heidelberg/New York, 2003, pp. 1 - 9.

12. A. HENZE, W. SCHRÖDER. "On the Influence of Thermal Boundary Conditions on Shock Boundary-Layer Interaction". DGLR-Paper 2000-175, 2000.

13. A. HENZE, W. SCHRÖDER, M. BLEILEBENS, H. OLIVIER. "Numerical and Experimental Investigations on the Influence of Thermal Boundary Conditions on Shock Boundary-Layer Interaction". Computational Fluid Dynamics J., Vol. 12, No. 2, 2003, pp. 401 - 407.

14. D. A. MILLER, R. M. WOOD. "An Investigation of Leading Edge Vortices at Supersonic Speeds". AIAA-Paper 83-1816, 1983.

15. W. STAUDACHER. "Die Beeinflussung von Vorderkantenwirbelsystemen schlanker Tragflügel (Influencing Leading-Edge Vortex Systems at Slender Wings)". Doctoral Thesis, Universität Stuttgart, Germany, 1992.

16. M. TOBAK, D. J. PEAKE. "Topology of Three-Dimensional Separated Flows". Annual Review of Fluid Mechnics, Vol. 14, 1982, pp. 61 - 85.

17. CH. MUNDT, F. MONNOYER, R. HÖLD. "Computational Simulation of the Aerothermodynamic Characteristics of the Reentry of HERMES". AIAA-Paper 93-5069, 1993.

18. H. W. LIEPMANN, A. ROSHKO. "Elements of Gasdynamics". John Wiley & Sons, New York/ London/ Sidney, 1966.

19. A. DURAND, B. CHANETZ, T. POT, D. CARTIGNY, E. SZÉCHÉNYI A. CHPOUN. "Experimental and Numerical Investigations on Mach Hysteresis". Int. Symp. on Shock Waves 23, Forth Worth, USA, 2001.
20. M. S. HOLDEN. "Viscous/Inviscid and Real-Gas Effects Associated with Hypersonic Vehicles". AGARD R-813, 1986, pp. 4-1 to 4-81.
21. J. L. STOLLERY. "Some Viscous Interactions Affecting the Design of Hypersonic Intakes and Nozzles". *J. J. Bertin, J. Periaux, J. Ballmann (eds.), Advances in Hypersonics, Vol. 1, Defining the Hypersonic Environment.* Birkhäuser, Boston, 1992, pp. 418 - 437.
22. J. DELERY. "Shock/Shock Interference Phenomena in Hypersonic Flows". Space Course 1993, Vol. 1, TU München, 1993, pp. 13-1 to 13-27.
23. D. S. DOLLING. "Fifty Years of Shock-Wave/Boundary-Layer Interaction Research: What Next?". AIAA J., Vol. 39, No. 8, 2001, pp. 1517 - 1531.
24. B. EDNEY. "Anomalous Heat Transfer and Pressure Distributions on Blunt Bodies at Hypersonic Speeds in the Presence of an Impinging Shock". FFA Rep. 115, 1968.
25. M. S. HOLDEN. "Two-Dimensional Shock Wave Boundary-Layer Interactions in High-Speed Flows. Part II, Experimental Studies on Shock Wave Boundary-Layer Interactions". AGARDograph No. 203, 1975, pp. 41 - 110.
26. M. MARINI. "Analysis of Hypersonic Compression Ramp Laminar Flows under Sharp Leading Edge Conditions". Aerospace Science and Technology, Vol. 5, 2001, pp. 257 - 271.
27. F. GRASSO, M. MARINI, G. RANUZZI, S. CUTTICA, B. CHANETZ. "Shock-Wave/Turbulent Boundary-Layer Interactions in Nonequilibrium Flows". AIAA J., Vol. 39, No. 11, 2001, pp. 2131 - 2140.
28. R. D. NEUMANN. "Defining the Areothermodynamic Methodology". *J. J. Bertin, R. Glowinski, J. Periaux (eds.), Hypersonics, Vol. 1, Defining the Hypersonic Environment.* Birkhäuser, Boston, 1989, pp. 125 - 204.
29. J. M. LONGO, R. RADESPIEL . "Flap Efficiency and Heating of a Winged Re-Entry Vehicle". J. Spacecraft and Rockets, Vol. 33, No. 2, 1996, pp. 178 - 184.
30. T. CORATEKIN, J. VAN KEUK, J. BALLMANN. "On the Performance of Upwind Schemes and Turbulence Models in Hypersonic Flows)". AIAA J., Vol. 42, No. 5, 2004, pp. 945 - 957.
31. T. CORATEKIN, A. SCHUBERT, J. BALLMANN. "Assessment of Eddy Viscosity Models in 2D and 3D Shock/Boundary-Layer Interactions". *W. Rodi, D. Laurence (eds.), Engineering Turbulence Modelling and Experiments 4.* Elsevier Science LTD., 1999, pp. 649 - 658.
32. G. T. COLEMAN, J. L. STOLLERY. "Heat Transfer from Hypersonic Turbulent Flow at a Wedge Compression Corner". J. Fluid Mechanics, Vol. 56, 1972, pp. 741 - 752.
33. G. SIMEONIDES, W. HAASE. "Experimental and Computational Investigations of Hypersonic Flow About Compression Corners". J. Fluid Mechanics, Vol. 283, 1995, pp. 17 - 42.

34. S. Brück. "Ein Beitrag zur Beschreibung der Wechselwirkung von Stössen in reaktiven Hyperschallströmungen (Contribution to the Description of the Interaction of Shocks in Reacting Hypersonic Flows)". Doctoral Thesis, Universität Stuttgart, Germany, 1998. Also DLR-FB 98-06, 1998.

35. G. Brenner. "Numerische Simulation von Wechselwirkungen zwischen Stössen und Grenzschichten in chemisch reagierenden Hyperschallströmungen (Numerical Simulation of Interactions between Shocks and Boundary Layers in Chemically Reacting Hypersonic Flows)". Doctoral Thesis, RWTH Aachen, Germany, 1994.

36. R. Schwane, J. Muylaert. "Design of the Validation Experiment Hyperboloid-Flare". ESA Doc. YPA/1256/RS, ESTEC, 1992.

37. G. Brenner, T. Gerhold, K. Hannemann, D. Rues. "Numerical Simulation of Shock/Shock and Shock-Wave/Boundary-Layer Interactions in Hypersonic Flows". Computers Fluids, Vol. 22, No. 4/5, 1993, pp. 427 - 439.

38. P. Perrier. "Concepts of Hypersonic Aircraft". *J. J. Bertin, J. Periaux, J. Ballmann (eds.), Advances in Hypersonics, Vol. 1, Defining the Hypersonic Environment.* Birkhäuser, Boston, 1992, pp. 40 - 71.

39. J. Fischer. "Selbstadaptive, lokale Netzverfeinerung für die numerische Simulation kompressibler, reibungsbehafteter Strömungen (Self-Adaptive Local Grid Refinement for the Numerical Simulation of Compressible Viscous Flows)". Doctoral Thesis, Universität Stuttgart, Germany, 1993.

40. A. R. Wieting, M. S. Holden. "Experimental Study of Shock Wave Interference Heating on a Cylindrical Leading Edge at Mach 6 and 8". AIAA-Paper 87-1511, 1987.

41. G. H. Klopfer, H. C. Yee. "Viscous Hypersonic Shock-on-Shock Interaction on Blunt Cowl Lips". AIAA-Paper 88-0233, 1988.

42. M. S. Holden, S. Sweet, J. Kolly, G. Smolinsky. "A Review of the Aerothermal Characteristics of Laminar, Transitional and Turbulent Shock/Shock Interaction Regions in Hypersonic Flows". AIAA-Paper 98-0899, 1988.

43. D. D'Ambrosio. "Numerical Prediction of Laminar Shock/Shock Interactions in Hypersonic Flow". J. Spacecraft and Rockets, Vol. 40, No. 2, 2003, pp. 153 - 161.

44. S. Brück. "Investigation of Shock-Shock Interactions in Hypersonic Reentry". *B. Sturtevant, J. E. Shepherd, H. G. Hornung (eds.), Proc. 20th Intern. Symposium on Shock Waves, Vol. I.* World Scientific, Pasadena, 1995, pp. 215 - 220.

45. W. D. Hayes, R. F. Probstein. "Hypersonic Flow Theory". Academic Press, New York/London, 1959.

46. E. H. Hirschel. "Hypersonic Flow of a Dissociated Gas over a Flat Plate". *L. G. Napolitano (ed.), Astronautical Research 1970.* North-Holland Publication Co., Amsterdam, 1971, pp. 158 - 171.

47. E. H. Hirschel. "Influence of the Accommodation Coefficients on the Flow Variables in the Viscous Interaction Region of a Hypersonic Slip-Flow Boundary Layer". Zeitschrift für Flugwissenschaften (ZFW), Vol. 20, No. 12, 1972, pp. 470 - 475.

48. S. RUDMAN, S. G. RUBIN. "Hypersonic Viscous Flow over Slender Bodies Having Sharp Leading Edges". Polytechnic Institute of Brooklyn, PIBAL Rep. No. 1018 (also AFOSR 67-1118), 1967.

49. M. BECKER. "Die ebene Platte in hypersonischer Strömung geringer Dichte. Experimentelle Untersuchungen im Stossformierungs- und Übergangsgebiet (The Flat Plate in Hypersonic Flow of Low Density. Experimental Investigations in the Shock Formation and the Transition Region". Doctoral Thesis, RWTH Aachen, Germany, 1970, also DLR FB 70-79, 1970.

50. L. TALBOT. "Criterion for Slip near the Leading Edge of a Flat Plate in Hypersonic Flow". AIAA J., Vol. 1, 1963, pp. 1169 - 1171.

51. J. H. KEMP. "Hypersonic Viscous Interaction on Sharp and Blunt Inclined Plates". AIAA-Paper 68-720, 1968.

52. D. R. CHAPMAN, M. W. RUBESIN. "Temperature and Velocity Profiles in the Compressible Laminar Boundary Layer with Arbitrary Distribution of Surface Temperature". Journal of the Aeronautical Sciences, Vol. 16, 1949, pp. 547 - 565.

53. L. LEES, R. F. PROBSTEIN. "Hypersonic Viscous Flow over a Flat Plate". Report No. 195, Dept. of Aero. Eng., Princeton University, 1952.

54. L. LEES. "On the Boundary-Layer Equations in Hypersonic Flow and their Approximate Solutions". Journal of the Aeronautical Sciences, Vol. 20, 1953, pp. 143 - 145.

55. G. KOPPENWALLNER. "Fundamentals of Hypersonics: Aerodynamics and Heat Transfer". VKI Short Course *Hypersonic Aerothermodynamics.* Von Kármán Institute for Fluid Dynamics, Rhode-Saint-Genese, Belgium, LS 1984-01, 1984.

56. J. L. STOLLERY. "Viscous Interaction Effects and Re-Entry Aerothermodynamics: Theory and Experimental Results". AGARD Lecture Series, LS 42, Vol. 1, 1972, pp. 10-1 to 10-28.

57. J. J. BERTIN. "General Characterization of Hypersonic Flows". *J. J. Bertin, R. Glowinski, J. Periaux (eds.), Hypersonics, Vol. 1, Defining the Hypersonic Environment.* Birkhäuser, Boston, 1989, pp. 1 - 65.

58. S. A. SCHAAF, P. L. CHAMBRÉ. "Flow of Rarefied Gases". Princeton University Press, Princeton, N.J., 1961.

59. V. P. SHIDLOVSKIY. "Introduction to the Dynamics of Rarefied Gases". American Elsevier Publ. Co., New York, 1967.

60. F. S. SHERMAN. "The Transition from Continuum to Molecular Flow". Annual Review of Fluid Mechnics, Vol. 1, 1969, pp. 317 - 340.

61. J. K. HARVEY. "Rarefied Gas Dynamics for Spacecraft". *J. J. Bertin, R. Glowinski, J. Periaux (eds.), Hypersonics, Vol. 1, Defining the Hypersonic Environment.* Birkhäuser, Boston, 1989, pp. 483 - 509.

62. H. K. CHENG. "Numerical and Asymptotic Analysis of Hypersonic Slip Flows". *Proc. First Conference on Numerical Methods in Gas Dynamics.* Novosibirsk, Russia, 1969.

63. M. BECKER, F. ROBBEN, R. CATTOLICA. "Velocity Distribution in Hypersonic Helium Flow near the Leading Edge of a Flat Plate". AIAA-Paper 73-0691, 1973.

64. E. H. Hirschel. "Untersuchung grenzschichtähnlicher Strömungen mit Druckgradienten bei grossen Anström-Machzahlen (Investigation of Boundary Layer Like Flows at Large Free-Stream Mach Numbers)". Inaugural Thesis, RWTH Aachen, Germany, 1975, also DLR FB 76-22, 1976.

65. R. G. Deissler. "An Analysis of Second Order Slip Flow and Temperature Jump Conditions for Rarefied Gases". Int. J. of Heat and Mass Transfer, Vol. 7, 1964, pp. 681 - 694.

66. G. Koppenwallner. "Rarefied Gas Dynamics". *J. J. Bertin, R. Glowinski, J. Periaux (eds.), Hypersonics, Vol. 1, Defining the Hypersonic Environment.* Birkhäuser, Boston, 1989, pp. 511 - 547.

67. R. T. Davis. "Numerical Solution of the Hypersonic Shock-Layer Equations". AIAA J., Vol. 8, No. 5, 1970, pp. 843 - 851.

68. W. L. Hendricks. "Slip Conditions with Wall Catalysis and Radiation for Multicomponent, Non-Equilibrium Gas Flow". Technical Report NASA-TM-X-64942, Marshall Space Flight Center, Huntsville, Alabama, 1974.

69. R. T. Davis. "Hypersonic Flow of a Chemically Reacting Binary Mixture past a Blunt Body". AIAA-Paper 70-0805, 1970.

70. R. K. Höld. "Viscous Shock-Layer Equations for the Calculation of Reentry Aerothermodynamics". *Proc. First European Symposium on Aerothermodynamics for Space Vehicles.* ESTEC, Noordwijk, The Netherlands, 1991, pp. 273 - 280.

10 Simulation Means

We give in this final chapter a short overview over the simulation means of aerothermodynamics. We do this with regard to scientific and to practical vehicle design aspects. The latter is getting a little more attention, because usually it is somewhat neglected.

We look first at some issues of flight vehicle design in order to provide a frame for this overview. This frame, of course, holds in a wider sense also for research activities. We treat the flight vehicle design phases, the task and tools of aerothermodynamics, the role changes of the simulation means due to the evolving discrete numerical methods of aerothermodynamics, the need to reduce the design margins, and give an assessment of the capabilities of the simulation means. A discussion of major issues of computational simulation, ground-facility simulation, and in-flight simulation follows next. The Oswatitsch independence principle is considered again and the issue of the so-called hot experimental technique is put forward.

The reader should be aware that simulation means are tools. A tool, however, is only as good as the person who wields it. In our context this means that only a sound knowledge of the physical phenomena, their mathematical models, their background, and their interrelations permits to use simulation tools successfully for the solution of a scientific or a design problem.

10.1 Some Notes on Flight Vehicle Design

Product Phases When speaking about flight vehicle design, it is useful to consider first the principal product phases. These phases are the same everywhere, also in hypersonic flight vehicle design. They may have other names, they may have different organizational support, and they will be somewhat different in the classical[1] and the hypersonic vehicle design, especially in the latter, if only one or very few specimens are to be produced.

[1] We define classical flight vehicles as aircraft flying in the low-speed, the subsonic, the transonic and the low supersonic domain. Hypersonic flight vehicles of course also operate in these domains. The low-speed domain is especially important in view of take-off (CAV-type vehicles) and landing (RV-type and CAV-type vehicles), and the transonic regime especially for CAV-type vehicles (drag divergence).

The principle product phases are, see, e.g., [1]:

1. Product definition.
   a) Conceptual design.
   b) Pre-design (preliminary design).
   c) Design.
2. Product development.
3. Product manufacturing.
4. Product support.

In the context of the present considerations product definition, phase 1, with its three sub-phases, is the most important one. It can extend over several years[2]. Conceptual design, sub-phase 1.a, defines the system "flight vehicle" together with its subsystems, however, in not very deep technical detail. Many concepts and variants are studied, and finally that one is frozen, which fulfills best the given specifications. In the pre-design phase, sub-phase 1.b, then the selected concept is refined until its feasibility is established. At the end of phase 1.b the concept is frozen, no major changes are expected in the following phases, unless serious problems show up. At the end of the pre-design phase also design sensitivities, design margins, further technology needs et cetera are established.

In the design phase, sub-phase 1.c, details of the flight vehicle are worked out, optimizations are made and data sets are produced. At the end of this phase the design is frozen. In the case of a go-ahead, the development phase, phase 2, which may last several years, begins. The airframe is being constructed, while the propulsion system and sub-systems of all kinds are developed and integrated. Finally after one or more prototypes have been built and tested, also in order to improve the data bases of the product, the manufacturing phase will begin[3]. In all definition sub-phases and in the development phase functions, properties, components, sub-systems et cetera are studied, defined and tested with mathematical/numerical models, in ground-simulation facilities and also in flight experiments.

In the early product definition phase the outer (aerodynamic) shape, the inner shape and the performance parameters of the flight vehicle are settled. In this phase also, what is very important, intentionally or unintentionally, the final manufacturing, operation and support costs are fixed to a large degree. In classical aircraft design this may regard up to 70 to 80 per cent of the life-cycle cost. Mistakes in this phase therefore must be avoided. Indeed,

---

[2] Of course, especially in hypersonic vehicle design, a forerunner technology development phase will exist over many years. It will, as a rule, overlap with phase 1. Very important is the observation, that all what is discussed here, also applies for an experimental vehicle, which may be designed, produced and flown during the forerunner technology development phase [2].

[3] We don't consider the manufacturing phase and the support phase, because they are of less importance to the purpose of the book.

it is a major problem to detect potential future cost drivers early enough. In general, the knowledge about the future product (performance, physical properties, functions) increases with the advancing product definition subphases. However, the freedom to change things decreases, because the cost to make changes rises. Therefore the general need is to rise the knowledge about the future product as high as possible as early as possible in the definition phase [3].

Tasks and Tools of Aerothermodynamic Design We have discussed shortly in Sub-Section 1.3 the different objectives of aerothermodynamics. We now have a look at the "tasks" and "tools" of aerothermodynamic design.

The tasks of the aerothermodynamic design, always in concert with the tasks of the other involved disciplines are, again simplified and idealized:

- the aerothermodynamic shape definition,
- the participation in multidisciplinary design problems,
- the verification of the aerothermodynamic design and of the demanded aerothermodynamic performance,
- the aerothermodynamic data-set generation (input also for performance determination, flight mechanics, et cetera),
- the problem diagnosis,
- and, if it applies, the participation in the flight-vehicle certification.

The tools or "simulation means", are:

- approximate (engineering) methods,
- discrete numerical methods for computational simulation and optimization,
- ground-facility simulation,
- in-flight simulation/experimentation with the help of either "passenger experiments" on other flight vehicles, on one or more dedicated experimental vehicles, or on a prototype.

This is very similar in classical aircraft design, see, e. g., [1].

Changing Roles of the Simulation Means The roles of the simulation means are:

- <u>computational simulation</u> with
  - approximate methods: aerothermodynamic shape definition, participation in multidisciplinary design problems,
  - discrete numerical methods: aerothermodynamic shape definition, participation in multidisciplinary design problems,
- <u>ground-facility simulation</u>: verification of aerothermodynamic design and performance, data-set generation,
- <u>ground-facility simulation and in-flight simulation</u>: overall design verification, problem diagnosis, and flight vehicle certification.

We use in the following sometimes the term "simulation-means triangle" which consists of computational simulation, ground-facility simulation and in-flight simulation. Computational simulation encompasses the simulation with approximate and sophisticated discrete numerical methods, but we contract now its meaning to the latter methods. Each of the simulation means has its potentials and deficits. However, since about a couple of years ago we have a change of the roles of the simulation means, which is due to the large advances in computational simulation, i. e., in the discrete numerical methods and in the computer technology. This is clearly the case in classical aircraft design [1], but it is naturally happening also in the design of hypersonic flight vehicles.

Discrete numerical methods increasingly substitute approximate methods in the flight-vehicle definition and development processes. This holds for both disciplinary and multidisciplinary simulation, and in the future also for the optimization[4]. Large computer power permits to use higher-level methods already in pre-design, phase 1.b. Parametric methods, based on results of Euler and Navier-Stokes/RANS simulations, have the potential to substitute simple potential and impact methods.

Ground-facility simulation on the other hand is increasingly less used in the definition and development processes of flight vehicles, but will remain for a long time the major tool in design verification and data-set generation. In classical flight vehicle design the high productivity of subsonic and transonic wind tunnels, i. e., the low cost and time per polar, today and in the near future by far cannot be matched by computational simulation, even if wind-tunnel model building and instrumentation is costly and time consuming. The transonic regime can now also be sufficiently covered by cryogenic wind tunnels, for instance the Transonic European Windtunnel (ETW), [4], [5]. However, the particularities of high Reynolds number aerodynamic design are not yet fully understood and mastered.

In hypersonic flight vehicle design ground-facility simulation will still remain for a long time to come the major tool in design verification and data-set generation, although there, as we will see in Section 10.3, the situation is not throughout so favorable. Moreover, ground-facility simulation must take over increasingly the very important task of data generation for model building and verification of the methods of computational simulation. This change in the use of ground-simulation facilities is an on-going process and keeps the need for such facilities alive, including the development of more and more sophisticated measurement devices.

In-flight simulation is and will remain the major means for the final design verification (systems identification, flight envelope opening). However, its other classical role as flight-experimental means is to be expanded towards high-accuracy data gathering for model building and verification of

[4] Not to be forgotten, bottlenecks of both numerical and approximate methods are flow-physics and partly also thermodynamic models.

the methods of computational simulation, and of ground-to- flight transfer methodologies.

The development of future aero-assisted space-transportation systems is driven by the need to attain high system cost efficiency. It has to take into account the weakening of Cayley's design paradigm, Section 1.1, and must make use of the second mathematization wave in science and engineering, [3]. Therefore it is very important to realize the potential of numerical disciplinary and multidisciplinary simulation and optimization methods. Here we are today in the phase of stepwise model building and verification.

High cost-efficiency of future space-transportation systems is expected from reusable or partly reusable aero-assisted single-stage or two-stage to orbit systems. Regarding the involved technologies it will probably not be reached by a single extraordinary break-through in one of the involved disciplines. Of large importance will be a highly integrated system definition and development approach with, for instance, a much better aerothermodynamic loads definition approach than we have today. As long as in the system definition phase, phase 1, the feasibility of a system is established in the classical way, which all too often does not care enough for the vehicle-system engineering problems of the later development phase, phase 2, and hence also does not identify all critical technologies, costly surprises will show up again and again. This view is only gradually being accepted in the community.

General Goal: Reduction of Design Margins None of the simulation means in the simulation triangle permits a full simulation of the flight vehicles properties and functions. It is the art of the engineer to arrive nevertheless at a viable design. In the following a short consideration of the three important entities in the definition and development processes, "sensitivities", "uncertainties", and "margins", is made [6].

The objectives of the definition and development processes in phase 1 and 2 are:

- the design of the flight vehicle with its performance, properties and functions according to the specifications,
- the provision of the describing data of the vehicle, the vehicle's data set(s).

Design sensitivities are sensitivities of the flight vehicle with regard to its performance, properties and functions. Hypersonic airbreathing flight vehicles are, for instance, sensitive with regard to vehicle drag (the "thrust minus drag" problem, where drag may turn out to be larger than the available thrust), and quite in general, to aerothermodynamic propulsion integration. We can state :

- small sensitivities permit rather large uncertainties in describing data (vehicle data set),
- large sensitivities demand small uncertainties in the describing data sets.

Uncertainties in describing data are due to deficits of the simulation means, i. e. the prediction and verification tools:

- computational simulation, there especially flow-physics and thermodynamics models,
- ground facility simulation,
- in-flight simulation.

Design margins finally allow for uncertainties in the describing data. The larger the uncertainties in design data, the larger are the design margins, which then are employed in the system design. They concern for instance flight performance, flight mechanics, flight dynamics, et cetera.

It is easy to imagine that design margins potentially give away performance. However, design margins anyway have the tendency to grow, not the least because vehicle weight has the tendency to grow. The general goal, which must be accomplished, is to reduce uncertainties in describing data (particularly where sensitivities are large) and hence to reduce design margins. Of course it is desirable to keep design sensitivities as small as possible, but demands of high performance and high cost-efficiency will always lead to large design sensitivities. Reduced uncertainties in describing data definitely will reduce design risks, cost and time, which is the essence of the Virtual Product philosophy [3], [1].

Assessment of Simulation Means We have noted the changing roles of the simulation means. We give now a short assessment of the capabilities of computational simulation and ground-facility simulation to treat potentially critical phenomena present on the flight vehicle configurations of the vehicle classes defined in Section 1.1 and their components [7]. Taken into account are phenomena, Table 10.1, which are typical for the hypersonic flight regime.

We see in Table 10.1 that the simulation of phenomena typical for viscosity-effects dominated CAV-type and ARV-type flight vehicles with turbulent flow has the largest deficiencies. In the design process especially the verification and data-set generation in ground-simulation facilities has the biggest shortcomings.

However, the message of Table 10.1 demands a very differentiated consideration. A major rule is that not each of the considered phenomena necessarily needs a simulation of high accuracy in the design of the given hypersonic vehicle. The accuracy demand is a function of the sensitivity of the vehicle or component on the respective phenomenon. If it is small, the data uncertainty must not be small, and the design margins are not influenced.

Take for instance the recovery temperature, which is not much affected by the laminar or turbulent state of the boundary layer, Fig. 7.9 in Section 7.3. If only this temperature would be of interest, or the structure and materials concept of the flight vehicle permits large uncertainties, the laminar-turbulent transition location does not play a deciding role. If then neither the drag nor other issues are critical, the flow can be assumed to be fully turbulent from

Table 10.1. Assessment of capabilities of simulation means to simulate potentially critical aerothermodynamic phenomena [7]. (...) indicates secondary criticality, → points to principle problems like the possibility/impossibility to attain Reynolds numbers large enough, the tripping problem, and the influence of the thermal state of the surface (TSS).

No.	Phenomenon	Vehicle class	Computational simulation	Ground facility simulation
1	attached laminar flow	RV, CAV, ARV, AOTV	good	good
2	laminar-turbulent transition	CAV, ARV, (RV)	poor	poor
3	attached turbulent flow	CAV, ARV, (RV)	fair	fair (?) → *Re*, tripping, TSS
4	thermal surface effects	CAV, ARV, RV, AOTV	good	not possible now
5	radiation cooling	RV, CAV, RV, AOTV	good	not possible
6	strong interaction: laminar	RV, AOTV	good	good → *Re*, TSS
7	strong interaction: turbulent	CAV, ARV, (RV)	poor	fair (?) → *Re*, tripping, TSS
8	low density effects	RV, ARV,	good	fair (?)
9	turbulent heat transfer on real-life surfaces	CAV, ARV, RV	fair (?)	fair → *Re*, tripping, TSS
10	turbulent mass transfer/ turbulent mixing	CAV, ARV	poor	fair (?) → *Re*, tripping, TSS
11	equilibrium real-gas effects	CAV, ARV, (RV)	good	good (?)
12	non-equilibrium real-gas effects	RV, ARV, AOTV, (CAV)	good (range?)	fair (?)
13	catalytic surface recombination	RV, ARV,	good (?)	fair (?) → TSS
14	plasma effects	AOTV	fair	fair (?)
15	micro-aerothermodynamics (gap flow, leakage flow, et cetera)	RV, CAV, RAV, AOTV	good (?)	fair (?)

the beginning, and, if possible at all for other reasons, in the ground-facility simulation boundary-layer tripping near the tip of the configuration would suffice.

If on the other hand surface radiation cooling, for instance at a CAV-type flight vehicle, would be necessary in order to employ a certain structure and materials concept, the cooling would affect the viscous drag[5] of a large part of the vehicle, the onset-flow of the propulsion-system inlet, but also the effectiveness of trim and control surfaces. Even these issues would not make an accurate determination of the transition location necessary from the beginning. Instead first it should be studied parametrically with numerical methods how the transition location affects the structure and materials concept, drag, et cetera. If all these are insensitive to the parametric changes, and the whole vehicle design is not affected, the flow can be considered as completely turbulent like in the recovery-temperature case. Only if one or more of the above would emerge to be sensitive to the laminar-turbulent transition location, the designer has a simulation problem. This concerns then not only item no. 2, but probably also no. 3 to 5, 7 and 9 of Table 10.1.

The table however also shows that in view of computational simulation large efforts are necessary to improve sufficiently flow-physics models and partly also thermodynamics models. Certainly also ground-simulation facilities and especially model and measurement techniques need further improvements, because they partly have several severe shortcomings [8], [2]. However, one has to realize, that a full simulation of reality is not possible in ground-simulation facilities, and *that it should not be attempted by trying to project fancy new hardware* (V. Y. Neyland, 1992, [8]). Instead disciplinary and multidisciplinary high-performance simulation and optimization, rooted firmly in the simulation triangle, should be pursued in, for instance, transfer model approaches [9], [2].

10.2 Computational Simulation

General Developments Numerical aerodynamics is now a more and more mature sub-discipline of aerodynamics and this holds also for numerical aerothermodynamics. This is due to the large advances in algorithm development for both simulation and optimization tasks, and, what is for industry very important, huge and cheap computer power. Computer power presently rises by a factor of ten every five years, and no saturation is discernible at this time. Computer architectures, which govern algorithm structures, appear to develop possibly in the direction of vector parallel rather than massive parallel architectures. This trend will be enhanced, if in about 2012 the 100.0 Gigaflops chip will be reached, which INTEL prognosticated in 1998. This will

[5] Possibly also the viscosity-effect induced pressure drag of the configuration.

affect also the developments in homogeneous or heterogeneous distributed (e. g., clusters of different kinds of PCs) computation.

Numerical aerodynamics/aerothermodynamics is now also the driver in multidisciplinary numerical simulation and optimization for all flight regimes, but especially for non-linear design regimes, [3]. Information technologies as they concern data storage technologies, data-base systems, networks and broad-band communications (both important for distributed computation), visualization tools, graphical user environments et cetera have also reached a high level of performance, reliability, and user friendliness.

Methods of Numerical Aerothermodynamics The flow regimes of interest here are the continuum and the slip-flow regime, Section 2.3. The Navier-Stokes equations, hence, are the basic governing equations, and Navier-Stokes methods[6] are the most general methods. In the limit of large Reynolds numbers they reduce to the Euler equations, Section 4.3. With Euler methods, however, only inviscid flows can be described. Together with the boundary-layer equations they permit to describe weakly interacting flow fields (coupled Euler/boundary-layer methods), Section 7.1.3. Viscous shock-layer methods model the flow situation at blunt bodies at large Mach numbers and moderate Reynolds numbers, Section 9.4.

The development of computer speed and storage permits increasingly, even in the early design phases of a flight vehicle, to employ discrete numerical methods. We concentrate therefore on the above mentioned methods and give a short overview[7].

The Navier-Stokes and the Euler equations are solved in the space-time domain with either time-asymptotic (time-marching) or time-accurate schemes [11], predominantly with (conservative) finite-volume discretization, [12]. The convective terms in Navier-Stokes solvers are treated like in Euler solvers [13], the viscous terms[8] are usually added with a central discretization approach. For the supersonic and hypersonic speed domain in many methods upwind schemes are used [14], [15], because they are naturally suited to describe wave propagation phenomena. According to Pandolfi and D'Ambrosio, [16], flux-difference splitting is the best method to deal with such phenomena. Solution acceleration schemes like the multigrid scheme, [17], have proven to be very effective also for hyperbolic problems, see, e. g., [18]. In fact, it is not

[6] When referring to Navier-Stokes methods, those for the computation of either laminar or turbulent flows, the latter being the Reynolds-averaged Navier-Stokes (RANS) methods, are meant.

[7] A detailed discussion of computational methods for hypersonic flow, although already partly somewhat outdated, can be found in [10].

[8] These are in general the components of the viscous stress tensor. In the "thin-layer" approximation only the boundary-layer terms in the x- and the z-momentum equation, and the term $\partial/\partial y(\mu \partial v/\partial y)$ in the y-momentum equation are included in order to reduce the algorithmic work.

only the sheer power of modern computers, but also the algorithm development, which has advanced the application of numerical methods.

We have discussed in Sub-Section 6.3.3 the treatment of shock waves in numerical methods and seen that shock-capturing is today the common method to treat shock waves in flow fields. This demands a sufficiently fine resolution of the computation domain, especially if the intersection of the bow-shock wave with downstream configuration parts has to be accurately described. Here local grid refinement is a desirable feature of a numerical scheme. If very high accuracy in special cases, for instance for scientific purposes, is needed, bow-shock fitting is still a viable alternative to bow-shock capturing. No review seems to be available regarding shock fitting methods, but the reader can find valuable information in, e. g., [19], [20].

Space-marching Navier-Stokes methods, see, e. g., [22], can be applied to sharp-nosed slender bodies, or at blunt bodies away from the blunt nose, if the equations are parabolized (PNS equations), Sub-Section 7.1.7, and at most flow separation in stream-wise direction, Fig. 9.1 b), occurs. At a blunt-nosed body, a time-marching scheme must be employed or a viscous shock-layer method, [21], [22], in order to get the necessary data for the initialization of the PNS solution.

The classical solution of the Euler equations for inviscid flow fields, the method of characteristics, [23], [24], is also a space-marching scheme, which demands purely supersonic (hyperbolic) flow. Space-marching Navier-Stokes (PNS) methods and the method of characteristics are now only seldom used[9].

Coupled Euler/second-order boundary-layer methods are a very efficient tool to compute the flow field at the windward side of, for instance, a flight vehicle of RV-type, [25], [26], [27]. More and more they give way to full Navier-Stokes methods, because with them one does not have to bother with the coupling procedure and moreover, one describes the flow past the whole configuration, including trim and control surfaces.

With the beginning of the application of the methods of numerical aerothermodynamics in flight vehicle design the need for test and validation data arose. The reader is referred in this regard to [28], [29], [30] and also to publications dealing with credible practices, e. g., [31], [32].

Grid Generation The problem of grid generation now appears to be solved in principle. For real-life flight vehicle configurations the generation of a block-structured grid, [33], for a RANS solution still needs several weeks or even months as it was about ten years ago [34]. For the generation of unstructured grids, [33], a few days now are needed. Structured codes run two to three times faster than unstructured ones. However, if one takes the total work time including grid generation, the latter are to be preferred, even if their accuracy is still to be improved somewhat. On parallel computer architectures a better

[9] It should be mentioned that basic concepts of the method of characteristics are applied in upwind methods [11].

load distribution can be made with unstructured codes, which is further in their favour, also in view that upwinding and multigrid acceleration is also possible with them.

A true breakthrough appears to happen presently with self-organizing hybrid Cartesian grid generation, see [35], but also [36] and [37]. For a real-life air vehicle configuration the creation of the initial Cartesian grid with an octree structure and a quasi-prism resolution of the boundary-layer domains and Kutta panels at trailing edges, takes, with minimal user intervention, about 30 *min* on a workstation, [35], [38]. After that the solution on a large-scale computer begins, during which the grid organizes itself in a way that it adapts to small-scale flow phenomena: the boundary layers, the vortex sheets behind the wings, stabilizers and so on, shock waves, et cetera.

The initial grid sets up on a CAD representation of the vehicle surface, [34]. The CAD surface of course must have been repaired, see, e. g., [39], and must be "watertight". The discretization of the surface, in form of a triangulation, is then made which takes into account the surface curvature of the configuration. Besides, the CAD representation is used in order to adapt later during the self-organizing solution process also the surface discretization/triangulation. With an automatic feature-based sampling technique, using a scan-line algorithm from computer graphics, [40], the CAD import, the sampling, and the surface triangulation takes about the same time as the creation of the initial grid [41]. In such a way after about at most 60 *min* a Navier-Stokes/RANS or a Euler solution can begin with an unstructured code with multigrid acceleration [38].

The self-organizing approach with hybrid Cartesian grids has the potential to treat also aerodynamic/structure couplings on moving and deforming surfaces, which are needed for transient simulations, of, for instance, control surface movements, where the configuration part or component moves simply through the background Cartesian grid. Here at present basis approaches begin to emerge. Not sufficiently resolved problems are still CAD repair (as for any grid generation) and also matters of re-importation of optimized shapes into CAD systems. Partly these are simply topics of process engineering.

Flow-Physics and Thermodynamics Modeling As was discussed in Chapter 8, a major problem for computational simulation lies in the flow-physics modeling, i. e. modeling of laminar-turbulent transition, turbulent shear flows, and turbulent separation on both hydraulically smooth surfaces, and real flight vehicle surfaces with surface irregularities. General, robust and versatile criteria and models are not available to the needed degree. This really limits the use of numerical aerodynamics/aerothermodynamics in aerodynamic design verification and data-set generation, which, with the growth of computer power, could otherwise become possible in about one decade.

Regarding laminar-turbulent transition mathematical/numerical models are needed, which do away with the dependence on empirical data. The capabilities of statistical turbulence models especially for separated flows remain to be defined, application classes should be built, and if and where necessary turbulence models should be newly devised or improved. Large eddy simulation (LES) of turbulent flow at high Reynolds numbers is still in its infancy, and definitely will be no way out of the turbulence modeling problem for realistic configurations for a long time to come. Maybe hybrid methods, which combine RANS methods locally with LES methods, will become available in the next decade for industrial applications. LES is, like direct numerical simulation (DNS), and together with ground-facility and flight experiments, a tool which should be employed in the efforts necessary to advance flow-physics modeling.

Thermodynamics modeling appears not to be as critical as flow-physics modeling. However, with growing accuracy demands of computational simulation, also here further progress is needed concerning accuracy and reliability of transport and thermo-chemical models.

Multidisciplinary Simulation and Optimization Aerothermodynamics is already multidisciplinary as the term suggests. Multidisciplinary design and optimization however encompasses here also the design of thermal protection systems[10] (TPS), and the - for flight vehicles - classical problems of aero-servoelasticity, which with hot primary structures can be dramatically enhanced. For airbreathing CAV-type vehicles with forebody pre-compression a very strong coupling of the flow past the elastic forebody and the performance of the propulsion system exists, see, e. g., [9].

Multidisciplinary design and optimization makes presently large strides in classical flight vehicle design and appears to take hold also in the design of hypersonic flight vehicles. A problem, which can sometimes be observed in both research and industry, is that the "leading" discipline has the tendency to sustain its own, usually reduced aerodynamic/aerothermodynamic capabilities - simple approximate methods - which then don't take advantage of the development of the methods of the mother discipline, in this case the methods of numerical aerothermodynamics [3]. However, the large potential of true multidisciplinary simulation and optimization methods can only be realized with the modern discrete numerical methods of aerothermodynamics.

[10] One has to keep in mind, that different structure and materials concepts, like the "classical" cold primary structure with thermal protection system for RV-type flight vehicles, the hot primary structure, or finally a mixture of both for CAV-type flight vehicles, pose different definition and development problems.

10.3 Ground-Facility Simulation

General Developments The only operational[11] hypersonic flight vehicle so far is the American Space Shuttle. This vehicle was designed and developed predominantly with the help of approximate methods and especially ground-facility simulation. At that time the capabilities of the discrete numerical methods of aerodynamics/aerothermodynamics were still very limited.

The first flight of the Space Shuttle was successful, however with regard to aerothermodynamics three major problems were identified [42]. The first, the hypersonic longitudinal stability problem, also called pitching-moment anomaly, was a very serious problem. The causes of it were connected to incomplete ground simulation of high-temperature real-gas effects and the Mach number[12], see, e. g., [43], [44].

The second problem was that the thermal loads on the TPS were overestimated. The temperatures on the TPS surface in flight were substantially smaller than predicted. The third problem were thermal loads larger than predicted on the OMS pod, which were due to "vortex scrubbing", i. e. attachment-line heating, see Sub-Section 3.2.3. One of the minor problems was that the lift to drag ratio (L/D) was underestimated in the lower Mach number regimes, see Section 3.1, which was also observed on the BURAN [45].

The above two major thermal-loads prediction problems and the L/D problem probably can be attributed to the fact that thermal surface effects have not been taken into account properly. They regard, as we have noted in Section 1.4, wall and near-wall viscous-flow and thermo-chemical phenomena. They are insufficiently or not at all captured in ground-facility simulation, if the Reynolds number is not representative and the usual cold-surface models are inadequate with regard to the boundary-layer properties but also with regard to catalytic wall recombination.

One has to be careful with the drawing of conclusions from these issues. It is not permitted to assume that with an eye on these problems and possible cures in ground-facility and computational simulation the aerothermodynamic design of RV-type flight vehicles is a mature discipline. One has to keep in mind that the configurational experience base is very small compared to that available in classical aircraft design. Therefore the risks and costs in the development of new vehicles of this kind are still high, especially if one strives for a much improved performance for instance regarding the ratio of vehicle mass to payload. Research for the clarification of principal configuration-related aerothermodynamic phenomena is necessary especially also in advanced ground-simulation facilities.

[11] We don't count re-entry capsules, experimental vehicles and missiles, target drones et cetera. The Russian BURAN flew only once.

[12] In the background is the Mach number independence principle.

Regarding CAV-type flight vehicles even less is known, when considering that these are viscosity-effects dominated, see, e. g. [46], [9]. Here the Reynolds number and the thermal state of the surface play a large role regarding viscous drag, forebody flow/inlet onset flow (aerothermodynamic propulsion integration), and control surface performance. These pose new challenges for ground-facility simulation, the role of which needs further clarification and concretion, and also new approaches regarding the embedding of ground-simulation facilities into the simulation-means triangle.

Hypersonic Ground-Simulation Facilities Overviews of ground-simulation facilities can be found in [47] (facilities worldwide), [48], [49] (facilities in Europe including Russia), and [50] (facility development in the last decade, worldwide). The references include partly also propulsion test facilities and give details about the model and measurement techniques available.

Information concerning hypersonic ground-facility simulation in general is given in, for instance, [51], [52], [53], [54]. Bushnell [55] discusses hypersonic ground test requirements with a very detailed classification of both civil and military hypersonic flight vehicles in the background. The (in)adequacy of existing facilities is seen rather critical, especially with regard to the simulation of high-temperature real-gas effects.

Among the relevant effects the dissociative behaviour is seen to be of importance in most cases, [56]. In order to duplicate the relevant Damköhler number(s), Section 5.4, the relevant product(s) characteristic density times characteristic length[13] (ρL) must be duplicated (binary scaling, [57], [58]), for instance $\rho_\infty R_s$, with R_s being the radius of the bow shock in front of the blunt forebody of the flight vehicle, [57]. The recombination behaviour scales with $\rho^2 L$.

With a small model in the ground-simulation facility this makes a very large density necessary. Modern shock-tube type facilities, [50], permit to obtain these densities, however, with small measurement times, not necessarily the true Mach number, and a very low productivity, i. e., only very few "shots" per day. That in a high-enthalpy facility the model onset flow may have been "frozen" in the nozzle, in general is not critical, Sub-Section 5.5.2, as long as it is properly defined and satisfies the simulation needs.

This all holds in principle for RV-type flight vehicles. Potentially critical issues of CAV-type flight vehicles are only insufficiently covered in these references. Thermal surface effects[14], for instance, cannot be duplicated, because the models have cold surfaces. We note that a full simulation of thermal surface effects, if the flight vehicle to be designed has a radiation cooled surface, is not possible, even if the model surface would be conditioned (heated) accordingly, [6], [59]. The problem of transition simulation, Section 8.1.6, which

[13] Stalker relates the binary scaling directly to the Mach number independence principle, [58].

[14] We remember that these can be critical also for RV-type flight vehicles when considering catalytic surface recombination.

is hampered by the noise radiation of the facility walls, also finds only little attention.

Implications and Use of the Oswatitsch Independence Principle
The hypersonic flight domain is usually considered to begin with $M_\infty \approx 5$, reaching up to $M_\infty \approx 25$. The argument for the lower bound is in general, that at that Mach number high-temperature real-gas effects begin to be felt. Indeed we have at this speed already substantial vibration excitation, Fig. 2.4. At $T = 1{,}000.0\ K$, for instance, the ratio of specific heats of air has already dropped to $\gamma \approx 1.33$, Fig. 5.2.

Another argument is connected to the Oswatitsch Mach number independence principle. We remember that Oswatitsch defines the hypersonic regime as that for which at a blunt body the shape of the bow-shock surface, the patterns of the streamlines, the sonic line, and the Mach lines in the supersonic part of the flow field, the density ratio, the surface pressure coefficient, and with the latter the force and moment coefficients become independent of M_∞, Section 6.8. We see from Fig. 6.37, that for a sphere the drag becomes independent of M_∞ at $M'_\infty \approx 5$. For the cone-cylinder configuration, lower curve in Fig. 6.37, this independence begins at $M_\infty = M'_\infty \approx 8$. The sphere certainly is representative for a RV-type configuration, whereas it is not clear whether the cone-cylinder is representative for a CAV-type configuration[15].

The observation made above, that in a shock-tube type facility the Mach number may not be the true Mach number, hence, is of no concern as long as it is large enough that the Mach-number independence principle holds ($M_\infty > M'_\infty$).

Another important conclusion can be drawn and has been drawn from the Mach number independence principle, viz., that it should be possible for the data-set generation of a hypersonic (RV-type) flight vehicle to make the measurements at a Mach number just, but sufficiently above M'_∞. This conclusion assumes that high-temperature real-gas effects on forces and moments are negligible.

This would allow to use basically a continuously running ground-simulation facility with a rather large cross section of the test section. This would have the benefit of sufficiently large and detailed models and a facility productivity which by far exceeds that of shock-tube type facilities, [51]. The latter is an important point, because hundreds of polars, each consisting of 10 to 20 data points, have to be produced for a typical data base.

This all was done during the development of the Space Shuttle, however not as simple as we sketch it here. The continuous AEDC tunnels B and C

[15] When considering the Mach-number independence principle with regard to ground-facility simulation, we always deal with that for blunt bodies. The principle for slender, sharp-nosed bodies has no relevance in that regard, because it says that some coefficients are proportional to some powers of the thickness ratio τ, Section 6.8. A ground-facility model will always have the same shape as the original configuration.

(M = 8 and 10), [53], were the basis, complemented by several other facilities. Pre-flight estimates, [61], had indicated that real-gas effects would lead to reduced loads (forces) and nose-up pitching moments at high altitudes and Mach numbers. These increments (in our above terminology "data uncertainties"), however, would not exceed the margins present in the overall aerodynamic design, [61]. As it then turned out during the first flight, all was correct except for the hypersonic longitudinal stability problem which we have mentioned above.

The lesson learned from this is that the use of a continuously running ground-simulation facility for the data-set generation, if desirable for productivity reasons, must be justified very carefully, and powerful extrapolation schemes must be devised. This can be done today with the methods of numerical aerothermodynamics. Otherwise large design margins must be provided. The use of special gas, for instance CF_4, facilities for the simulation of high-temperature real-gas effects, even if interesting results can be obtained with them, see, e. g., [62], is no way out of the problem.

In closing we remark that this all holds for hypersonic RV-type flight vehicles. How and how far Oswatitsch's Mach number independence principle is applicable to CAV-type vehicles is not clear. In any case it seems to be advisable to reconsider the ground-facility simulation problem for both vehicle types starting from this principle, and to establish in more detail its strengths and limitations[16].

Hot-Experimental Technique This topic concerns thermal surface effects to be investigated in either ground-facility or in-flight simulation, [6]. We address the hot model technique in ground-simulation facilities and the hot measurement technique in both ground-facility and in-flight simulation.

In ground-facility simulation we work in general with "cold" models, which are actually "uncontrolled cold" models with regard to thermal surface effects of both viscous and thermodynamic phenomena, for instance laminar-turbulent transition, strong interacting flows and catalytic surface recombination. It is conceivable to condition models or model surfaces not only in the sense that they are cooled, but that they are heated. First investigations of this kind have already been performed, see, e.g, Chapter 9. The type of the facility, either continuous or short duration/intermittent, plays a role. However, even if it would be desirable, a true simulation of radiation cooling is not possible, [6], [59], although some thermal surface effects can be simulated to a certain degree, see, e. g., [60]. The hot model topic basically is a research

[16] We have remarked in Section 6.8 that the Mach-number independence principle is valid (for a given γ) on the windward side of a body, i. e. in the portions of the flow past a body, where the bow-shock surface lies close to the body surface. It certainly is not valid on the lee-side of a body. Nothing is known about transition regimes between windward side and lee-side surface portions, and also regarding shape particularities like boat-tailing of the lower side aft portion, et cetera.

topic, but is of importance especially also for the design of CAV-type flight vehicles, and of RV-type vehicles [6].

Regarding the hot measurement technique we have a short look at necessary improvements of surface gauges only. Opto-electronic and surface coating techniques are not considered. Take for instance the need to determine during flight exactly the thermal state of the surface, if boundary-layer instability and transition phenomena are to be measured and correlated. To determine the heat flux in the gas at the wall, q_{gw}, Fig. 3.1, we need to measure the wall temperature T_w, and the heat flux into the wall, q_w. q_{gw} is then found with eq. (3.12), provided the thermal emissivity ϵ of the gauge surface is known. This, however, implies, that the gauge has the same thermal emissivity as the surrounding vehicle surface has.

We put it in general terms: any insert into a surface (temperature and heat flux measurement gauges, pressure measurement gauges) must not locally falsify the surface emissivity and catalycity, and the material-in-depth heat transmission and capacity properties. Otherwise the measurements represent the situation at the gauge insert, and not necessarily that of the surrounding vehicle surface.

The need for surface instrumentation of such kind exists for in-flight measurements, but to a certain extent also for ground-facility simulation with hot model surfaces. Innovative solutions of the falsification problem of gauges are therefore demanded.

10.4 In-Flight Simulation

We don't consider hypersonic experimental flight vehicle design and operation problems, but concentrate on five basic issues of in-flight simulation of aerothermodynamic phenomena, see, e. g., [9], [2]. This holds for both passenger experiments and experimentation on dedicated flight vehicles.

Size of the flight vehicle. The flight vehicle must have a sufficiently large size such that the phenomena in question are present, that they can be isolated sufficiently, and that the related flow and thermodynamic parameters are large enough to be measured. A dedicated experimental flight vehicle will be smaller than the reference vehicle. In general it will not be possible to scale down the reference vehicle linearly (flying wind tunnel model), because the surface temperature of the radiation cooled vehicle surface portions is scale dependent, Section 3.2.5.

Instrumentation and data acquisition. It is necessary to put high demands on in-flight experimentation. On the one hand we need new data, which correspond to the present state of knowledge and to the conceived design problems, on the other hand in-flight simulation is very costly. Hence the measurements must give data with sufficient accuracy, data rates, and repeatability. Above we have considered shortly the needs of the hot experimental technique, namely the falsification problem of surface gauges, which must be overcome to a sufficient degree. The instrumentation must not aim for the maximum possible, but for the necessary in view of the phenomenon to be investigated and quantified. Data acquisition rates must be such that the flight speed and atmospheric inhomogenities are covered. The latter is important especially at extended flight in the stratosphere, Section 2.1.

Air data and vehicle system behaviour. Without air data measurements, i. e., the determination of the velocity vector $\underline{v_\infty}$ of the flight vehicle relative to the surrounding atmosphere, and free-stream density ρ_∞ and temperature T_∞, the measured data cannot be correlated and are essentially worthless. A special problem is that on the flight trajectory the actual density of the atmosphere can deviate from the assumed atmosphere model, a topic which we touched already in Section 2.1. New developments in opto-electronics make it likely that, in future, air data can be measured ahead of the bow shock, i. e., directly in the undisturbed atmosphere. Vehicle system behaviour (flight quality, transient states, aero-servoelasticity of the airframe) must be taken sufficiently into account already during the mission definition.

Pre-flight analysis. Pre-flight analysis, like post-flight analysis, must employ the methods of numerical aerothermodynamics. Necessary, and possible today, is a complete and high-fidelity simulation of the aerothermodynamic environment in which the phenomena in question are to be simulated in flight[17]. This concerns the vehicle configuration, the flight trajectory - in fact the pre-flight analysis also serves to define the flight trajectory - and the instrumentation. The determination of the thermal state of the surface must take into account the heat flux into the wall, if the radiation-adiabatic situation does not reflect accurately enough the actual surface situation[18].

Programmatical embedding. In-flight simulation and experimentation is not conceivable without a firm embedding into a larger technology or development effort. The scope, the relation to a reference vehicle, or several reference vehicles, the pre-flight and the post-flight analysis are costly and time consuming and will be justifiable only in the frame of a larger effort.

References

1. J. B. Vos, A. Rizzi, D. Darracq, E. H. Hirschel. "Navier-Stokes Solvers in European Aircraft Design". Progress in Aerospace Sciences, Vol. 38, 2002, pp. 601 - 697.
2. E. H. Hirschel, F. G. J. Kremer. "Technology Development and Verification Plan - Final Report, Slice D". FESTIP FSS-SCT-RP-0068, 1998.

[17] Pre-flight analysis is based on the nominal flight trajectory - and some variations of it - and a nominal atmosphere (standard atmosphere). Post-flight analysis is made with the actual flight trajectory and the actual atmosphere (air data).

[18] We remember that in flight the radiation-adiabatic temperature is a realistic estimate of the wall temperature, Section 3.1. In general the actual surface temperature will be (somewhat) lower or higher.

3. E. H. HIRSCHEL. "Towards the Virtual Product in Aircraft Design?" *J. Periaux, M. Champion, J.-J. Gagnepain, O. Pironneau, B. Stoufflet, P. Thomas (eds.), Fluid Dynamics and Aeronautics New Challenges.* CIMNE Handbooks on Theory and Engineering Applications of Computational Methods, Barcelona, Spain 2003, pp. 453 - 464.

4. J. A. VAN DER BLIEK. "ETW, a European Resource for the World of Aeronautics. The History of ETW in the Context of European Research and Development Cooperation". ETW, Cologne, Germany, 1996.

5. D. SCHIMANSKI, J. QUEST. "Tools and Techniques for High Reynolds Number Testing, Status and Recent Improvements of ETW". AIAA-Paper 2003-0755, 2003.

6. E. H. HIRSCHEL. "Thermal Surface Effects in Aerothermodynamics". *Proc. Third European Symposium on Aerothermodynamics for Space Vehicles, Noordwijk, The Netherlands, November 24 - 26, 1998.* ESA SP-426, 1999, pp. 17 - 31.

7. E. H. HIRSCHEL. "Hypersonic Aerodynamics". Space Course 1993, Vol. 1, Technical University München, 1993, pp. 2-1 to 2-17.

8. V. Y. NEYLAND. "Scientific and Engineering Problems of Preflight Development of Orbiter". TsAGI Central Aerohydrodynamic Institute, Moscow, Russia, 1992.

9. E. H. HIRSCHEL. "The Technology Development and Verification Concept of the German Hypersonics Technology Programme". AGARD R-813, 1986, pp. 12-1 to 12-15.

10. J. J. BERTIN, J. PERIAUX, J. BALLMANN (EDS.). "Advances in Hypersonics, Vol. 3, Computing Hypersonic Flows". Birkhäuser, Boston, 1992.

11. A. EBERLE, A. RIZZI, E. H. HIRSCHEL. "Numerical Solutions of the Euler Equations for Steady Flow Problems". *Notes on Numerical Fluid Mechanics, NNFM 34.* Vieweg, Braunschweig/Wiesbaden, 1992.

12. C. HIRSCH. "Numerical Computation of Internal and External Flows". Vol. 1, Fundamentals of Numerical Discretization, J. Wiley and Sons, New York, 1988.

13. R. LÖHNER. "Applied Computational Fluid Dynamics Techniques". J. Wiley and Sons, Chichester, U. K., 2001.

14. M. Y. HUSSAINI, B. VAN LEER , J. VAN ROSENDALE (EDS.). "Upwind and High-Resolution Schemes". Springer-Verlag, Berlin/Heidelberg/New York, 1997.

15. M. PANDOLFI, D. D'AMBROSIO. "Numerical Instabilities in Upwind Methods: Analysis and Cures fo the "Carbuncle" Phenomenon". J. of Computational Physics, Vol. 166, 2001, pp. 271 - 301.

16. M. PANDOLFI, D. D'AMBROSIO. "A Critical Analysis of Upwinding". *J. Periaux, M. Champion, J.-J. Gagnepain, O. Pironneau, B. Stoufflet, P. Thomas (eds.), Fluid Dynamics and Aeronautics New Challenges.* CIMNE Handbooks on Theory and Engineering Applications of Computational Methods, Barcelona, Spain 2003, pp. 161 - 177.

17. U. Trottenberg, C. Oosterlee, A. Schüller. "Multigrid". Academic Press, 2001.

18. R. Radespiel, R. C. Swanson. "Progress with Multigrid Schemes for Hypersonic Flow Problems". J. of Computational Physics, Vol. 116, 1995, pp. 103 - 122.

19. J.-A. Désidéri, R. Glowinski, J. Periaux (eds.). "Hypersonic Flows for Re-Entry Problems". Vol. II and III, Springer-Verlag, Berlin/Heidelberg/New York, 1992.

20. S. Menne, C. Weiland, D. D'Ambrosio, M. Pandolfi. "Validation of Real Gas Simulations Using Different Non-Equilibrium Methods". Computers and Fluids, Vol. 24, 1995, pp. 189 - 208.

21. R. T. Davis. "Numerical Solution of the Hypersonic Shock-Layer Equations". AIAA J., Vol. 8, No. 5, 1970, pp. 843 - 851.

22. J. J. Bertin. "State-of-the-Art Engineering Approaches to Flow Field Computations". *J. J. Bertin, R. Glowinski, J. Periaux (eds.), Hypersonics, Vol. 2, Computation and Measurement of Hypersonic Flows.* Birkhäuser, Boston, 1989, pp. 1 - 91.

23. K. Oswatitsch. "Gas Dynamics". Academic Press, New York, 1956.

24. H. W. Liepmann, A. Roshko. "Elements of Gasdynamics". John Wiley & Sons, New York/ London/ Sidney, 1966.

25. Ch. Mundt, M. Pfitzner, M. A. Schmatz. "Calculation of Viscous Hypersonic Flows Using a Coupled Euler/Second-Order Boundary-Layer Method". Notes of Numerical Fluid Mechanics, NNFM 29, Vieweg, Braunschweig/Wiesbaden, 1990, pp. 422 - 433.

26. B. Aupoix, J. Ph. Brazier, J. Cousteix, F. Monnoyer. "Second-Order Effects in Hypersonic Boundary Layers". *J. J. Bertin, J. Periaux, J. Ballmann (eds.), Advances in Hypersonics, Vol. 3, Computing Hypersonic Flows.* Birkhäuser, Boston, 1992, pp. 21 - 61.

27. Ch. Mundt, F. Monnoyer, R. Höld. "Computational Simulation of the Aerothermodynamic Characteristics of the Reentry of HERMES". AIAA-Paper 93-5069, 1993.

28. S. Borrelli, F. Grasso, M. Marini, J. Periaux (eds.). "Proceedings of the First Europe-US High Speed Flow Field Database Workshop, Naples, Italy". Published with permission by AIAA, 1998.

29. M. S. Holden (supervisor). "Experimental Database from CUBRC Studies in Hypersonic Laminar and Turbulent Interacting Flows Including Flowfield Chemistry". *Prepared for RTO Code Validation of DSMC and Navier-Stokes Code Validation Studies.* Calspan-University at Buffalo Research Center, Buffalo, NY, 2000.

30. J.-A. Désidéri, M. Marini, J. Periaux. "Validation Databases in Fluid Mechanics: from the European Space Shuttle Program HERMES to the European Thematic Network FLOWNET". *J. Periaux, M. Champion, J.-J. Gagnepain, O. Pironneau, B. Stoufflet, P. Thomas (eds.), Fluid Dynamics and Aeronautics New Challenges.* CIMNE Handbooks on Theory and Engineering Applications of Computational Methods, Barcelona, Spain 2003, pp. 535 - 546.

31. J. G. MARVIN. "A CFD Validation Roadmap for Hypersonic Flows". AGARD CP-514, 1992, pp. 17-1 to 17-16.
32. U. B. MEHTA. "Guide to Credible Computer Simulations of Fluid Flows". J. Prop. and Power, Vol. 12, No. 5, 1996, pp. 940 - 948.
33. J. F. THOMPSON, B. K. SONI , N. P. WEATHERILL (EDS.). "Handbook of Grid Generation". CRC Press, Boca Raton/London/New York/Washington D. C., 1999.
34. E. H. HIRSCHEL, W. SCHWARZ. "Mesh Generation for Aerospace CFD Applications". J. on Surveys on Mathematics for Industry, Vol. 4, 1995, pp. 249 - 265.
35. F. DEISTER. "Selbstorganisierendes hybrid-kartesisches Netzverfahren zur Berechnung von Strömungen um komplexe Konfigurationen (Self-Organizing Hybrid Cartesian Grid Method for the Computation of Flows past Complex Configurations)". Doctoral Thesis, Universität Stuttgart, Germany, 2002.
36. M. J. AFTOSMIS, M. J. BERGER, G. ADOMAVICIUS. "A Parallel Multilevel Method for Adaptively Refined Cartesian Grids with Embedded Boundaries". AIAA-Paper 2000-0808, 2000.
37. M. DELANAYE, A. PATEL, K. KOVALEV, B. LÉONARD, CH. HIRSCH. "From CAD to Adapted Solution for Error Controlled CFD Simulations". *J. Periaux, M. Champion, J.-J. Gagnepain, O. Pironneau, B. Stoufflet, P. Thomas (eds.), Fluid Dynamics and Aeronautics New Challenges.* CIMNE Handbooks on Theory and Engineering Applications of Computational Methods, Barcelona, Spain 2003, pp. 465 - 477.
38. F. DEISTER, E. H. HIRSCHEL. "Self-Organizing Hybrid Cartesian Grid/Solution System with Multigrid". AIAA-Paper 2002-0112, 2002.
39. A. A. MEZENTSEV, TH. WÖHLER. "Methods and Algorithms of Automated CAD Repair for Incremental Surface Mehing". Proc. 8th International Meshing Roundtable, South Lake Tahoe, Cal., October 10 to 13, 1999, pp. 299 - 307.
40. J. D. FOLEY, A. VAN DAM, S. K. FEINER, J. F. HUGHES. "Computer Graphics". Addison-Wesley Publ. Comp., Reading, Mass. 1996.
41. F. DEISTER, U. TREMEL, E. H. HIRSCHEL, H. RIEGER. "Automatic Feature-Based Sampling of Native CAD Data for Surface Grid Generation". *C. Breitsamter, B. Laschka, H.-J. Heinemann, R. Hilbig (eds.), New Results in Numerical and Experimental Fluid Mechanics IV.* Notes on Numerical Fluid Mechanics and Multidisciplinary Design, NNFM 87, Springer, Berlin/Heidelberg/New York, 2004, pp. 374 - 381.
42. J. P. ARRINGTON, J. J. JONES (EDS.). "Shuttle Performance: Lessons Learned". NASA CP-2283, 1983.
43. B. F. GRIFFITH, J. R. MAUS, J. T. BEST. "Explanation of the Hypersonic Longitudinal Stability Problem - Lessons Learned". *J. P. Arrington, J. J. Jones (eds.), Shuttle Performance: Lessons Learned.* NASA CP-2283, Part 1, 1983, pp. 347 - 380.
44. B. F. GRIFFITH, J. R. MAUS, J. T. BEST. "Explanation of the Hypersonic Longitudinal Stability Problem - Lessons Learned". *D. A. Throckmorton (ed.), Shuttle Performance: Lessons Learned.* NASA CP-3248, Part 1, 1995, pp. 347 - 380.

45. V. Y. NEYLAND. "The Convergence of the Orbiter BURAN Flight Test and Preflight Study Results, and the Choice of a Strategy to Develop a Second Generation Orbiter". AIAA-Paper 1989-5019, 1989.

46. J. L. HUNT. "Hypersonic Airbreathing Vehicle Design (Focus on Aero-Space Plane)". *J. J. Bertin, R. Glowinski, J. Periaux (eds.), Hypersonics, Vol. 1, Defining the Hypersonic Environment.* Birkhäuser, Boston, 1989, pp. 205 - 262.

47. J. ARNOLD, J. F. WENDT "Test Facilities". AGARD-AR-319, Vol. I, 1996, pp. 8-1 to 8-27.

48. W. KORDULLA, J. MUYLAERT "Inventory of Aerothermodynamic Capabilities in Europe". *IAF Conference 2003, Bremen, Germany.* IAC-Paper-03-V.5.07, 2003.

49. J. C. TRAINEAU, C. PELISSIER, A. M. KHARITONOV, V. M. FOMIN, V. I. LAPYGIN, V. A. GORELOV. "Review of European Facilities for Space Aerothermodynamics". ONERA, Techn. Rep. RT 1/06302 DMAE, 2003.

50. F. K. LU, D. E. MARREN (EDS.). "Advanced Hypersonic Test Facilities". Vol. 198, Progress in Astronautics and Aeronautics, AIAA, Washington, DC, 2002.

51. R. D. NEUMANN. "Defining the Aerothermodynamic Methodology". *J. J. Bertin, R. Glowinski, J. Periaux (eds.), Hypersonics, Vol. 1, Defining the Hypersonic Environment.* Birkhäuser, Boston, 1989, pp. 125 - 204.

52. J. J. BERTIN, R. GLOWINSKI, J. PERIAUX (EDS.). "Hypersonics, Vol. 2, Computation and Measurement of Hypersonic Flows". Birkhäuser, Boston, 1989.

53. R. K. MATTHEWS. "Hypersonic Wind Tunnel Testing". *J. J. Bertin, J. Periaux, J. Ballmann (eds.), Advances in Hypersonics, Vol. 1, Defining the Hypersonic Environment.* Birkhäuser, Boston, 1992, pp. 72 - 108.

54. N. N. "Hypersonic Experimental and Computational Capability, Improvement and Validation". AGARD-AR-319, Vol. I, 1996, Vol. II, 1998.

55. D. M. BUSHNELL. "Hypersonic Ground Test Requirements". *F. K. Lu, D. E. Marren (eds.), Advanced Hypersonic Test Facilities.* Vol. 198, Progress in Astronautics and Aeronautics, AIAA, Washington, DC, 2002, pp. 1 - 15.

56. F. K. LU, D. E. MARREN. "Principles of Hypersonic Test Facility Development". *F. K. Lu, D. E. Marren (eds.), Advanced Hypersonic Test Facilities.* Vol. 198, Progress in Astronautics and Aeronautics, AIAA, Washington, DC, 2002, pp. 17 - 27.

57. W. E. GIBSON, P. V. MARRONE. "A Similitude for Non-Equilibrium Phenomena in Hypersonic Flight". AGARDograph 68, 1964, pp. 105 - 131.

58. R. J. STALKER. "Hypervelocity Aerodynamics with Chemical Nonequilibrium". Annual Review of Fluid Mechnics, Vol. 21, 1989, pp. 37 - 60.

59. A. HENCKELS, F. MAURER. "Hypersonic Wind Tunnel Testing with Simulation of Local Hot Wall Boundary Layer and Radiation Cooling". Zeitschrift für Flugwissenschaften und Weltraumforschung (ZFW), Vol. 18, 1994, pp. 160 - 166.

60. G. SIMEONIDES, X. GIBERGY, X. DE LA CASA, J. M. CHARBONNIER. "Combined Convective Heating - Radiation Cooling Analysis for Aerodynamic Surfaces in Hypersonic Flow and Experimental Simulation of Temperature Viscous

Effects". *Proc. EUROTHERM SEMINAR 55, "Heat Transfer in Single Phase Flows 5"*. NTUAthens, Greece, 1997.

61. W. C. Woods, J. P. Arrington, H. H. Hamilton II. "A Review of Preflight Estimates of Real-Gas Effects on Space Shuttle Aerodynamic Characteristics". *J. P. Arrington, J. J. Jones (eds.), Shuttle Performance: Lessons Learned.* NASA CP-2283, Part 1, 1983, pp. 309 - 346.
62. D. K. Prabhu, P. E. Papadopoulos, C. B. Davies, M. J. Wright, R. D. McDaniel. "Shuttle Orbiter Contingency Abort Aerodynamics, II: Real-Gas Effects and High Angles of Attack". AIAA-Paper 2003-1248, 2003.

11 The RHPM-Flyer

The Rankine-Hugoniot-Prandtl-Meyer- (RHPM-) flyer - an infinitely thin flat plate - is the highly simplified configuration of a hypersonic flight vehicle. It approximates for $M_\infty > 1$ at small angles of attack a CAV-type flight vehicle (RHPM-CAV-flyer), and at large angle of attack a RV-type vehicle (RHPM-RV-flyer). Aerodynamic control surfaces or inlet ramps can be simulated by attachment of a separate flat plate element with a deflection or ramp angle.

The parameters of the inviscid flow on the windward (w) side of the RHPM-flyer - found with the Rankine-Hugoniot relations - and on its lee (l) side - found with the Prandtl-Meyer relation - are each constant, Fig. 11.1, also on the possible attachments (control surfaces or inlet ramps). They are functions only of the free-stream Mach number M_∞, the assumed effective ratio of specific heats γ_{eff}, and of the angle of attack α, which of course must be smaller than α_{max}, at which the shock wave at the windward side becomes detached, eq. (6.116). If on the lee side the expansion limit is reached, $M_l \to \infty$ is obtained as result and all thermodynamic variables are set to zero.

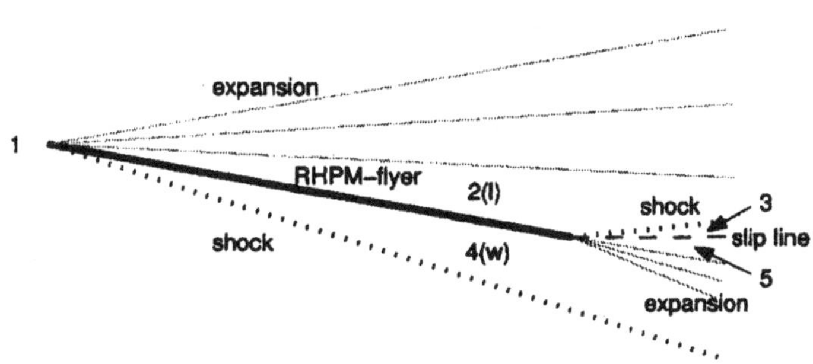

Fig. 11.1. Sketch of the RHPM-flyer [1] at $M_\infty = 6$, $\alpha = 10°$, $\gamma = 1.4$. True scale, "1" ≡ "∞".

The RHPM-flyer serves as a teaching and demonstration flight vehicle. It permits a fast and cheap illustration and approximate quantification of configurational, inviscid, viscous and thermal phenomena based on the perfect-gas/γ_{eff} assumption. Of course all phenomena related to blunt configuration parts, especially to the blunt vehicle nose, cannot be treated. This holds in particular for the windward side of RV-type vehicles, where the subsonic pocket at large angles of attack extends far downstream on the windward side. Nevertheless, also in that case is the RHPM-flyer an acceptable approximation.

Treated can be configurational phenomena like the dependence of forces and moments on Mach number, angle of attack and effective ratio of specific heats, the Mach number independence, the hypersonic shadow effect, vehicle planform effects (inclusion of the third dimension with the help of a lateral extension of the flat plate), and also the boundary-layer development and the thermal state of the surface.

The different behaviour of phenomena on the windward and the lee side at, for instance, RV-type vehicles can be demonstrated. Figs. 11.2 and 11.3 show as example the ratio 'unit Reynolds number on the windward and on the lee side of the RHPM-flyer' to 'unit Reynolds number at infinity' as function of the angle of attack and the free-stream Mach number. The ratios are each constant along the flyer's surface. At the windward side they rise initially with the angle of attack above the free-stream value and then drop below it, see also Section 6.6. On the lee side the unit Reynolds number is smaller than that at infinity for all angles of attack.

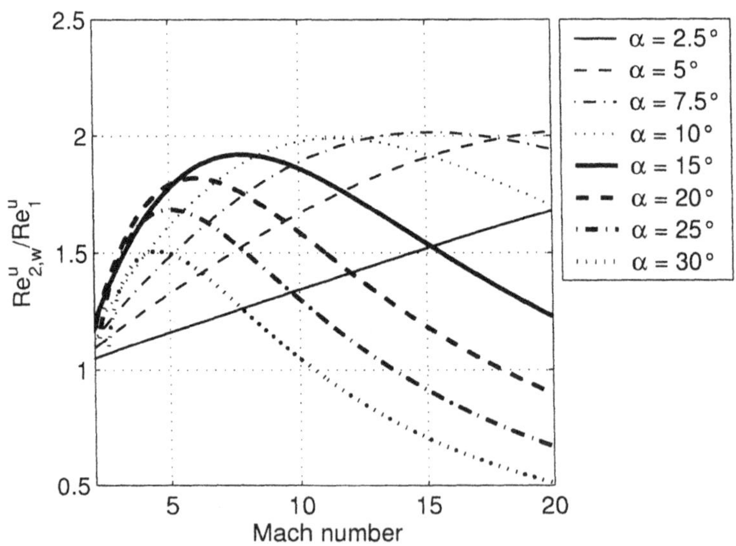

Fig. 11.2. $Re^u_{2,w}/Re^u_1$ at the windward side (w) of the RHPM-flyer as function of the angle of attack α and the free-stream Mach number ("1" $\equiv$ "∞") [1], perfect gas.

Regarding the inviscid flow parameters the RHPM-flyer is employing simple shock-expansion theory, [2], which results in constant values along the

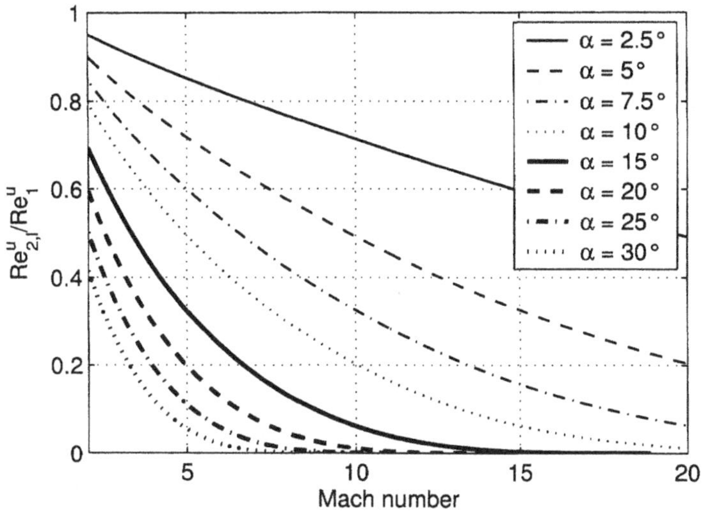

Fig. 11.3. $Re^u_{2,l}/Re^u_1$ at the lee side (l) of the RHPM-flyer as function of the angle of attack α and the free-stream Mach number ("1" ≡ "∞") [1], perfect gas.

surface. Viscous and thermal phenomena (laminar and turbulent flow) are described with the approximate relations with reference-temperature extensions given in Section 7.2. They depend, of course, on the boundary-layer running length, i. e., the distance from the plate's leading edge or the effective origin of the turbulent boundary layer, when laminar-turbulent transition occurs a finite distance downstream of the leading edge. The use of a power-law approximation for the viscosity is an appropriate and acceptable simplification.

The RHPM-flyer was defined and applied in [1], [3]. It is programmed in MATLAB 6.1 and contains the Rankine-Hugoniot relations and the Prandtl-Meyer relation, [4], as well as boundary-layer approximations. Where it applies, also inverse operations can be performed. The ratio of specific heats can deliberately be chosen (γ_{eff} approach), such that high-temperature equilibrium real-gas effects can be approximated.

The programmes can be obtained from the authors of [1], [3]: `bastian.thorwald@unibw-muenchen.de` and `mharchi@web.de`, or at `http://www.unibw-muenchen.de/campus/LRT/LRT10/web_home.htm`.

References

1. B. Thorwald. "Demonstration of Configurational Phenomena Exemplified by the RHPM-Hypersonic-Flyer". Diploma Thesis, University Stuttgart, Germany, 2003.

2. H. W. LIEPMANN, A. ROSHKO. "Elements of Gasdynamics". John Wiley & Sons, New York/London/Sidney, 1966.
3. M. MHARCHI. "Demonstration of Hypersonic Thermal Phenomena and Viscous Effects with the RHPM-Flyer". Diploma Thesis, University Stuttgart, Germany, 2003.
4. AMES RESEARCH STAFF. "Equations, Tables, and Charts for Compressible Flow". NACA R-1135, 1953.

12 Governing Equations for Flow in General Coordinates

We collect the transport equations for a multi-component, multi-temperature[1] non-equilibrium flow. All transport equations have been discussed in Chapter 4, as well as in Chapters 5, 6, 7 (see also [1], [2], [3]). We write the equations in (conservative) flux-vector formulation, which we have used already for the energy equation in Sub-Section 4.3.2, and for three-dimensional Cartesian coordinates:

$$\frac{\partial \underline{Q}}{\partial t} + \frac{\partial(\underline{E} + \underline{E}_{visc})}{\partial x} + \frac{\partial(\underline{F} + \underline{F}_{visc})}{\partial y} + \frac{\partial(\underline{G} + \underline{G}_{visc})}{\partial z} = \underline{S}. \tag{12.1}$$

$\underline{Q}$ is the conservation vector, $\underline{E}$, $\underline{F}$, $\underline{G}$ are the convective (inviscid) and $\underline{E}_{visc}$, $\underline{F}_{visc}$, $\underline{G}_{visc}$ the viscous fluxes in x, y, and z direction. $\underline{S}$ is the source term of mass and of vibration energy.

The conservation vector $\underline{Q}$ has the form:

$$\underline{Q} = [\rho_i, \rho u, \rho v, \rho w, \rho e_t, \rho_m e_{vibr,m}], \tag{12.2}$$

where ρ_i are the partial densities of the involved species i, Section 2.2, ρ is the density, u, v, w are the Cartesian components of the velocity vector $\underline{V}$, $e_t = e + 1/2\ V^2$ is the mass-specific total energy ($V = |\underline{V}|$), Sub-Section 4.3.2, and $\rho e_{vibr,m}$ the mass-specific vibration energy of the molecular species m.

The convective and the viscous fluxes in the three directions read:

$$\underline{E} = \begin{bmatrix} \rho_i u \\ \rho u^2 + p \\ \rho u v \\ \rho u w \\ \rho u h_t \\ \rho_m u e_{vibr,m} \end{bmatrix}, \ \underline{E}_{visc} = \begin{bmatrix} j_{i_x} \\ \tau_{xx} \\ \tau_{xy} \\ \tau_{xz} \\ q_x - \sum_i j_{i,x} h_i + u\tau_{xx} + v\tau_{xy} + w\tau_{xz} \\ -k_{vibr,m}\frac{\partial T_{vibr,m}}{\partial x} + j_{m_x} e_{vibr,m} \end{bmatrix}, \tag{12.3}$$

---

[1] We include only the transport of non-equilibrium vibration energy. Equations for other entities, like rotation energy, can be added.

$$\underline{F} = \begin{bmatrix} \rho_i v \\ \rho u v \\ \rho v^2 + p \\ \rho v w \\ \rho v h_t \\ \rho_m v e_{vibr,m} \end{bmatrix}, \ \underline{F}_{visc} = \begin{bmatrix} j_{i_y} \\ \tau_{yx} \\ \tau_{yy} \\ \tau_{yz} \\ q_y - \sum_i j_{i,y} h_i + u\tau_{yx} + v\tau_{yy} + w\tau_{yz} \\ -k_{vibr,m} \frac{\partial T_{vibr,m}}{\partial y} + j_{m_y} e_{vibr,m} \end{bmatrix}, \tag{12.4}$$

$$\underline{G} = \begin{bmatrix} \rho_i w \\ \rho u w \\ \rho v w \\ \rho w^2 + p \\ \rho w h_t \\ \rho_m w e_{vibr,m} \end{bmatrix}, \ \underline{G}_{visc} = \begin{bmatrix} j_{i_z} \\ \tau_{zx} \\ \tau_{zy} \\ \tau_{zz} \\ q_z - \sum_i j_{i,x} h_i + u\tau_{zx} + v\tau_{zy} + w\tau_{zz} \\ -k_{vibr,m} \frac{\partial T_{vibr,m}}{\partial z} + j_{m_z} e_{vibr,m} \end{bmatrix}. \tag{12.5}$$

The convective flux vectors $\underline{E}$, $\underline{F}$, $\underline{G}$ represent from top to bottom the transport of mass, Sub-Section 4.3.3, momentum, Sub-Section 4.3.1, of total energy, Sub-Section 4.3.2, and of non-equilibrium vibration energy[2], Section 5.4. In the above $h_t = e_t + p/\rho$ is the total enthalpy, eq. (5.7).

In the viscous flux vectors $\underline{E}_{visc}$, $\underline{F}_{visc}$, $\underline{G}_{visc}$, the symbols j_{i_x} et cetera represent the Cartesian components of the diffusion mass-flux vector $\underline{j}_i$ of the species i, Section 4.3.3, and τ_{xx}, τ_{xy} et cetera the components of the viscous stress tensor $\underline{\underline{\tau}}$, eqs. (7.10) to (7.15) in Section 7.1.3. In the fifth line each we have the components of the energy-flux vector $\underline{q}_e$, eq. (4.58), Sub-Section 4.3.2, which we have summarized as generalized molecular heat-flux vector in eqs. (4.61) and (4.62), the latter being the equation for the heat flux in the gas at the wall in presence of slip flow. In the sixth line, finally, we find the terms of molecular transport of the non-equilibrium vibration energy in analogy to the terms of eq. (4.59).

The components of the molecular heat flux vector $\underline{q}$ in the viscous flux vectors in eqs. (12.3) to (12.4) read, due to the use of a multi-temperature model with m vibration temperatures:

$$\begin{pmatrix} q_x \\ q_y \\ q_z \end{pmatrix} = -k_{trans} \nabla T_{trans} + \sum_m k_{vibr,m} \nabla T_{vibr,m}. \tag{12.6}$$

The source term $\underline{S}$ finally contains the mass sources Sm_i of the species i due to dissociation and recombination, eq. (4.84) in Sub-Section 4.3.3, and energy sources $Q_{k,m}$:

[2] We have not given the equation(s) for the transport of vibrational energy in Section 5.4, but have referred instead to [4].

$$\underline{S} = \begin{bmatrix} Sm_i \\ 0 \\ 0 \\ 0 \\ 0 \\ \sum_k \sum_m Q_{k,m} \end{bmatrix}. \tag{12.7}$$

The terms $Q_{k,m}$ represent the k mechanisms of energy exchange of the m vibration temperatures, for instance between translation and vibration, vibration and vibration, et cetera, Sub-Section 5.4.

To compute the flow past configurations with general geometries, the above equations are transformed from the physical space x, y, z into the computation space ξ, η, ζ:

$$\begin{aligned} \xi &= \xi(x, y, z), \\ \eta &= \eta(x, y, z), \\ \zeta &= \zeta(x, y, z). \end{aligned} \tag{12.8}$$

This transformation, which goes back to Viviand [5], regards only the geometry, and not the velocity components. This is in contrast to the approach for the general boundary-layer equations, Sub-Section 7.1.3, where both are transformed. ξ usually defines the main-stream direction, η the lateral direction, and ζ the wall-normal direction, however in general not in the sense of locally monoclinic coordinates, Sub-Section 7.1.3.

The transformation results in:

$$\frac{\partial \underline{\widehat{Q}}}{\partial t} + \frac{\partial(\underline{\widehat{E}} + \underline{\widehat{E}}_{visc})}{\partial \xi} + \frac{\partial(\underline{\widehat{F}} + \underline{\widehat{F}}_{visc})}{\partial \eta} + \frac{\partial(\underline{\widehat{G}} + \underline{\widehat{G}}_{visc})}{\partial \zeta} = \underline{\widehat{S}}, \tag{12.9}$$

which is the same as the original formulation, eq. (12.1).

The transformed conservation vector, the convective flux vectors, and the source term are now:

$$\begin{aligned} \underline{\widehat{U}} &= J^{-1}\underline{U}, \\ \underline{\widehat{E}} &= J^{-1}[\xi_x \underline{E} + \xi_y \underline{F} + \xi_z \underline{G}], \\ \underline{\widehat{F}} &= J^{-1}[\eta_x \underline{E} + \eta_y \underline{F} + \eta_z \underline{G}], \\ \underline{\widehat{G}} &= J^{-1}[\zeta_x \underline{E} + \zeta_y \underline{F} + \zeta_z \underline{G}], \\ \underline{\widehat{S}} &= J^{-1}\underline{S}, \end{aligned} \tag{12.10}$$

with J^{-1} being the Jacobi determinant of the transformation. The transformed viscous flux vectors have the same form as the transformed convective flux vectors.

The fluxes, eqs. (12.3) to (12.5), are transformed analogously, however we don't give the details, and refer instead to, for instance, [6], [7].

References

1. R. B. Bird, W. E. Stewart, E. N. Lightfoot. "Transport Phenomena". John Wiley, New York and London/Sydney, 2nd edition, 2002.
2. W. G. Vincenti, C. H. Kruger. "Introduction to Physical Gas Dynamics". John Wiley & Sons, New York/London/Sydney, 1965. Reprint edition, Krieger Publishing Comp., Melbourne, Fl., 1975
3. K. A. Hoffmann, S. T. L. Chiang, M. S. Siddiqui, M. Papadakis. "Fundamental Equations of Fluid Mechanics". Engineering Education System,Wichita, Ca., 1996.
4. C. Park. "Nonequilibrium Hypersonic Flow". John Wiley & Sons, New York, 1990.
5. H. Viviand. "Conservative Forms of Gas Dynamic Equations". La Recherche Aerospatiale, No. 1974-1, 1974, pp. 65 - 68.
6. T. H. Pulliam, J. L. Steger. "Implicit Finite-Difference Simulations of Three-Dimensional Compressible Flows". AIAA J., Vol. 18, No. 2, 1980, pp. 159 - 167.
7. C. Hirsch. "Numerical Computation of Internal and External Flows". Vol. 1, Fundamentals of Numerical Discretization, J. Wiley & Sons, New York, 1988.

13 Constants, Functions, Dimensions and Conversions

The dimensions are in general the SI units (Système International d'unités), see [1], [2], where also the constants can be found. The basic units, the derived units, and conversions[1] to US units are given in Section 13.2.

13.1 Constants and and Air Properties

Molar universal gas constant R_0 $= 8.314472{\cdot}10^3\ kg\,m^2/s^2\,kg\text{-}mol\,K$
$= 4.97201{\cdot}10^4\ lb_m\,ft^2/s^2\,lb_m\text{-}mol\,{}^\circ R$

Stephan-Boltzmann constant σ $= 5.670400{\cdot}10^{-8}\ W/m^2\,K^4$
$= 1.7123{\cdot}10^{-9}\ Btu/hr\,ft^2\,{}^\circ R^4$

Table 13.1. Molecular weights, gas constants, and intermolecular force parameters of air constituents for the low temperature domain, [3], [4]. * is the U. S. standard atmosphere value, + the value from [4].

Gas	Molecular weight M $[kg/kg\text{-}mol]$	Specific gas constant R $[m^2/s^2K]$	Collision diameter $\sigma \cdot 10^{10}$ [m]	2nd Lennard-Jones parameter ϵ/k $[K]$
air	28.9644* (28.97+)	287.06	3.617	97.0
N_2	28.02	296.73	3.667	99.8
O_2	32.00	259.83	3.430	113.0
NO	30.01	277.06	3.470	119.0
N	14.01	593.47	2.940	66.5
O	16.00	519.65	2.330	210.0
Ar	39.948	208.13	3.432	122.4
He	4.003	2077.06	2.576	10.2

[1] Details can be found, for instance, at
`http://www.chemie.fu-berlin.de/chemistry/general/si_en.html` and at
`http://physics.nist.gov/cuu/Units/units/html`.

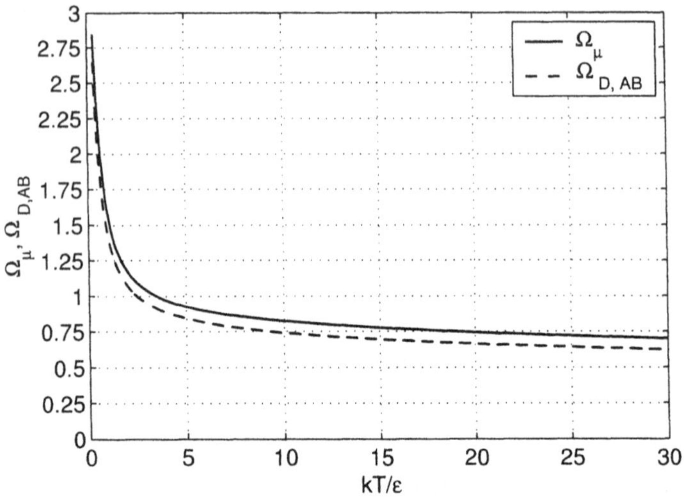

Fig. 13.1. Dimensionless collision integrals $\Omega_\mu = \Omega_k$, and $\Omega_{D_{AB}}$ of air as function of kT/ϵ or kT/ϵ_{AB}, [6], [4].

Table 13.2. Characteristic rotational, vibrational, and dissociation temperatures of air molecules, [5].

Gas:	N_2	O_2	NO
Θ_{rot} $[K]$	2.9	2.1	2.5
Θ_{vibr} $[K]$	3,390.0	2,270.0	2,740.0
Θ_{diss} $[K]$	113,000.0	59,500.0	75,500.0

13.2 Dimensions and Conversions

SI basic units and SI derived units are listed of the major flow, transport, and thermal entities. In the left column name and symbol are given and in the right column the unit (dimension), with → the symbol used in Chapter 14, and in the line below its conversion.

SI Basic Units

length, L — $[m]$, → $[L]$
$1.0\ m = 100.0\ cm = 3.28084\ ft$
$1{,}000.0\ m = 1.0\ km$

mass, m — $[kg]$, → $[M]$
$1.0\ kg = 2.20462\ lb_m$

time, t — $[s]$ (= $[sec]$), → $[t]$

temperature, T	$[K]$, $\rightarrow [T]$ $1.0\ K = 1.8\ ^{\circ}R$
amount of substance, *mole*	$[kg\text{-}mol]$, $\rightarrow [mole]$ $1.0\ kg\text{-}mol = 2.20462\ lb_m\text{-}mol$

SI Derived Units

area, A	$[m^2]$, $\rightarrow L^2]$ $1.0\ m^2 = 10.76391\ ft^2$
volume, V	$[m^3]$, $\rightarrow L^3]$ $1.0\ m^3 = 35.31467\ ft^3$
speed, velocity, v, u	$[m/s]$, $\rightarrow [M/t]$ $1.0\ m/s = 3.28084\ ft/s$
force, F	$[N] = [kg\, m/s^2]$, $\rightarrow [M\, L/t^2]$ $1.0\ N = 0.224809\ lb_f$
pressure, p	$[Pa] = [N/m^2]$, $\rightarrow [M/L\, t^2]$ $1.0\ Pa = 10^{-5}\ bar = 9.86923 \cdot 10^{-6}\ atm =$ $= 0.020885\ lb_f/ft^2$
density, ρ	$[kg/m^3]$, $\rightarrow [M/L^3]$ $1.0\ kg/m^3 = 0.062428\ lb_m/ft^3$
(dynamic) viscosity, μ	$[Pa\, s] = [N\, s/m^2]$, $\rightarrow [M/L\, t]$ $1.0\ Pa\, s = 0.020885\ lb_f\, s/ft^2$
kinematic viscosity, ν	$[m^2/s]$, $\rightarrow [M^2/t]$ $1.0\ m^2/s = 10.76391\ ft^2/s$
shear stress, τ	$[Pa] = [N/m^2]$, $\rightarrow [M/L\, t^2]$ $1.0\ Pa = 0.020885\ lb_f/ft^2$
energy, enthalpy, work, quantity of heat	$[J] = [N\, m]$, $\rightarrow [M\, L^2/t^2]$ $1.0\ J = 9.47813 \cdot 10^{-4}\ BTU =$ $= 23.73036\ lb_m ft^2/s^2 = 0.737562\ lb_f/s^2$
(mass specific) internal energy, enthalpy, e, h	$[J/kg] = [m^2/s^2]$, $\rightarrow [L^2/t^2]$ $1.0\ m^2/s^2 = 10.76391\ ft^2/s^2$

(mass) specific heat, c_v,c_p	$[J/kg\,K] = [m^2/s^2\,K]$, $\rightarrow [L^2/t^2\,T]$
specific gas constant, R	$1.0\ m^2/s^2\,K = 5.97995\ ft^2/s^2\,{}^\circ R$
power, work per unit time	$[W] = [J/s] = [N\,m/s]$, $\rightarrow [M\,L^2/t^3]$ $1.0\ W = 9.47813{\cdot}10^{-4}\ BTU/s =$ $= 23.73036\ lb_m\,ft^2/s^3$
thermal conductivity, k	$[W/m\,K] = [N/s^2\,K]$, $\rightarrow [M\,L/t^3\,T]$ $1.0\ W/m\,K = 1.60496{\cdot}10^{-4}\ BTU/s\,ft\,R =$ $= 4.018342\ lb_m\,ft/s^3\,R$
heat flux, q	$[W/m^2] = [J/m^2\,s]$, $\rightarrow [M/t^3]$ $1.0\ W/m^2 = 0.88055{\cdot}10^{-4}\ BTU/s\,ft^2 =$ $= 2.204623\ lb_m/s^3$
(binary) mass diffusivity, D_{AB}	$[m^2/s]$, $\rightarrow [L^2/t]$ $1.0\ m^2/s = 10.76391\ ft^2/s$
thermo diffusivity, D_A^T	$[kg/m\,s]$, $\rightarrow [M/L\,t]$ $1.0\ kg/m\,s = 0.67197\ lb_m/ft\,s$
diffusion mass flux, j	$[kg/m^2\,s]$, $\rightarrow [M/L^2\,t]$ $1.0\ kg/m^2\,s = 0.20482\ lb_m/ft^2\,s$

References

1. B. N. TAYLOR, ED.. "The International System of Units (SI)". US Dept. of Commerce, National Institute of Standards and Technology, NIST Special Publication 330, 2001, US Government Printing Office, Washington, D. C., 2001.
2. B. N. TAYLOR. "Guide for the Use of the International System of Units (SI)". US Dept. of Commerce, National Institute of Standards and Technology, NIST Special Publication 811, 1995, US Government Printing Office, Washington, D. C., 1995.
3. J. O. HIRSCHFELDER, C. F. CURTISS, R. B. BIRD. "Molecular Theory of Gases and Liquids". John Wiley, New York, corrected printing, 1964.
4. R. B. BIRD, W. E. STEWART, E. N. LIGHTFOOT. "Transport Phenomena". John Wiley & Sons, New York/London/Sydney, 2nd edition, 2002.
5. W. G. VINCENTI, C. H. KRUGER. "Introduction to Physical Gas Dynamics". John Wiley, New York and London/Sydney, 1965. Reprint edition, Krieger Publishing Comp., Melbourne, Fl., 1975
6. P. D. NEUFELD, A. R. JANSEN, R. A. AZIZ. J. Chem. Phys., Vol. 57, 1972, pp 1100 - 1102.

14 Symbols

Only the important symbols are listed. If a symbol appears only locally or infrequent, it is not included. In general the page number is indicated, where a symbol appears first or is defined. Dimensions are given in terms of the SI basic units: length $[L]$, time $[t]$, mass $[M]$, temperature $[T]$, and amount of substance $[mole]$, Chapter 13. For actual dimensions and their conversions see Section 13.2.

14.1 Latin Letters

A	amplitude, p. 271
A	surface, $[L^2]$
a	speed of sound, p. 142, $[L/t]$
C	Chapman-Rubesin factor, p. 340, $[-]$
C_D	drag coefficient, p. 182, $[-]$
C_L	lift coefficient, p. 182, $[-]$
C_R	resultant (total) force coefficient, p. 183, $[-]$
c	molar density, p. 20, $[mole/L^3]$
c	phase velocity, p. 271, $[L/t]$
c_i	molar concentration of species i, p. 20, $[mole/L^3]$
c_p	pressure coefficient, p. 145, $[-]$
c_p	(mass) specific heat at constant pressure, p. 106, $[L^2/t^2T]$
c_v	(mass) specific heat at constant volume, p. 106, $[L^2/t^2T]$
D	diameter, $[L]$
D	drag, $[ML/t^2]$
$DAM1$	first Damköhler number, p. 109, $[-]$
$DAM2$	second Damköhler number, p. 110, $[-]$
D_{AB}	mass diffusivity coefficient of a binary system, p. 78, $[L^2/t]$
D_A^T	thermo-diffusion coefficient of species A, p. 95, $[M/Lt]$
E	Eckert number, p. 91, $[-]$
e	internal (mass-specific) energy, p. 87, $[L^2/t^2]$
F	force, p. 195, $[ML/t^2]$
f	degree of freedom, p. 105, $[-]$
f	frequency, p. 271, $[1/t]$
G	Görtler parameter, p. 290, $[-]$

H	altitude, p. 2, $[L]$
H	shape factor, p. 229, $[-]$
h	(mass-specific) enthalpy, p. 105, $[L^2/t^2]$
h_t	total enthalpy, p. 29, $[L^2/t^2]$
h^*	reference enthalpy, p. 217, $[L^2/t^2]$
$\underline{j}_i$	diffusion mass-flux vector of species i, p. 94, $[M/L^2t]$
K	hypersonic similarity parameter, p. 194, $[-]$
K	acceleration parameter, p. 291, $[-]$
K_c	equilibrium constant, p. 113, $[(L^3/mole)^{\nu''-\nu'}]$
Kn	Knudsen number, p. 22, $[-]$
k	thermal conductivity, p. 76, $[ML/t^3T]$
k	roughness height, surface parameter, p. 287, $[L]$
k	turbulent energy, p. 294, $[L^2/t^2]$
k_{fr}	forward reaction rate, p. 113, $[(L^3/mole)^{\nu'-1}/t]$
k_{br}	backward reaction rate, p. 113, $[L^3/mole)^{\nu''-1}/t]$
$k_{w_{ia}}$	catalytic recombination rate, p. 123, $[L/t]$
L	characteristic length, $[L]$
L	lift, $[ML/t^2]$
Le	Lewis number, p. 91, $[-]$
M	Mach number, p. 83, $[-]$
M	molecular weight, p. 20, $[M/mole]$
M	moment, $[ML^2/t^2]$
M_N	Mach number normal (locally) to the shock wave, p. 153, $[-]$
M_i	molecular weight of species i, p. 20, $[M/mole]$
M_∞	flight Mach number, $[-]$
M_*	critical Mach number, p. 143, $[-]$
Pe	Peclét number, p. 90, $[-]$
Pr	Prandtl number, p. 90, $[-]$
p	pressure, p. 20, $[M/Lt^2]$
p_e	pressure at the boundary-layer edge, p. 184, $[M/Lt^2]$
p_i	partial pressure of species i, p. 20, $[M/Lt^2]$
q	dynamic pressure, p. 144, $[M/Lt^2]$
q_∞	free-stream dynamic pressure, $[M/Lt^2]$
q	heat flux, p. 74, $[M/t^3]$
q_{gw}	heat flux in the gas at the wall, p. 30, $[M/t^3]$
q_w	heat flux into the wall, p. 30, $[M/t^3]$
q_{rad}	thermal radiation heat flux, p. 35, $[M/t^3]$
q'	disturbance amplitude, p. 270
R	gas constant, p. 20, $[L^2/t^2T]$
R	radius, $[L]$
R_0	universal gas constant, p. 20, $[ML^2/t^2moleT]$
Re	Reynolds number, p. 84, $[-]$
Re^u	unit Reynolds number, p. 18, $[1/L]$
Re_θ	contamination Reynolds number, p. 283, $[-]$

r	recovery factor, p. 31, [–]
r_i	volume fraction of species i, p. 21, [–]
S	source term, p. 385, [–]
Sc	Schmidt number, p. 95, [–]
Sm_i	species source term, p. 94, 112, $[M/L^3t]$
Sr	Strouhal number, p. 73, [–]
St	Stanton number, p. 30, [–]
s	entropy, p. 142, $[L^2/t^2T]$
T	temperature, p. 20, $[T]$
T_{gw}	temperature of the gas at the wall, p. 10, $[T]$
T_{ra}	radiation-adiabatic temperature, p. 31, $[T]$
T_t	total temperature, p. 30, $[T]$
T_w	wall temperature, p. 10, $[T]$
T_r	recovery temperature, p. 30, $[T]$
Tu	level of free-stream turbulence, p. 293, [–]
T^*	reference temperature, p. 217, $[T]$
u, v, w	Cartesian velocity components, p. 70, $[L/t]$
u', v', w'	non-dimensional Cartesian velocity components, p. 190, $[L/t]$
$\widetilde{u}, \widetilde{v}$	velocity components normal and tangential to an oblique shock wave, p. 152, $[L/t]$
u_∞, v_∞	free-stream velocity, flight speed, p. 17, 18, $[L/t]$
u_{vs}	viscous sub-layer edge-velocity, p. 225, $[L/t]$
u^+	non-dimensional velocity, p. 226, [–]
u_τ	friction velocity, p. 225, $[L/t]$
V_m	maximum speed, p. 143, $[L/t]$
V_1, V_2	resultant velocities ahead and behind an oblique shock wave, p. 152, $[L/t]$
$\underline{V}$	velocities vector, p. 70, $[L/t]$
$V, \overline{V}$	interaction parameters, p. 343, [–]
v^i, v^{*i}	contravariant and physical velocity components, p. 202, $[1/t]$, $[L/t]$
v_0	surface suction or blowing velocity, p. 227, $[L/t]$
x, y, z	Cartesian coordinates, p. 70, $[L]$
x^i	surface-oriented locally monoclinic coordinates, p. 203, [–]
x_i	mole fraction of species i, p. 20, [–]
y^+	non-dimensional wall distance, p. 226, [–]
Z	real-gas factor, p. 102, [–]

14.2 Greek Letters

α	angle of attack, p. 6, [°]
α	thermal accommodation coefficient, p. 93, [–]
α	thermal diffusivity, p. 90, $[L^2/t]$

α	wave number, p. 271, $[1/L]$
γ	ratio of specific heats, p. 106, [–]
γ_{eff}	effective ratio of specific heats, p. 165, [–]
γ_{i^a}	recombination coefficient of atomic species, p. 122, [–]
Δ	characteristic boundary-layer length, p. 36, $[L]$
Δ_0	shock stand-off distance, p. 163, $[L]$
δ	flow (ordinary) boundary-layer thickness, p. 85, $[L]$
δ	ramp angle, p. 152, [°]
δ	shock stand-off distance ($\delta \equiv \Delta_0$), p. 164, $[L]$
δ_{flow}	flow (ordinary) boundary-layer thickness ($\delta_{flow} \equiv \delta$), p. 85, $[L]$
δ_M	mass-concentration boundary-layer thickness, p. 96, $[L]$
δ_T	thermal boundary-layer thickness, p. 92, $[L]$
δ_{sc}	turbulent scaling thickness, p. 226, $[L]$
δ_{vs}	viscous sub-layer thickness, p. 225, $[L]$
δ_1	boundary-layer displacement thickness ($\delta_1 \equiv \delta^*$), p. 227, $[L]$
δ_2	boundary-layer momentum thickness ($\delta_2 \equiv \theta$), p. 227, $[L]$
ε	emissivity coefficient, p. 35, [–]
ε	density ratio, p. 163, [–]
ε_f	fictitious emissivity coefficient, p. 46, [–]
Θ_{diss}	characteristic dissociation temperature, p. 113, $[T]$
Θ_{rot}	characteristic rotational temperature, p. 109, $[T]$
Θ_{vibr}	characteristic vibrational temperature, p. 111, $[T]$
θ	flow angle, p. 70, [°]
θ	shock angle, p. 152, [°]
θ_*	sonic shock angle, p. 155, [°]
κ	bulk viscosity, p. 82, $[M/Lt]$
λ	mean free path, p. 22, $[L]$
λ	wave length, p. 271, $[L]$
μ	viscosity, p. 75, $[M/Lt]$
μ	Mach angle, p. 159, [°]
ν	kinematic viscosity, p. 211, $[L^2/t]$
ν	Prandtl-Meyer angle, p. 175, [°], []
ν'_{ir}, ν''_{ir}	stoichiometric coefficients, p. 112, [–]
ρ	density, p. 20, $[M/L^3]$
ρ_i	partial density of species i, p. 20, $[M/L^3]$
ρ_t	total density, p. 142, $[M/L^3]$
ρ_i^*	fractional density of species i, p. 21, $[M/L^3]$
ρ_0	surface suction or blowing density, p. 227, $[M/L^3]$
σ	reflection coefficient, p. 86, [–]
τ	relaxation time, p. 111, $[t]$
τ	thickness ratio, p. 194, $[t]$
$\underline{\underline{\tau}}$	viscous stress tensor, p. 82, $[M/t^2L]$
τ_{xx}, τ_{xy}, ...	components of the viscous stress tensor, p. 205, $[M/t^2L]$
τ_w	skin friction, wall shear stress, p. 238, $[M/t^2L]$

Φ	angle, p. 332, [°]
Φ	transported entity, p. 71
Φ_{ij}	term in Wilke's mixing formula, p. 80
φ	characteristic manifold, p. 210
φ	sweep angle of leading edge or cylinder, p. 187, [°]
χ	cross-flow Reynolds number, p. 284, [-]
χ, $\overline{\chi}$	viscous interaction parameter, p. 341, [−]
Ψ'	disturbance stream function, p. 272, [$1/t$]
ψ	angle, p. 186, [°]
Ω	vorticity content vector, p. 202, [L/t]
Ω_k	dimensionless thermal conductivity collision integral, p. 77, [−]
Ω_μ	dimensionless viscosity collision integral, p. 75, [−]
ω	circular frequency, p. 271, [$1/t$]
ω_i	mass fraction of species i, p. 20, [−]
ω_k	exponent in the power-law equation of thermal conductivity (ω_k, ω_{k1}, ω_{k2}), p. 77, [−]
ω_μ	exponent in the power-law equation of viscosity (ω_μ, $\omega_{\mu1}$, $\omega_{\mu2}$), p. 75, [−]
$\underline{\omega}$	vorticity vector, p. 202, [$1/t$]

14.3 Indices

14.3.1 Upper Indices

i	$i = 1, 2, 3$, general coordinates and contravariant velocity components
$*i$	physical velocity component
T	thermo-diffusion
u	unit
$+$, $-$	ionized
$+$	dimensionless sub-layer entity
$*$	reference-temperature/enthalpy value

14.3.2 Lower Indices

A, B	species of binary gas
br	backward reaction
c	compressible
$corr$	corrected
cr	critical
D	drag
e	boundary-layer edge, external (flow)

$elec$	electronic
eff	effective
$equil$	equilibrium
exp	experiment
fp	flat surface portion
fr	forward reaction
gw	gas at the wall
Han	Hansen
i	imaginary part
i, j	species
ic	incompressible
k	thermal conductivity
L	lift
l	lee side
lam	laminar
M	mass concentration
$molec$	molecule
ne	non-equilibrium
R	resultant
r	real part
r	reaction
r	recovery
ra	radiation adiabatic
rad	radiation
ref	reference
res	residence
rot	rotational
SL	sea level
$Suth$	Sutherland
s	shock
s	stagnation point
sc	turbulent scaling
scy	swept cylinder
sp	sphere
T	thermal
t	total
tr,l	transition, lower location
tr,u	transition, upper location
$trans$	translational
$turb$	turbulent
$vibr$	vibrational
vs	viscous sub-layer
w	wall
w	windward side

μ	viscosity
τ	friction velocity
0	reference
1	ahead of the shock wave
2	behind the shock wave
∞	infinity
$*$	critical

14.4 Other Symbols

$O()$	order of magnitude
$'$	non-dimensional and stretched
$'$	fluctuation entity
$\underline{v}$	vector
$\underline{\underline{t}}$	tensor

14.5 Acronyms

Indicated is the page where the acronym is used for the first time.

AOTV	aeroassisted orbital transfer vehicle, p. 3
ARV	ascent and re-entry vehicle, p. 3
BDW	blunt delta wing, p. 54
CAD	computer aided design, p. 367
CAV	cruise and acceleration vehicle, p. 3
DNS	direct numerical solution, p. 279
FESTIP	Future European Space Transportation Investigations Programme, p. 3
GETHRA	general thermal radiation, p. 46
HALIS	high alpha inviscid solution, p. 125
LES	large eddy simulation, p. 294
OMS	orbital maneuvering system, p. 43
PNS	parabolized Navier-Stokes, p. 366
RANS	Reynolds-averaged Navier-Stokes, p. 46
RHPM	Rankine-Hugoniot-Prandtl-Meyer, p. 5
RV	re-entry vehicle, p. 3
SSTO	single stage to orbit, p. 3
TSTO	two stage to orbit, p. 3
TPS	thermal protection system, p. 8

Name Index

Subject Index

Permissions

Figures reprinted with permission:

Fig. 2.5, by W. G. Vincenti and C. H. Kruger
Fig. 3.3, by S. Wüthrich
Figs. 3.7, 3.8, by ASME
Fig. 6.7, by the American Institute of Aeronautics and Astronautics, Inc.
Figs. 6.11, 9.7, 9.8, by Elsevier
Fig. 6.16, by H. W. Liepmann and A. Roshko
Fig. 6.21, by WIT Press
Fig. 6.37, by J. Zierep

The permissions to reprint the figures provided directly from the authors, see the Acknowledgements at the beginning of the book, were given in all cases by the authors.

Papier | Fördert gute Waldnutzung
FSC® C083411

Kolibri 360 GmbH
Ferdinand-Jühlke-Straße 7
99095 Erfurt, Deutschland
produktsicherheit@kolibri360.de